BIOCHEMICAL SOCIETY SYMPOSIA

No. 51

# GENES AND PROTEINS
# IN IMMUNITY

BIOCHEMICAL SOCIETY SYMPOSIUM No. 51
held at the University of Oxford, July 1985

# Genes and Proteins in Immunity

ORGANIZED AND EDITED BY

## J. KAY, M. A. KERR, A. F. WILLIAMS AND K. B. M. REID

1986
LONDON: THE BIOCHEMICAL SOCIETY

The Biochemical Society,
7 Warwick Court,
London WC1R 5DP, U.K.



ISBN: 0 904498 18 2
ISSN: 0067-8694



*Printed in Great Britain by the
University Press, Cambridge*

# List of Contributors

O. Acuto (*Laboratory of Immunobiology, Dana-Farber Cancer Institute, and the Departments of Medicine and Pathology, Harvard Medical School, Boston, MA 02115, U.S.A.*)

G. Alcaraz (*Section on Chemical Immunology, Arthritis and Rheumatism Branch, National Institute of Arthritis, Diabetes, and Digestive and Kidney Diseases, National Institutes of Heath, Bethesda, MD 20892, U.S.A.*)

C. A. Alper (*The Center for Blood Research, 800 Huntington Avenue, Boston, MA 02115, U.S.A.*)

D. C. Anderson (*Baylor College of Medicine and Texas Children's Hospital, 6621 Fannin, Houston, TX 77030, U.S.A.*)

J. T. August (*Department of Pharmacology and Experimental Therapeutics, The Johns Hopkins University School of Medicine, Baltimore, MD 21205, U.S.A.*)

Z. Awdeh (*The Center for Blood Research, 800 Huntington Avenue, Boston, MA 02115, U.S.A.*)

A. N. Barclay (*MRC Cellular Immunology Unit, Sir William Dunn School of Pathology, University of Oxford, South Parks Road, Oxford OX1 3RE, U.K.*)

K. T. Belt (*Department of Biochemistry, University of Oxford, South Parks Road, Oxford OX1 3QU, U.K.*)

D. R. Bentley (*MRC Immunochemistry Unit, Department of Biochemistry, University of Oxford, South Parks Road, Oxford OX1 3QU, U.K.*)

C. Berek (*Medical Research Council Laboratory of Molecular Biology, Hills Road, Cambridge CB2 2QH, U.K.*)

G. Blobel (*Rockefeller University, 1230 York Avenue, New York, NY 10021, U.S.A.*)

W. Born (*Department of Medicine, National Jewish Hospital, and University of Colorado Health Sciences Center, Denver, CO 80202, U.S.A.*)

R. D. Campbell (*MRC Immunochemistry Unit, Department of Biochemistry, University of Oxford, South Parks Road, Oxford OX1 3QU, U.K.*)

T. J. Campen (*Laboratory of Immunobiology, Dana-Farber Cancer Institute, and the Departments of Medicine and Pathology, Harvard Medical School, Boston, MA 02115, U.S.A*)

M. C. Carroll (*Department of Pediatrics, Harvard Medical School, Children's Hospital, Boston, MA 02115, U.S.A.*)

E. R. Cebra (*Department of Biology, University of Pennsylvania, Philadelphia, PA 19104, U.S.A.*)

LIST OF CONTRIBUTORS

J. J. Cebra (*Department of Biology, University of Pennsylvania, Philadelphia, PA 19104, U.S.A.*)

G. L. Chen (*University of California, Berkeley, CA 94720, U.S.A.*)

J. W. Chen (*Department of Pharmacology and Experimental Therapeutics, The Johns Hopkins University School of Medicine, Baltimore, MD 21205, U.S.A.*)

Y.-H. Chien (*Department of Medical Microbiology, Stanford University School of Medicine, Stanford, CA 94305, U.S.A.*)

M. J. Clark (*MRC Cellular Immunology Unit, Sir William Dunn School of Pathology, University of Oxford, South Parks Road, Oxford OX1 3RE, U.K.*)

H. R. Colten (*Department of Pediatrics, Harvard Medical School, Children's Hospital, Boston, MA 02115, U.S.A.*)

M. M. Davis (*Department of Medical Microbiology, Stanford University School of Medicine, Stanford, CA 94305, U.S.A.*)

A. J. Day (*MRC Immunochemistry Unit, Department of Biochemistry, University of Oxford, South Parks Road, Oxford OX1 3QU, U.K.*)

S. B. Dowton (*Department of Pediatrics, Harvard Medical School, Children's Hospital, Boston, MA 02115, U.S.A.*)

M. P. D'Souza (*Department of Pharmacology and Experimental Therapeutics, The Johns Hopkins University School of Medicine, Baltimore, MD 21205, U.S.A.*)

R. A. Dwek (*Department of Biochemistry, University of Oxford, South Parks Road, Oxford OX1 3QU, U.K.*)

J. Even (*Medical Research Council Laboratory of Molecular Biology, Hills Road, Cambridge CB2 2QH, U.K.*)

M. Fabbi (*Laboratory of Immunobiology, Dana-Farber Cancer Institute, and the Departments of Medicine and Pathology, Harvard Medical School, Boston, MA 02115, U.S.A.*)

A. Farr (*Department of Biological Structure, University of Washington, Seattle, WA 98195, U.S.A.*)

M. Friedlander (*Rockefeller University, 1230 York Avenue, New York, NY 10021, U.S.A.*)

Y. Fukuoka (*Department of Immunology, IMM18, Scripps Clinic and Research Foundation, 10666 North Torrey Pines Road, La Jolla, CA 92037, U.S.A.*)

N. R. J. Gascoigne (*Department of Medical Microbiology, Stanford University School of Medicine, Stanford, CA 94305, U.S.A.*)

I. Gigli (*Department of Medicine/Dermatology, University of California, San Diego, School of Medicine, San Diego, CA 92103, U.S.A.*)

D. Givol (*Department of Chemical Immunology, Weizmann Institute of Science, Rehovot 76100, Israel*)

C. Goodnow (*Department of Medical Microbiology, Stanford University School of Medicine, Stanford, CA 94305, U.S.A.*)

P. D. Hoeprich, Jr. (*Department of Immunology, IMM18, Scripps Clinic and Research Foundation, 10666 North Torrey Pines Road, La Jolla, CA 92037, U.S.A.*)

S. W. Homans (*Department of Biochemistry, University of Oxford, South Parks Road, Oxford OX1 3QU, U.K.*)

R. Huey (*Department of Immunology, IMM18, Scripps Clinic and Research Foundation, 10666 North Torrey Pines Road, La Jolla, CA 92037, U.S.A.*)

T. E. Hugli (*Department of Immunology, IMM18, Scripps Clinic and Research Foundation, 10666 North Torrey Pines Road, La Jolla, CA 92037, U.S.A.*)

J. Kappler (*Department of Medicine, National Jewish Hospital, and University of Colorado Health Sciences Center, Denver, CO 80202, U.S.A.*)

J.-P. Kinet (*Section on Chemical Immunology, Arthritis and Rheumatism Branch, National Institute of Arthritis, Diabetes, and Digestive and Kidney Diseases, National Institutes of Health, Bethesda, MD 20892, U.S.A.*)

D. B. Kotloff (*Department of Biology, University of Pennsylvania, Philadelphia, PA 19104, U.S.A.*)

D. A. Lebman (*Department of Biology, University of Pennsylvania, Philadelphia, PA 19104, U.S.A.*)

T. Lindsten (*Department of Medical Microbiology, Stanford University School of Medicine, Stanford, CA 94305, U.S.A.*)

V. Malhotra (*MRC Immunochemistry Unit, Department of Biochemistry, University of Oxford, South Parks Road, Oxford OX1 3QU, U.K.*)

P. Marrack (*Department of Medicine, National Jewish Hospital, and University of Colorado Health Sciences Center, Denver, CO 80202, U.S.A.*)

G. W. McCaughan (*MRC Cellular Immunology Unit, Sir William Dunn School of Pathology, University of Oxford, South Parks Road, Oxford OX1 3RE, U.K.*)

H. Metzger (*Section on Chemical Immunology, Arthritis and Rheumatism Branch, National Institute of Arthritis, Diabetes, and Digestive and Kidney Diseases, National Institutes of Health, Bethesda, MD 20892, U.S.A.*)

K. J. Micklem (*Nuffield Department of Pathology, Level 1, John Radcliffe Hospital 1, Headington, Oxford OX1 9DU, U.K.*)

C. Milstein (*Medical Research Council Laboratory of Molecular Biology, Hills Road, Cambridge CB2 2QH, U.K.*)

K. E. Mostov (*Whitehead Institute, Nine Cambridge Center, Cambridge, MA 02142, U.S.A.*)

T. L. Murphy (*Department of Pharmacology and Experimental Therapeutics, The Johns Hopkins University School of Medicine, Baltimore, MD 21205, U.S.A.*)

A. Palsdottir (*Department of Biochemistry, University of Oxford, South Parks Road, Oxford OX1 3QU, U.K.*)

LIST OF CONTRIBUTORS

R. B. Parekh (*Department of Biochemistry, University of Oxford, South Parks Road, Oxford OX1 3QU, U.K.*)

R. Quarto (*Section on Chemical Immunology, Arthritis and Rheumatism Branch, National Institute of Arthritis, Diabetes, and Digestive and Kidney Diseases, National Institutes of Health, Bethesda, MD 20892, U.S.A.*)

T. W. Rademacher (*Department of Biochemistry, University of Oxford, South Parks Road, Oxford OX1 3QU, U.K.*)

D. Ramarli (*Laboratory of Immunobiology, Dana-Farber Cancer Institute, and the Departments of Medicine and Pathology, Harvard Medical School, Boston, MA 02115, U.S.A.*)

D. Raum (*The Center for Blood Research, 800 Huntington Avenue, Boston, MA 02115, U.S.A.*)

E. L. Reinherz (*Laboratory of Immunobiology, Dana-Farber Cancer Institute, and the Departments of Medicine and Pathology, Harvard Medical School, Boston, MA 02115, U.S.A.*)

J. Ripoche (*MRC Immunochemistry Unit, Department of Biochemistry, University of Oxford, South Parks Road, Oxford OX1 3QU, U.K.*)

N. Roehm (*Department of Medicine, National Jewish Hospital, and University of Colorado Health Sciences Center, Denver, CO 80202, U.S.A.*)

H. D. Royer (*Laboratory of Immunobiology, Dana-Farber Cancer Institute, and the Departments of Medicine and Pathology, Harvard Medical School, Boston, MA 02115, U.S.A.*)

P. A. Schweitzer (*Department of Biology, University of Pennsylvania, Philadelphia, PA 19104, U.S.A.*)

D. A. Shackleford (*The Salk Institute for Biological Studies, Post Office Box 85800, San Diego, CA 92138, U.S.A.*)

R. D. Shahin (*Department of Biology, University of Pennsylvania, Philadelphia, PA 19104, U.S.A.*)

E. Sim (*Wellcome Toxicology Unit, Department of Pharmacology, University of Oxford, South Parks Road, Oxford OX1 3QU, U.K.*)

R. B. Sim (*MRC Immunochemistry Unit, Department of Biochemistry, University of Oxford, South Parks Road, Oxford OX1 3QU, U.K.*)

T. A. Springer (*Dana-Farber Cancer Institute and Harvard Medical School, 44 Binney Street, Boston, MA 02115, U.S.A.*)

I. S. Trowbridge (*The Salk Institute for Biological Studies, Post Office Box 85800, San Diego, CA 92138, U.S.A.*)

C.-Y. Yu (*Department of Biochemistry, University of Oxford, South Parks Road, Oxford OX1 3QU, U.K.*)

E. J. Yunis (*The Centre for Blood Research, 800 Huntington Avenue, Boston, MA 02115, U.S.A.*)

# Preface

This Symposium volume was intended initially to be a celebration of Rod Porter's outstanding career in immunology. Sadly, however, it must now appear as a memorial tribute as a consequence of his tragic death in a car accident on 6th September 1985. He was due to retire on 1st October 1985 from the Whitley Chair of Biochemistry at Oxford, a post he had held since 1967. Rod would have been 68 at that time, but in view of his very successful and active role as Honorary Director of the Medical Research Council Immunochemistry Unit (1967–1985), he had been invited by the MRC to continue as Director until 1989.

The Symposium title 'Genes and Proteins in Immunity' was Rod's own choice and it was typical of him to select a title 'concise and to the point' in preference to other more convoluted suggestions which were considered. It was also characteristic that he agreed that the meeting should be open to the whole scientific community and that speakers should be chosen from among his own co-workers and others whose work was of current interest. By this means, a series of talks presenting up-to-date material was arranged. The meeting was largely centred on new developments and dealt extensively with the genes of the complement components and with receptors and cell surface antigens of relevance to immunity. In some cases, speakers switched from their initially proposed topics in order to present their most recent data and the final blend that appears in this volume shows a merging of immunology with biochemistry. This not only reflects current trends but is also appropriate in that Rod Porter considered himself as a biochemist who worked on the immune system rather than an uncompromising specialist in immunology.

From 1946 to 1949 Rod Porter worked with F. Sanger at Cambridge and, after obtaining his Ph.D. for a study on the determination of free amino groups using the DNP method, he joined the MRC Scientific Staff at Mill Hill in 1949. It was during this period (1949–1960) that he acquired a great knowledge of the art of protein purification from A. J. P. Martin and a background in immunological techniques by collaboration with J. H. Humphrey. In 1958 he described the limited proteolysis of rabbit antibody IgG by papain and the separation of the two Fab fragments with retained specificity of antigen from the third fragment (Fc) which was crystallizable. This was followed by a very productive period (1960–1967) as Pfizer Professor of Immunology at St Mary's Hospital in London where his initial proposals for the structure of IgG were confirmed and amplified and the concept of variable and constant regions was reinforced. At Oxford (1967 onwards) he was keen to develop new themes and work was started on the complement system and on molecular approaches to lymphocyte surface molecules. His own interest increasingly centred on complement and developed into a study of the various components, control proteins and receptors of the system. By the early 1980s, his group was involved in the cloning of many of these complement proteins with particular emphasis on the C2, factor B and C4 components which are encoded within the human histocompatibility complex.

Rod Porter was awarded many prizes and honours, which included the Nobel Prize (1972) and the Companion of Honour (1985). However, one honour which gave him particular pleasure was to be appointed a Curator of the Botanic Gardens at Oxford. This interest in gardening, bee-keeping and other rural pursuits, along with climbing and walking, occupied much of his time away from the laboratory. He and his wife Julia had a knack of making the members of his Unit and visitors feel like part of one big family. The warmth of this feeling was tangible thoughout the Symposium, which was attended by a high proportion of those who had worked with Rod and enjoyed his unique companionship.

J. KAY
M. A. KERR
A. F. WILLIAMS
K. B. M. REID

*Cardiff, Dundee and Oxford*

# Contents

Page

CONTENTS

# Abbreviations

| | |
|---|---|
| bp | Base pairs |
| C4bp | C4b-binding protein |
| CIC | Circulating immune complex |
| CR | Complement receptor |
| CTL | Cytotoxic effector lymphocyte |
| DAF | Decay accelerating factor |
| EBV | Epstein–Barr virus |
| EGS | Ethylene glycol bis(succinimidyl succinate) |
| FACS | Fluorescence-activated cell sorting |
| FLU | Fluorescein |
| hPMN | Human polymorphonuclear leukocyte |
| IAP | Intracisternal A-particle |
| IFN | Interferon |
| Ig | Immunoglobulin |
| IL | Interleukin |
| In | Inulin |
| kb | Kilobases (or base pairs) |
| LAMP | Lysosome associated membrane protein |
| LFA | Lymphocyte function associated antigen |
| LINE | Long interdispersed element |
| LPS | Lipopolysaccharide |
| LTR | Long terminal repeat |
| MAb | Monoclonal antibody |
| MAC | Membrane attack complex |
| MHC | Major histocompatibility complex |
| MLC | Mixed lymphocyte culture |
| 21-OH | 21-Hydroxylase (deficiency) |
| PC | Phosphocholine |
| pIg | Polymeric immunoglobulin |
| PMA | Phorbol myristate acetate ( = TPA) |
| SC | Secretory component |
| SDS | Sodium dodecyl sulphate |
| sIg | Surface immunoglobulin |
| SINE | Short interdispersed element |
| RER | Rough endoplasmic reticulum |
| RFLP | Restriction fragment length polymorphism |
| TD | Thymus dependent |
| TPA | 12-*O*-Tetradecanoylphorbol 13-acetate ( = PMA) |
| TVT | Transmissible venereal tumour |

The nomenclature of complement components is as recommended by the World Health Organisation [*Bull. W.H.O.* (1968) **39**, 935–936; (1981) **59**, 489–490].

*Biochem. Soc. Symp.* **51**, 1–6
*Printed in Great Britain*

# Introduction to Contemporary Complement Research

IRMA GIGLI

*Department of Medicine/Dermatology, University of California, San Diego, School of Medicine,
San Diego, CA 92103, U.S.A.*

'**Even if complement appears to have little relevance to immunity, its
importance remains unimpaired at the laboratory level as a reagent in
complement fixation tests.**' (**M. F. Burnet**)

The strides made in complement research during the years that followed this
comment have clearly dispelled this notion. A fair measure of credit for these
advances should go to Rodney Porter and his research group. The structural
elucidation of complement proteins has progressed rapidly on the protein and
DNA level [1–4]. Intricate genetic complexities have been uncovered, particularly
in the case of C4 [5]. Cell surface complement receptors and their role in host
defense [6–8] and immunoregulation [9] are subjects of intense basic and clinical
investigation. A group of interrelated regulatory binding proteins that share
functional properties has been identified. This group includes several plasma and
cell membrane proteins [10–12]. The cell killing mechanism of complement has
been largely unravelled and is now being compared with that of lymphocytes
[13]. Complement research also extends into clinical investigation of genetic
deficiencies, infectious disease, autoimmune disease and such novel disease
entities as acquired immunodeficiency syndrome (AIDS) [14–16].

The complement system is old phylogenetically and appears early in ontogeny
[17]. The proteins in the system are synthesized by cells of differing origin, such
as hepatocytes, macrophages, lymphocytes and fibroblasts [18,19]. Together
with the 20 plasma proteins recognized in the system, at least eight distinct cell
surface receptors have evolved on a number of cells, particularly on inflammatory
cells and cells of the immune system [18]. These receptors interact with fragments
of complement proteins which are generated during their activation.

Functionally, it is now recognized that complement is an integral part both
of host defense against infections and of the inflammatory process. The activities
in both processes are derived from proteolytic fragments of precursor proteins
and from the fusion of multiple proteins into macromolecular organizations.
This knowledge is derived from biochemical investigation of the proteins, their
reaction products, their interaction with cellular receptors, as well as from the
study of genetic deficiencies in humans and the design and exploration of
experimental disease models.

Most of the actions of complement are directed toward cell membranes. We
now recognize three modes of action: (1) cytolysis or disruption of the lipid
compartment of cell membranes due to the formation of transmembrane protein

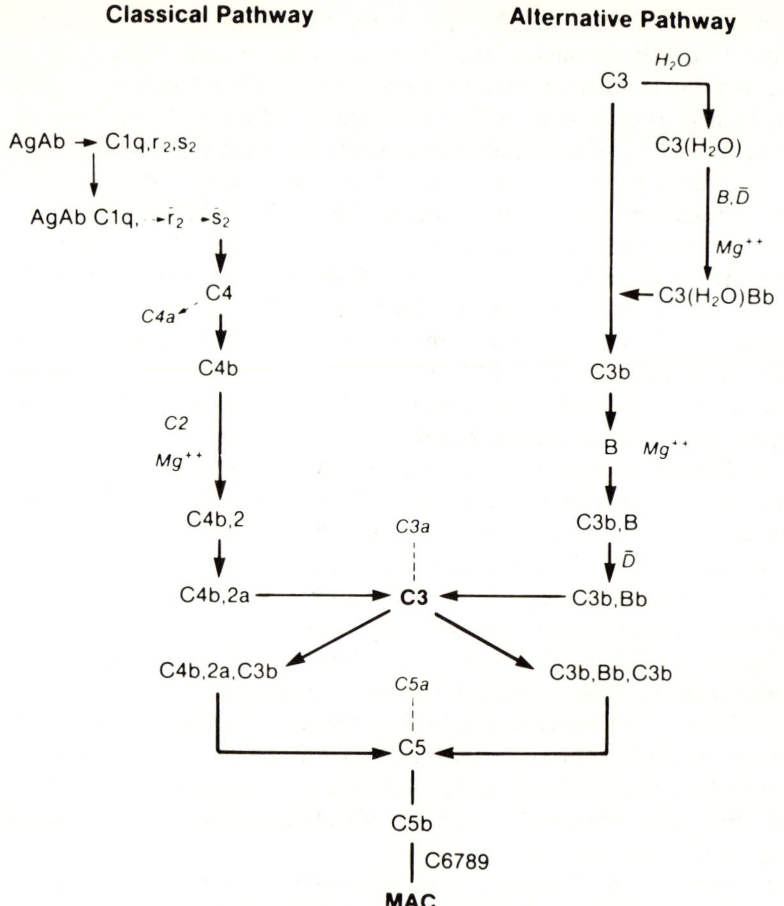



Fig. 1. *Representation of the molecular organization of the complement system*

The left side of the scheme represents the classical pathway, the right side the alternative pathway. Only the fragments with anaphylotoxin activity have been identified.

channels [20]; (2) covalent binding of some of the complement proteins to hydroxyl or amino groups on a cell surface, a mechanism by which complement proteins are transferred from the plasma to a target membrane during activation [21]; and (3) specific interaction of complement fragments with complement receptors [5]. These receptors permit fragments of complement proteins to elicit and regulate cellular responses [9]. They also participate in the processing of circulating immune complexes [12]. To generate fragments with such diversified functions the complement system is organized in two activation pathways, the classical and alternative pathways (Fig. 1), which resemble each other in several respects.

The classical pathway is initiated by binding of the first component, C1, to antigen–antibody complexes. C1 is a macromolecule consisting of two weakly interacting subunits, C1q and $C1r_2$, $C1s_2$. C1 undergoes spontaneous activation by an intramolecular autocatalytic mechanism which is physiologically controlled by the C1-inhibitor [22]. However, on binding to immune complexes through C1q, the subunits, of the C1 macromolecule become firmly associated and activation is initiated, even in the presence of the C1 inhibitor. Conformationally activated pro-C1r$_2$ is the initial enzyme of the pathway. Cleaved C1r$_2$ then activates C1s$_2$ which catalyses the assembly of the C3 convertase, C4b,2a, from the native proteins C2 and C4. The enzyme is either formed in the fluid phase or is covalently attached, through C4b, to the surface of target particles, which the catalytic site is located in the C2a subunit. The target-bound C3 convertase becomes the C5 convertase, C4b,2a,3b, on activation of C3 and binding of C3b, which modulates the substrate C5 for cleavage by C4b,2a. Cleavage of C5 and generation of metastable C5b initiates the assembly of the membrane attack complex (MAC) [20].

Polysaccharides are thought to be critical in the activation of the alternative pathway; however, antibodies may play an important part in this process [23]. This pathway is initiated spontaneously by hydrolysis of the internal thioester of native C3 in a reaction with a molecule of water [24]. C3 so modified [C3(H$_2$O)] becomes a structural subunit of the initial C3 convertase, C3(H$_2$O),Bb, when it reacts with Factor B and Factor D. The initial enzyme of the alternative pathway is Factor D [25], which is always present in plasma in active and uninhibited form, but cannot act on its substrate Factor B unless it is modulated by C3(H$_2$O) or C3b. C3(H$_2$O),Bb is a fluid-phase enzyme, its active site resides in the subunit Bb. The action of the enzyme on native C3 produces metastable C3b which can covalently attach to the surface of biological particles. C3b in the presence of Mg$^{2+}$ binds Factor B and Factor D then catalyses the formation of the particle-bound C3 convertase, C3b,Bb, by cleaving and activating Factor B [24]. With the addition of a molecule of C3b the enzyme becomes capable of acting on C5.

Considering that C3 and C4 are regarded as related in evolution, the molecular organizations of the two C3 and the two C5 convertases are indeed very similar. Both C3 convertases are genetically linked to the major histocompatibility complex (MHC) [26]. C2, Factor B and C4 are products of the class III MHC genes. C2 and Factor B probably arose by gene duplication, since in man their genes are adjacent on chromosome 6. They are precursors of unusual serine proteinases, and the ultrastructure of their active fragments C2a and Bb is very similar, exhibiting a two-domain arrangement [27]. Both convertases are under negative regulation by a number of plasma and membrane proteins: Factor I, C4-bp, Factor H, CR1 (C3b/C4b receptor) and decay accelerating factor (DAF) [28]. The alternative pathway C3 convertase is under positive regulation by properdin [29] (Fig. 2); its counterpart in the classical pathways has not been yet observed.

Both pathways eventuate in cleavage and activation of C5 and thus in assembly of the membrane attack complex (MAC). Through the metastable membrane-binding site in C5b MAC binds firmly to target membranes through hydrophobic interaction with the lipid bilayer. The final events are the polymerization and unfolding of C9, and the formation of a transmembrane poly(C9) channel. The functional result is the labilization of the structure of cell membranes [20]. This mechanism of action of MAC is capable of lysing

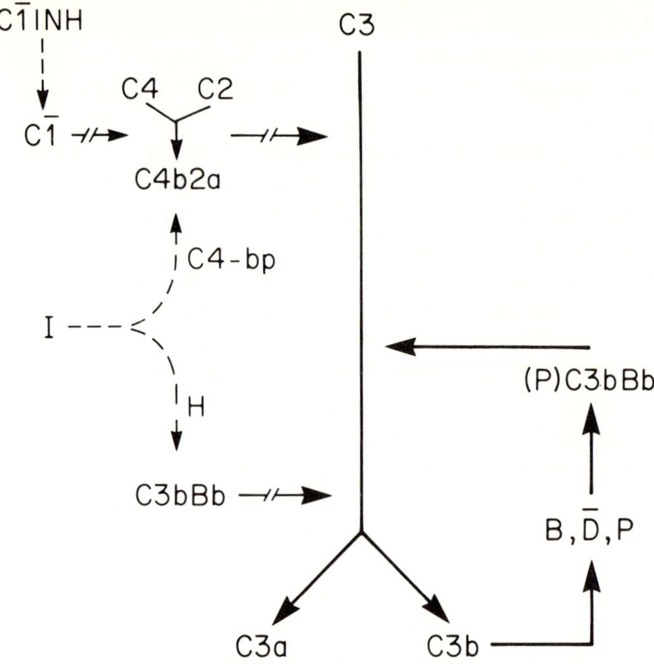

Fig. 2. *Representation of the regulatory proteins present in normal plasma and their primary site of action*

For the sake of clarity the complement receptor with regulatory function (CR1) and the membrane protein, decay accelerating factor (DAF), have not been included.

enveloped viruses, virus-infected cells [30] and susceptible bacteria [31], and thereby may be operative in host defense.

The protein C3 is pivotal in the organization and function of the complement system. It is the precursor of a number of physiological fragments which exert their biological functions by binding to cellular complement receptors or by interacting with other complement proteins. The molecule exposes during activation at least ten distinct binding sites, namely for CR1, CR2, CR3, C3a receptor, Factor B, properdin, Factor H, C5, conglutinin, and for attachment to membranes. The last named corresponds to the internal thioester. The structural analysis of C3 revealed a 29% identity in amino acid sequence between C3 and C4 [3]. Considerable homology also exists between C3 and C5 [4]. Therefore it is not surprising that fragments generated during activation of these proteins share functional activities.

The anaphylatoxins are highly potent mediators of inflammation derived from C3, C4, and C5 [32]. They are hormone-like messenger molecules as much as they bind to specific cell surface receptors with high affinity. The peptide–receptor complex is then internalized and the cell is functionally regulated. The cellular responses to the anaphylatoxins include release of histamine, serotinin, hydrolytic enzymes, platelet activating factor, interleukin-1, arachidonic acid metabolites and active oxygen species, chemotactic migration, cell adhesion, and smooth muscle contraction. The possibility was raised recently that C3a and C5a may be regulators of the immune response. Cells capable of responding to the stimuli imparted by the anaphylatoxins include polymorphs, monocytes, macrophages, mast cells, smooth muscle cells, and presumably certain subsets of T lymphocytes [32].

Cell surface receptors for complement have become increasingly important, particularly those with specificity for bound fragments of C3. The C3b receptor (CR1) [33] was the first to be isolated [34]. The C3d receptor, CR2 [35], has now been identified with the Epstein–Barr virus receptor on B lymphocytes [36,37] and the C3bi receptor, CR3, has been shown to be structurally related to the lymphocyte surface antigen LFA-1 [38]. By and large, C3 receptors aid in the clearance of potentially harmful substances from the circulation. Recent studies in primates suggest that, under normal conditions, infused immune complexes bind to erythrocyte CR1 and are delivered to the liver where they are processed by hepatic macrophages [39]. When the animals were depleted of complement, immune complexes could not bind to the CR1, and although they disappeared from the circulation they could not be found in either liver, spleen or gut. It was suggested that they were deposited in tissues prone to circulating immune complex (CIC)-mediated injury. It appears that, compared with erythrocytes, the role of the leukocytes in clearing CIC is limited [40]. Clinical support for this notion is provided by studies in patients with autoimmune diseases such as systemic lupus erythematosus, who show a significant decrease in the number of CR1 molecules on the surface of their erythrocytes [41–43]. More recently, we have shown that patients with certain infectious diseases produced by bacteria (lepromatous leprosy) or viruses (immunodeficiency syndrome, AIDS), in whom there is marked impairment of cellular immunity, also have a marked reduction in the number of CR1 per erythrocyte [13,44]. Furthermore, patients with AIDS-related complex, but not healthy homosexuals, also have a decreased number of CR1, suggesting that CR1 depletion could be an early prognostic indicator of disease activity.

The origin of this defect has been a matter of controversy. Wilson *et al.* [42] conducted studies which supported the concept of an inherited defect of CR1 expression in patients with systemic lupus erythematosus. Based on these studies and those on normal individuals they concluded that two codominant alleles determined the low levels of E-CR1 in their lupus patients, as well as in their relatives. They proposed that individual expression of CR1 falls into three genetically determined groups, high (HH), intermediate (HL) and low (LL). Other studies have shown that low CR1 in patients with systemic lupus erythematosus was influenced by factors related to the activity of the disease rather than genetic background. Our study reveals that family members of AIDS patients have a mean CR1 which does not differ from that of the normal propulation. Furthermore, family studies indicate that the low number of CR1 in AIDS patients could not be attributed to a codominant inherited pattern. We conclude that the defect in AIDS is probably acquired rather than inherited. It could be the result of either shedding of the receptors following their binding CIC or of micro-organisms bearing bound C3b or the result of inhibition of CR1 synthesis in nucleated erythrocyte precursors in the bone marrow.

In this introduction I have attempted to highlight a number of aspects in molecular biology, genetics, biochemistry, cell biology, immunology and medicine which are presently reached by research in complement. This, as well as the presentations to follow, clearly show that the complement system has long left the laboratory as a reagent in the performance of complement fixation tests. The contributions of this complex system to inflammation and multiple aspects of host defence can no longer be overlooked.

## References


1.  Bentley, D. R. & Porter, R. R. (1984) *Proc. Natl. Acad. Sci. U.S.A.* **81**, 1212–1215
2.  Campbell, D. R. & Porter, R. R. (1983) *Proc. Natl. Acad. Sci. U.S.A.* **80**, 4464–4468
3.  de Bruijin, M. H. L. & Fey, G. H. (1985) *Proc. Natl. Acad. Sci. U.S.A.* **82**, 708–712
4.  Lundwall, A. B., Wetsel, R. A., Kristensen, T. *et al.* (1985) *J. Biol. Chem.* **260**, 2108–2112
5.  Carroll, M. D. & Porter, R. R. (1983) *Proc. Natl. Acad. Sci. U.S.A.* **80**, 264–267
6.  Fearon, D. T. (1983) *Springer Semin. Immunopathol.* **6**, 159–173
7.  Gigli, I. & Nelson, R. A. (1968) *Exp. Cell Res.* **51**, 45–67
8.  Schreiber, R. D. (1984) *Springer Semin. Immunopathol.* **7**, 221–249
9.  Weigle, O. W., Goodman, M. G., Morgan, E. L. *et al.* (1983) *Springer Semin. Immunopathol.* **6**, 173–194
10. Gigli, I., Fujita, T. & Nussenweig, V. (1979) *Proc. Natl. Acad. Sci. U.S.A.* **76**, 6596–6600
11. Fearon, D. T. (1979) *Proc. Natl. Acad. Sci. U.S.A.* **76**, 5867–5871
12. Nicholson-Weller, A., Burge, J. & Austen, K. F. (1981) *J. Immunol.* **127**, 2035–2039
13. Zalman, L. S., Brothers, M. A., Chiu, F. J. *et al.* (1985) *Complement* **2**, 90(A)
14. Alper, C. A. & Rosen, F. S. (1984) *Springer Semin. Immunopathol.* **7**, 251–261
15. Agnello, V. (1978) *Medicine (Baltimore)* **57**, 1–23
16. Tausk, F., Hoffman, T., Schreiber, R. D. *et al.* (1985) *J. Invest. Dermatol.* **85**, 58s–61s
17. Gigli, I. & Austen, K. F. (1971) *Annu. Rev. Microbiol.* **25**, 309–332
18. Colten, H. R. (1977) in *Biological Amplification Systems in Immunology* (Day, N. K. & Good, R. A., eds.), pp. 47–67. Plenum Medical Book Co.
19. Hartung, H. P. & Hadding, U. (1983) *Springer Semin. Immunopathol.* **6**, 283–326
20. Muller-Eberhard, H. (1984) *Springer Semin. Immunopathol.* **7**, 93–141
21. Tack, B. F. (1983) *Springer Semin. Immunopathol.* **6**, 259–282
22. Ziccardi, R. J. (1983) *Springer Semin. Immunopathol.* **6**, 213–230
23. Ratnoff, W. D., Fearon, D. T. & Austen, K. F. (1983) *Springer Semin. Immunopathol.* **6**, 361–372
24. Pangburn, M. K. & Muller-Eberhard, H. J. (1984) *Springer Semin. Immunopathol.* **7**, 163–192
25. Fearon, D. T. & Austen, K. F. (1977) *J. Exp. Med.* **146**, 22–23
26. Colten, H. R. (1983) *Springer Semin. Immunopathol.* **6**, 149–158
27. Smith, C. A., Vogen, C. W. & Muller-Eberhard, H. J. (1983) *J. Exp. Med.* **159**, 324–329
28. Ross, G. D. & Medof, M. E. (1985) *Adv. Immunol.* **37**, 217–267
29. Fearon, D. T. & Austen, K. F. (1975) *J. Exp. Med.* **142**, 856–863
30. Cooper, N. R. & Nemerow, G. R. (1983) *Springer Semin. Immunopathol.* **6**, 327–348
31. Brown, E. J., Joiner, K. A. & Frank, M. M. (1983) *Springer Semin. Immunopathol.* **6**, 349–360
32. Hugli, T. E. (1984) *Springer Semin. Immunopathol.* **7**, 193–220
33. Nelson, R. A. (1956) *Proc. Soc. Exp. Biol. Med.* **49**, 55–61
34. Fearon, D. T. (1980) *J. Exp. Med.* **152**, 20–30
35. Weis, J. J., Tedder, T. F. & Fearon, D. T. (1984) *Proc. Natl. Acad. Sci. U.S.A.* **81**, 881–885
36. Nemerow, G. R. & Cooper, N. R. (1984) *Proc. Natl. Acad. Sci. U.S.A.* **81**, 4955–4959
37. Fingeroth, J. D., Weis, J. J., Tedder, T. F. *et al.* (1984) *Proc. Natl. Acad. Sci. U.S.A.* **81**, 4510–4515
38. Ross, G. D., Jarowski, C., Rabellino, E. M. *et al.* (1978) *J. Exp. Med.* **147**, 730–738
39. Cornacoff, J. F., Hebert, L. A. & Smead, W. L. (1983) *J. Clin. Invest.* **71**, 236–247
40. Waxman, L. A., Hebert, L. A. & Cornacoff, J. B. (1984) *J. Clin. Invest.* **74**, 1329–1340
41. Iida K., Mornaghi, R. & Nussenzweig, V. (1982) *J. Exp. Med.* **155**, 1427–1438
42. Wilson, J. C., Wong, W. W., Shur, P. H. *et al.* (1982) *N. Engl. J. Med.* **307**, 981–986
43. Ross, G. D., Walport, M. J., Parker, C. J. *et al.* (1984) *Arthritis Rheum.* **27**, S28
44. Tausk, F. A., Schreiber, R. D. & Gigli, I. (1984) *Trans. Assoc. Am. Physicians* **97**, 346–352

*Biochem. Soc. Symp.* **51**, 7–18
*Printed in Great Britain*

# C2 and Factor B: Structure and Genetics

DAVID R. BENTLEY AND R. DUNCAN CAMPBELL

*MRC Immunochemistry Unit, Department of Biochemistry, South Parks Road,
Oxford OX1 3QU, U.K.*

## Synopsis

Complement components C2 and factor B are novel types of serine protease
that are encoded by single loci in the major histocompatibility complex on
human chromosome 6. The two proteins share 39% homology, or 50% taking
into account conservative amino acid replacements. The catalytic chains, C2a
(509 residues) and Bb (505 residues) show homology in their $C$-terminal domains
to the catalytic polypeptides of other serine proteases. The non-catalytic chains,
C2b (223 residues) and Ba (234 residues) both contain three tandem repeats of
approx. 60 amino acids each, which are homologous to the repeats in C4b-binding
protein and factor H, and also the repeats in the non-complement protein
$\beta_2$-glycoprotein I. Molecular mapping and DNA sequence analysis has shown
that the factor B gene is 6 kb in length and contains 18 exons, while the C2 gene
is 18 kb in length; 425 bp separates the 3′ end of the C2 gene from the 5′ end
of the factor B gene. C2 and factor B are polymorphic and structural variants
have been detected at the protein level by differences in charge. The degree of
polymorphism at the factor B locus has been defined by DNA sequence analysis
of the two common alleles F and S. In addition restriction fragment length
polymorphisms have been detected in the C2 gene. These DNA polymorphisms
subdivide the common allelic variant of C2 (C2C) and reveal that there is much
greater variability at the C2 locus than that detected by protein typing.

## Introduction

C2 and factor B constitute a unique class of plasma serine protease that is
involved in activation of the complement system, the major effector mechanism
in humoral immunity. They share homology with the classical serine proteases,
but are unique in having catalytic chains with much-extended $N$-termini and a
mechanism of activation that is probably different from that of other serine
proteases [1–4]. The genes encoding C2 and factor B, and also the C4 genes, have
been mapped to the major histocompatibility complex in man (HLA) on the
short arm of chromosome 6 [5]. They lie in between the class I and class II loci,
and no crossovers have been observed between the complement genes [6]. C2
and factor B are synthesized primarily in the liver [7] and also in mononuclear
phagocytes [8]. The levels of expression of the C2 and factor B genes differ on
the basis of the observed levels of the two proteins in blood; C2 is present at

much lower concentrations (15 mg/l) [9] than factor B (150 mg/l) [10]. The two genes also appear to be under independent control, as the kinetics of C2 induction by $\gamma$-interferon in primary cultures of human monocytes are different from those of factor B [11].

Genetic deficiency of C2 is the most common complement deficiency in man [12–14], with homozygotes occurring at a frequency in 1 in 10000 [15]. Approx. 40% of these are associated with immune complex-related diseases [12,16]. C2 deficiency is an allele at the C2 structural locus [13], but no major gene deletion has been detected by Southern blot analysis [17]. The absence of detectable C2 mRNA in peripheral blood monocytes of C2-deficient individuals has suggested a pretranslational defect in C2 gene expression [17]. Total deficiency of factor B has not been observed, and may therefore be a lethal condition, but there are reports of heterozygous deficiency [18–20]. Structural polymorphism of both C2 and factor B have been detected, and the occurrence of particular variants may be associated with susceptibility to certain disease states [21].


**Structure of C2 and Factor B**

C2 and factor B circulate in the blood as single-chain polypeptides of $M_r$ 102000 and $M_r$ 90000 respectively. Activation of C2 yields the $N$-terminal non-catalytic polypeptide C2b ($M_r$ 30000) and the $C$-terminal catalytic chain C2a ($M_r$ 70000); factor B activation yields corresponding $N$-terminal (Ba, $M_r$ 30000) and $C$-terminal (Bb, $M_r$ 60000) chains. During activation of the classical pathway of complement, the zymogen form of C2 associates with activated C4 (C4b) bound to immune aggregates; this interaction is $Mg^{2+}$-dependent [22] and is believed to involve primarily the $N$-terminal domain of C2 (C2b) [23,24]. C2 is then activated by $\overline{C1}$ in a single cleavage event to form the classical pathway C3 convertase, $\overline{C4b2a}$; C2b is no longer essential for activity [24]. A similar series of events takes place in the alternative pathway: the zymogen form of factor B associates with bound C3b in a $Mg^{2+}$-dependent reaction [25] which is presumed to involve the $N$-terminal domain of factor B (Ba) [26,27]. Activation of factor B by factor D then results in the formation of the alternative pathway C3 convertase, $\overline{C3bBb}$. The activities of both the classical and alternative pathway C3 convertases are subsequently modified by binding of additional C3b to become C5 convertases [28,29] which initiate activation of the late components C5–C9 leading to the lysis of cellular antigens [30].

Inactivation of both C3 convertases is rapid. Dissociation of C2a from C4b is accelerated by C4b binding protein (C4BP), which competes with C2a for the binding sites on C4b [31]. C4BP also acts as a cofactor to render C4b susceptible to cleavage by the serine protease factor I. [32,33]. Similarly, the alternative pathway C3 convertase is rapidly inactivated by dissociation of Bb from the complex; factor H has an analogous role to C4BP [31] both in accelerating this dissociation, and in acting as an essential cofactor for cleavage of C3b by factor I [34,35]

Structural analysis of human C2 and factor B was initially carried out by purification of the proteins from serum and peptide sequence analysis. Using this approach, the complete sequences of Bb [36,37] and more recently of Ba

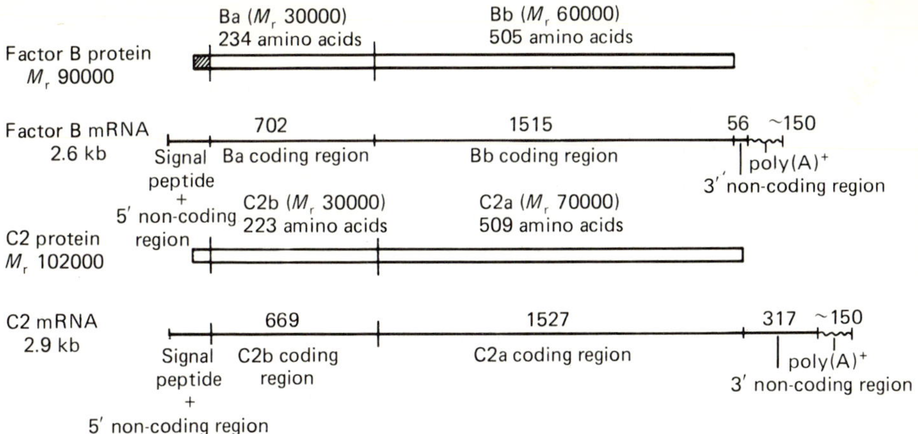

Fig. 1. *Organization of C2 and factor B mRNA and the encoded polypeptides*

The proteins are divided into three domains by vertical lines. The domains correspond to the signal peptide, the *N*-terminal chain, and the *C*-terminal catalytic chain.

[38] were obtained. The lower levels of C2 present in the serum, and the susceptibility of C2 to proteolysis during isolation [1,39], however, made it possible to obtain only partial amino acid sequence data for C2 [3]. The available data for C2 and also factor B were therefore used in the design of mixed synthetic oligonucleotides for use as hybridization probes to obtain cDNA clones for factor B [40,41] and C2 [42,43]. Complete cDNA sequences of factor B [44,45] and C2 [46] have now been obtained. The lengths of the C2 and factor B mRNAs were determined as 2.9 kb and 2.6 kb respectively by Northern blot analysis [42,44]. It has also been suggested in an independent analysis that there may be additional minor C2-specific RNA species [47]. The organization of the mRNAs and the encoded polypeptide chains of C2 and factor B is shown in Fig. 1; the extra length of the C2 mRNA is accounted for mostly in the 3′ non-coding region, which is 317 nucleotides in C2 compared with 56 nucleotides in factor B. The proteins are very similar in length, and the homology between them, which is approx. 39% (identical amino acids), or 50% including conservative amino acid replacements, extends over the entire polypeptide chain. C2 contains 15.9% carbohydrate, factor B contains 8.6% [48]. Most of the extra carbohydrate found in C2 is in C2a, based on the differences in $M_r$ values of the fragments of C2 and factor B (see above and Fig. 1), and on the observation that C2a contains six potential sites for asparagine-linked oligosaccharides, compared with only two in Bb [46]. The sequences of C2b and Ba indicate the presence of two potential glycosylation sites in each polypeptide [38,44,46].

The *N*-terminal chains Ba (234 amino acids) and C2b (223 amino acids) both contain the structural feature of a triple direct repeat of approx. 60 amino acids. In Ba, the homology between the three repeats is 27% (I versus II), 29% (I versus III) and 47% (II versus III) [44] (Fig. 2 shows their organization). The homology between the repeat units suggests that they arose by repeated duplication of an ancestral DNA segment; this postulate is supported by the gene structure encoding Ba, which shows that each repeat is exactly encoded in a separate exon

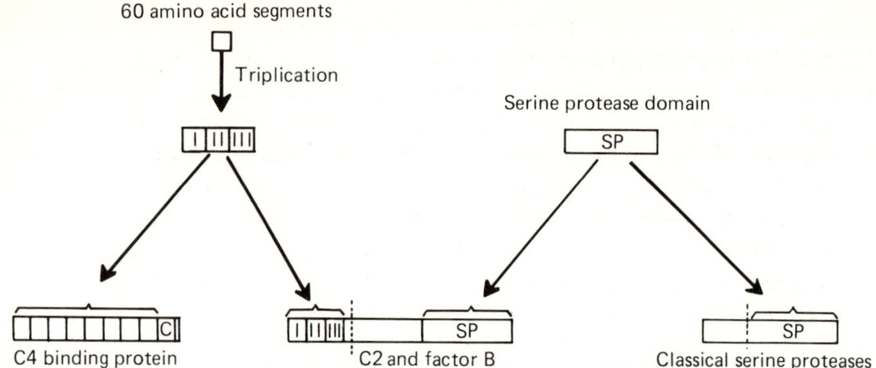


Fig. 2. *Structural and evolutionary relationships between the novel class of serine proteases C2 and factor B, the classical serine proteases, and the regulatory complement protein C4BP*

Factor H is also a member of the C4BP family, as it is composed almost entirely of 60 amino acid repeats [50,51]. It is proposed that C2 and factor B have evolved by fusion of gene segments derived during the evolution of the two different families of proteins.

and therefore defines the limits of the evolutionary segment at the DNA level [45]. The 60 amino acid repeat has also been discovered in the primary structures of the two regulatory proteins C4BP [49] and factor H [50,51], and also in the non-complement serum protein $\beta_2$-glycoprotein I [52]. C4BP has a polypeptide chain of 549 amino acids, composed of eight tandem repeats of approx. 60 amino acids each, plus an unrelated *C*-terminal domain of 58 residues (see Fig. 2) [49]. The homology between C4BP and C2b can be correlated with their common function in binding to C4b; this suggests a role for the 60 amino acid repeats in acting as a C4b-binding domain. By analogy, the corresponding relationship between factor H and Ba would suggest that the 60 amino acid repeats in these two polypeptides have a common role in binding C3b.

$\beta_2$-Glycoprotein I contains five 60 amino acid repeats [52]; the function of this protein is not known, and no interaction with complement components has been reported. The occurrence of the 60 amino acid repeat in this apparently functionally unrelated protein might therefore indicate that this sequence has been conserved as a structural domain in the evolution of many serum proteins.

The *C*-terminal halves of C2a and Bb share homology with the catalytic domains of other serine proteases (see Fig. 2). The majority of half-cystine residues are conserved in position [37,46], indicating that this part of C2a and Bb probably has a similar three-dimensional structure to that already determined for bovine trypsin [53,54] and other serine protease catalytic chains. C2 and factor B also show a trypsin-like specificity in that both cleave specifically after an arginine residue, and this suggests that the configuration of C2 and factor B should closely resemble trypsin in the region of the active site. This is confirmed by the sequences of C2 and factor B [37,42,46], which are highly conserved in the vicinity of the activity site serine and histidine residues, and also around the secondary substrate binding pocket (Ser-Trp-Gly 214–216 in trypsin is also found in both C2 and factor B). There is no direct correspondent to the Asp-189 of trypsin, which lies at the bottom of the substrate binding pocket and interacts with the positively charged arginyl and lysinyl side chains [53,54]. The nearest aspartic acid conserved in C2 and factor B corresponds to position 187 in trypsin [37,46]. The configurations of C2 and factor B might therefore differ from that of trypsin to accommodate the fact that the aspartic acid is two residues further away from the active site serine than in trypsin. Another alternative is that the ionic interaction required for trypsin activity is no longer part of the requirement for substrate binding in C2 and factor B.

In contrast to trypsin, C2 and factor B display absolute specificity in recognition of the correct substrate: they cleave only C3 and C5, and only at specific sites. Regions responsible for this substrate specificity may be the extra loops in the catalytic domains of C2 and factor B that are not present in trypsin: these include residues 594–627 which are encoded by a separate exon in factor B with no counterpart in other serine proteases [55], and also residues 712–721. Alternatively, the *N*-terminal regions of the catalytic chains of C2 and factor B, which are not homologous to any other protein domains so far analysed, may be important in this role.

The common evolutionary origin of the serine proteases of blood coagulation, fibrinolysis, and also other systems is now well established (see [56,57] for reviews), and recent studies have confirmed that the homology extends to the level of gene structure ([55] and references therein). The novel serine proteases of the complement system conform to this evolutionary model; but they are also related to another family of proteins, defined by the presence of the 60 amino acid repeat structure. As summarized in Fig. 2, C2 and factor B are therefore examples of the fusion of at least two gene segments of distinct evolutionary origin to form a novel class of gene product.

### Genetics of C2 and Factor B

The genes for C2 and factor B have been aligned in the class III HLA region along with the two C4 genes, C4A and C4B [58] and two cytochrome *P*-450-21-hydroxylase coding sequences, 21-OHase A and 21-OHase B [59, 60] (see Fig. 3). The orientation of the class III genes with respect to the class I and II loci is unknown, but it has been suggested that the C4 genes are closer to the HLA-B locus than the factor B gene [61]. Detailed mapping and DNA sequence analysis has shown that the C2 gene is 18 kb in length, in contrast to the adjacent factor B gene of 6 kb [62]. The lengths of the encoded mRNAs are similar (C2, 2.9 kb; factor B, 2.6 kb) and the difference in size of the two loci must therefore occur in length of non-coding (i.e. intron) sequences. The evolution of the class III loci therefore appears to have involved tandem duplication of the C4, 21-OHase coding region, and also a duplication to generate C2 and factor B followed by marked divergence of the intron lengths between C2 and factor B.

Polymorphism of C2 and factor B has been shown at the protein level. In the case of C2, isoelectric focusing followed by haemolytic overlay detected three variants: a common variant C2C (gene frequency 97%) and two rare variants, the acidic C2A (gene frequency < 1%) and the basic C2B (gene frequency 2%) [63–65]. Factor B polymorphism detected by agarose gel electrophoresis at alkaline pH followed by immunofixation showed two common variants F (gene

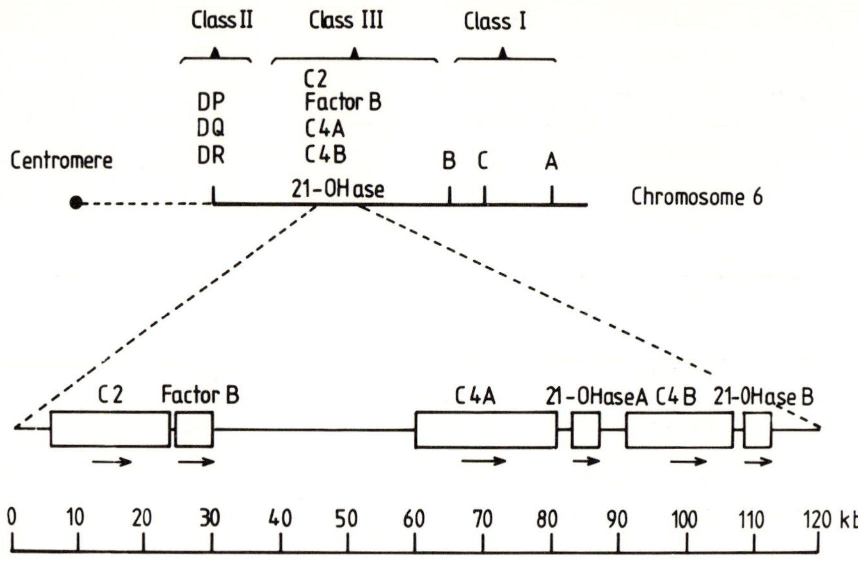


Fig. 3. *Map of the major histocompatibility complex in man (HLA) on the short arm of chromosome 6*

The class III region is expanded to show the alignment of the complement genes C2, factor B, C4A, C4B and also the *P*-450 21-hydroxylase coding sequences, 21-OHase A and 21-OHase B.

frequency 28%) and S (gene frequency 70%), and many rare variants [18,66], although the gene frequencies vary in different ethnic groups [67]. More recently isoelectric focusing has been used to resolve the different forms of factor B [68–70]. This method enabled subdivision of the F allele into two classes $F^a$ (gene frequency 11% in the total population) and $F^b$ (gene frequency 17%) [71].

The availability of DNA probes has allowed the study of C2 and factor B polymorphism to be extended to the level of the gene. Four restriction fragment length polymorphisms (RFLPs) have been detected so far in this region of the genome [62,72,73] (see Fig. 4). The *Taq*I polymorphism corresponds to the presence or absence of a single *Taq*I site in the 3′ region of the C2 gene. The two allelic forms show Mendelian inheritance and the absence of the *Taq*I site (in 11% of the haplotypes in the population) was always associated with the factor B F allele [73]. More recently, it has been shown that this fraction of the F alleles corresponds to the $F^b$ subclass identified by isoelectric focusing [74].

The *Bam*H1 polymorphism [62,72] corresponds to the presence or absence of a single *Bam*H1 restriction site in the middle of the C2 gene; the *Bam*H1 and *Taq*I polymorphisms show strong linkage disequilibrium, which is to be expected as the two sites are only approx. 3 kb apart [62], assuming that the two polymorphisms arose at similar times.

The *Sst*I polymorphism is multiallelic and maps to the 5′ region of the C2 gene. Four patterns have been identified in Southern blots using a 300 bp C2 genomic probe which hybridized to a single *Sst*I restriction fragment of 2.7, 2.65, 2.6 or 2.4 kb [62]. Length variation was also observed in Southern blots using other restriction enzymes [72] (D. R. Bentley, R. D. Campbell & S. J. Cross, unpub-

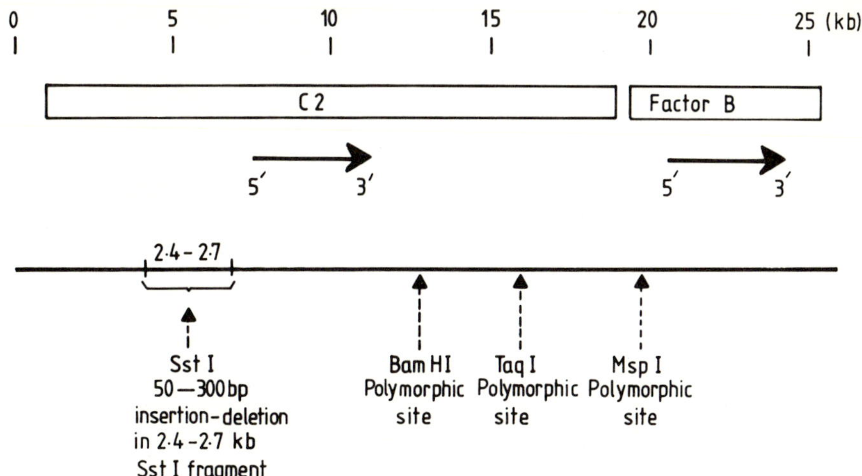



Fig. 4. *Map of the region encoding C2 and factor B in the HLA class III region, showing the location of the four restriction fragment length polymorphisms discovered to date*

In the case of the *Sst*I polymorphism, the boundaries of the variable *Sst*I fragment are shown (size range 2.4–2.7 kb), as the precise location of the proposed 50–300 bp deletion/insertion has not been established.

lished work) indicating that the *Sst*I RFLP is probably the result of insertion and/or deletion of short sequences in this region of the genome, and not caused by the presence or absence of specific restriction sites. Strong allelic association has so far been observed between the *Sst*I polymorphism and the *Bam*H1 and *Taq*I RFLPs; haplotypes containing the polymorphic *Bam*H1 and *Taq*I sites were associated with one of the shorter *Sst*I fragments, 2.65, 2.6 or 2.4 kb. The *Sst*I polymorphism is particularly useful as a genetic marker in this region as it subdivides previously indistinguishable haplotypes carrying the C2C and factor B F alleles. This haplotype occurs in approx. 28% of the population; on the basis of the *Sst*I polymorphism, it is subdivided into four classes. The haplotype C2C, Bf:F, *Sst*I:2.7 kb has a frequency of 0.16 in the total population; the haplotype C2:C, Bf:F, *Sst*I:2.4 kb has a frequency of 0.09, and two rare haplotypes C2:C, Bf:F, *Sst*I:2.65 kb and C2:C, Bf:F, *Sst*I:2.6 kb together have a frequency of about 0.03 [62].

The multiallelic polymorphism observed in the *Sst*I analysis of the C2 gene has also been observed for example in the human ζ-globin genes [75–77] and the insulin gene [78]. The resolution of restriction fragments in the range 2.4–2.7 kb in the *Sst*I polymorphism permitted the definition of four distinct size classes, but does not exclude the possibility of finding further variation at this position if Southern blot analysis is carried out using finer resolution techniques. This was possible in the analysis of the human ζ-globin gene polymorphism, in which high resolution of short restriction fragments (in the range 1.3–2.0 kb) demonstrated the presence of at least ten different variants [77]. The discovery of a highly polymorphic marker in the HLA class III region is significant in that it will be very useful in genetic characterization of the human genome, particularly in relation to susceptibility to diseases associated with the HLA.

Initial work on the DNA polymorphism of the factor B locus did not yield any differences between different individuals [72,73]. The difference in mobility between the fast (F) and slow (S) forms of the protein on agarose gels can be accounted for by a single charge difference which has previously been proposed to arise in the N-terminal peptide Ba [66]. A direct approach was therefore taken to identify the structural basis of the polymorphism between the F and S forms of factor B. Cosmid clones of genomic DNA from individuals carrying either the Bf:S, TaqI: + (TaqI: + denotes the presence of the polymorphic TaqI site) or the Bf:F, TaqI: − alleles were isolated and the nucleotide sequence of the regions encoding Ba in each case was determined. The results identified two nucleotide differences in coding sequences between the two alleles. One was a silent change; the other was a G→A transition in the second nucleotide of the codon for amino acid 7 of the mature polypeptide (see Fig. 5b). This would result in an amino acid change from arginine in the S allele to glutamine in the F allele (Fig. 5b), which could account for the observed mobility difference. It is necessary to determine the coding sequence of Bb in both these alleles to identify any additional differences which could affect electrophoretic mobility of either allele in agarose gels.

The nucleotide difference between the sequences of the F and S alleles determines two altered restriction sites. In the S allele, the sequence 5′-CCGGCC-3′ contains an MspI site (5′-CCGG-3′; underlined in Fig. 5b) and also a HaeIII site (5′-GGCC-3′). In the corresponding sequence of the F allele, 5′-CCAGCC-3′, neither restriction site is present. MspI was therefore used in Southern blot analysis to verify that the polymorphism defined by sequence analysis was present in the population, and also to extend the analysis to other individuals typed for factor B. The MspI restriction map of this part of the factor B gene is shown in Fig. 5(a), and the results of the Southern blot analysis of nine individuals, using a 400 bp DdeI genomic fragment as a 5′ factor B probe are shown in Fig. 6. In the FF homozygous individuals (lanes 2, 3, 5 and 9 in Fig. 6), the only band observed is 706 bp. This indicates the absence of the polymorphic MspI site, and the band corresponds to the sum of the 396 and 310 bp fragments shown in the map in Fig. 5(a). In the SS homozygotes (lanes 1 and 8 in Fig. 6), and in the $F_1S$ and $SS_1$ individuals (lanes 4 and 6 respectively) the bands oberved correspond to the 396 and 310 bp fragments, and the polymorphic MspI site is therefore present. The FS heterozygote in lane 7 shows all three bands; this is interpreted as being heterozygous for the MspI polymorphism, the 706 bp fragment being associated with the F allele.

The RFLP defined using MspI and the factor B probe is informative for F and S alleles. The polymorphic MspI site is absent from all the F alleles, as both Bf:F, TaqI: + and Bf:F, TaqI: −haplotypes were present in the above analysis (see legend to Fig. 6 for details). The MspI site is present in all other factor B alleles so far analysed, which included examples of S, $S_1$ and $F_1$ types.

To date, the identification of polymorphic forms of C2 and factor B proteins relies on the ability to separate the different allelic products on the basis of mobility in agarose gel electrophoresis or isoelectric focusing. It remains possible that novel variants without altered net charge are also present, which would not show altered mobility. It is not possible at present to test for functional

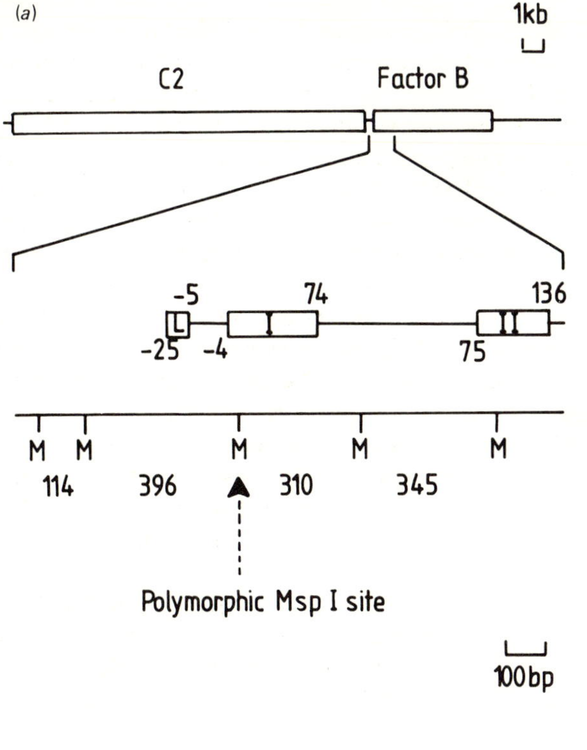


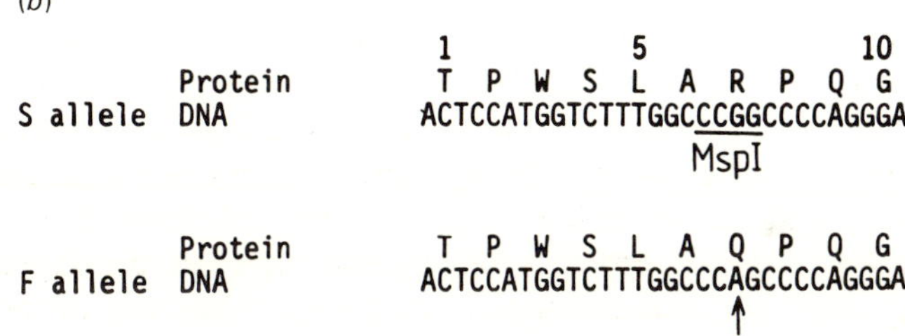


Fig. 5. *The MspI polymorphism.*

(a) Map of the region encoding C2 and factor B, with a small section corresponding to the 5′ end of the factor B gene shown expanded. L, leader coding sequence [45]; I and II, exons I and II, encoding the first and second 60 amino acid repeats, respectively [45]. The numbers below each box refer to the first amino acid encoded in each box, and the numbers above refer to the last amino acid encoded by each box; the first amino acid of the mature zymogen form of factor B = 1. The map of the *Msp*I sites in this region is also shown, with lengths of restriction fragments (114 bp, 396 bp, 310 bp, 345 bp), and the position of the polymorphic *Msp*I site is arrowed. Loss of this site in the F allele yields a 706 bp *Msp*I fragment (b) Sequence of the Bf:S, *Taq*I+ and Bf:F, *Taq*I:– alleles, showing the region encoding the first ten amino acids of Ba in each case. The *Msp*I site is underlined in the sequence of the S allele and the single base change is arrowed in the sequence of the F allele.

differences between either the known different structural forms of C2, or possible new protein variants which might be defined by the DNA polymorphisms described above. No functional difference has been observed between the F and S forms of factor B in plasma [79], although there may be differences in serum

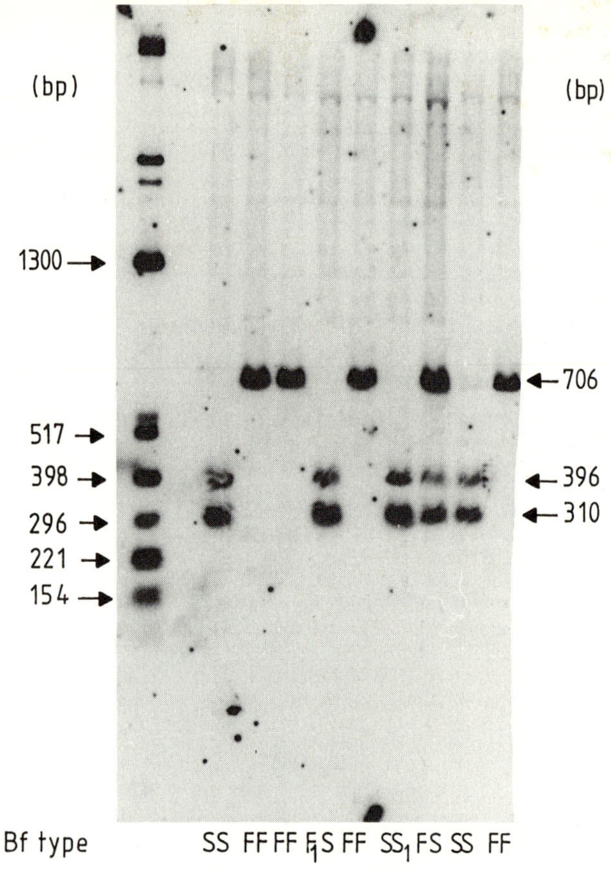



Fig. 6. *Southern blot analysis of factor B typed individuals with the restriction endonuclease MspI and a 400 bp factor B genomic DdeI fragment as a probe*

The Bf types are shown below each lane. Examples of the F allele included both *Taq*I: + and *Taq*I: − genotypes, as follows. Lane 2 is second from the left and is *Taq*I: + /*Taq*I: + ; lane 3, *Taq*I: − /*Taq*I: − ; lane 5, *Taq*I: + /*Taq*I: − ; lane 9, *Taq*I: + /*Taq*I: −. The FS heterozygote in lane 7 carries a Bf: F, *Taq*I: − allele. The size markers are shown on the left. All F alleles lack the polymorphic *Msp*I site, as demonstrated by the appearance of a 706 bp band in all individuals with a Bf: F allele.

concentration of the different forms in different individuals [79,80]. The identification of functionally altered forms of C2 and factor B would provide a basis for possible causative associations between these loci and certain disease states. Alternatively the DNA polymorphisms identified in this region will be of value in distinguishing chromosomes which otherwise appear identical, and in defining possible linked causative disease factors which are not yet identified.

# References


1. Reid, K. B. M. & Porter, R. R. (1981) *Annu. Rev. Biochem.* **50**, 433–464
2. Christie, D. L., Gagnon, J. & Porter, R. R. (1980) *Proc. Natl. Acad. Sci. U.S.A.* **77**, 4923–4927
3. Gagnon, J. (1984) *Philos. Trans. R. Soc. London Ser. B* **306**, 301–309
4. Gigli, I. (1986) *Biochem. Soc. Symp.* **51**, 1–6
5. Weitkamp, L. R. & Lamm, L. U. (1982) *Cytogenet. Cell. Genet.* **32**, 130–143

 6. Alper, C. A., Raum, D., Karp, S., Awdeh, Z. L. & Yanis, E. T. (1983) *Vox Sang.* **45**, 62–67
 7. Morris, K. M., Aden, D. P., Knowles, B. B. & Colten, H. R. (1982) *J. Clin. Invest.* **70**, 906–13
 8. Colton, H. R. (1982) *Mol. Immunol.* **19**, 1279–1285
 9. Porter, R. R. & Reid, K. B. M. (1979) *Adv. Protein Chem.* **33**, 1–71
10. Curman, B., Sanberg-Trägardh, L. & Pederson, P. A. (1977) *Biochemistry* **16**, 5368–5375
11. Colton, H. R. & Dowton, S. B. (1986) *Biochem. Soc. Symp.* **51**, 37–46
12. Glass, D., Raum, D., Gibson, D., Stillman, J. S. & Schur, P. H. (1976) *J. Clin. Invest.* **58**, 853–861
13. Parisier, K. M., Raum, D., Berkman, E. M., Alper, C. A. & Agnello, V. (1978) *J. Immunol.* **121**, 2580–2581
14. Mortensen, J. P., Buskjaer, L. & Lamm, L. U. (1980) *Immunology* **39**, 541–549
15. Agnello, V. (1978) *Medicine* **57**, 1–23
16. Rynes, R. I. (1982) *Clin. Rheum. Dis.* **8**, 29–47
17. Cole, F. S., Whitehead, A. S., Auerbach, H. S., Lint, T., Zeity, H. J., Kilbridge, P. & Colten, H. R. (1985) *New Engl. J. Med.* **313**, 11–16
18. Mauff, G., Hauptmann, G., Hitzeroth, H. W., Gauchel, F. Z. & Scherz, R. (1978) *Z. Immun. Forsch.* **154**, 115–120
19. Suciu-Foca, N., O'Neill, G. & Rubenstein, P. (1980) in *Histocompatibility Testing* (Terasaki, P. I., ed.) p. 935, UCLA, Los Angeles
20. Tokunaga, K., Omoto, K., Yukiyama, Y., Sakurai, M., Saji, H. & Maruya, E. (1984) *Hum. Genet.* **67**, 449
21. Rittner, C. & Bertrams, J. (1981) *Hum. Genet.* **56**, 235–247
22. Sitomer, G., Stroud, R. M. & Mayer, M. M. (1966) *Immunochemistry* **3**, 57–69
23. Nagasawa, S. & Stroud, R. M. (1977) *Proc. Natl. Acad. Sci. U.S.A.* **74**, 2998–3001
24. Kerr, M. A. (1980) *Biochem. J.* **189**, 173–181
25. Vogt, W., Dames, W., Schmidt, G. & Dieminger, L. (1977) *Immunochemistry* **14**, 201–205
26. Götze, O. & Müller-Eberhard, H. J. (1971) *J. Exp. Med.* **134**, 90s–108s
27. Hunsicker, L. G., Ruddy, S. & Austen, K. F. (1973) *J. Immunol.* **110**, 128–138
28. Vogt, W., Schmidt, G., von Buttlar, B. & Dieminger, L. (1978) *Immunology* **34**, 29–40
29. Medicus, R. G., Götze, O. & Müller-Eberhard, H. J. (1976) *Scand. J. Immunol.* **5**, 1049–1055
30. Lachmann, P. J. & Thompson, R. A. (1970) *J. Exp. Med.* **131**, 643–651
31. Gigli, I., Fujita, T. & Nussenzweig, V. (1979) *Proc. Natl. Acad. Sci. U.S.A.* **76**, 6596–6600
32. Fujita, T., Gigli, I. & Nussenzweig, V. (1978) *J. Exp. Med.* **148**, 1044–1051
33. Fujita, T. & Nussenzweig, V. (1979) *J. Exp. Med.* **150**, 267–276
34. Whaley, K. & Ruddy, S. (1976) *Science* **193**, 1011–1013
35. Weiler, J. M., Daha, M. R., Austen, F. & Fearon, D. T. (1976) *Proc. Natl. Acad. Sci. U.S.A.* **73**, 3268–3272
36. Gagnon, J. & Christie, D. L. (1983) *Biochem. J.* **209**, 51–60
37. Christie, D. L. & Gagnon, J. (1983) *Biochem. J.* **209**, 61–70
38. Mole, J. E., Anderson, J. K., Davison, E. A. & Woods, D. E. (1984) *J. Biol. Chem.* **259**, 3407–3412
39. Kerr, M. A. (1979) *Biochem. J.* **183**, 615–622
40. Campbell, R. D. & Porter, R. R. (1983) *Proc. Natl. Acad. Sci. U.S.A.* **80**, 4464–4468
41. Woods, D. E., Markham, A. F., Ricker, A. T., Goldberger, G. & Colten, H. R. (1982) *Proc. Natl. Acad. Sci. U.S.A.* **79**, 5661–5665
42. Bentley, D. R. & Porter, R. R. (1984) *Proc. Natl. Acad. Sci. U.S.A.* **81**, 1212–1215
43. Woods, D. E., Edge, M. D. & Colten, H. R. (1984) *J. Clin. Invest.* **74**, 634–638
44. Morley, B. J. & Campbell, R. D. (1984) *EMBO. J.* **3**, 153–157
45. Campbell, R. D., Bentley, D. R. & Morley, B. J. (1984) *Philos. Trans. R. Soc. London Ser. B* **306**, 367–378
46. Bentley, D. R. (1986) *Biochem. J.*, submitted
47. Perlmutter, D. M., Colten, H. R., Grossberger, D., Strominger, J., Seidman, J. G. & Chaplin, D. H. (1985) *J. Clin. Invest.*, in the press
48. Tomana, M., Niemann, M., Garner, C. & Volanakis, J. E. (1985) *Mol. Immunol.* **22**, 107–111
49. Chung, L. P., Bentley, D. R. & Reid, K. B. M. (1985) *Biochem. J.* **230**, 133–141
50. Kristensen, T., Wetsel, R. A. & Tack, B. F. (1985) *Fed. Proc. Fed. Am. Soc. Exp. Biol.* **44**, 1531 (abstr)
51. Sim, R. B., Malhotra, V., Ripoche, J., Day, A. J., Micklem, K. J. & Sim, E. (1986) *Biochem. Soc. Symp.* **51**, 83–96
52. Lozier, J., Takahashi, N. & Putnam, F. W. (1984) *Biochemistry* **81**, 3640–3644
53. Stroud, R. M., Kay, L. M. & Dickerson, R. E. (1974) *J. Mol. Biol.* **83**, 185–208
54. Hüber, R., Kukla, D., Bode, W., Schwager, P., Bartels, K., Deisenhofer, J. & Steigemann, W. (1974) *J. Mol. Biol.* **89**, 73–101
55. Campbell, R. D. & Bentley, D. R. (1985) *Immunol. Rev.* **87**, 19–37

56. Patthy, L. (1985) *Cell* **41**, 657–663
57. Young, C. L., Barker, W. C., Tomaselli, C. M. & Dayhoff, M. O. (1978) in *Atlas of Protein Sequence and Structure* (Dayhoff, M. O., ed.), pp. 73–93, National Biomedical Research Foundation, Silver Spring, MD
58. Carroll, M. C., Campbell, R. D., Bentley, D. R. & Porter, R. R. (1984) *Nature (London)* **307**, 237–241
59. Carroll, M. C., Campbell, R. D. & Porter, R. R. (1985) *Proc. Natl. Acad. Sci. U.S.A.* **82**, 521–525
60. White, P. C., Grossberger, D., Onufer, B. J., Chaplin, D. D., New, M. I., Dupont, B. & Strominger, J. L. (1985) *Proc. Natl. Acad. Sci. U.S.A.* **82**, 1089–1093
61. Olaisen, B., Teisberg, P., Jonassen, R., Thorsby, E. & Gedde-Dahl, T. (1983) *Ann. Hum. Genet.* **47**, 285–292
62. Bentley, D. R., Campbell, R. D. & Cross, S. J. (1985) *Immunogenetics* **22**, 377–390
63. Alper, C. A. (1976) *J. Exp. Med.* **144**, 1111–1115
64. Meo, T., Atkinson, J., Bernoco, M., Bernoco, D. & Ceppellini, R. (1976) *Eur. J. Immunol.* **6**, 916–919
65. Meo, T., Atkinson, J., Bernoco, M., Bernoco, D. & Ceppellini, R. (1977) *Proc. Natl. Acad. Sci. U.S.A.* **74**, 1672–1675
66. Alper, C. A., Boenisch, T. & Watson, L. (1972) *J. Exp. Med.* **135**, 68–80
67. Alper, C. A. (1981) in *The Role of the Major Histocompatibility Complex in Immunobiology* (Dorf, M. E., ed.), pp. 173–220, Garland, New York
68. Teng, Y. S. & Tan, S. G. (1982) *Hum. Hered.* **32**, 362
69. Geserick, G., Palzelt, D., Schröder, H. & Nagai, T. (1983) *Vox Sang* **44**, 178–182
70. David, V., Fauchet, F., Phengsavath, H., Guenet, L. & LeGall, J.Y. (1983) *Hum. Genet.* **64**, 189–190
71. Abbal, M., Thomsen, M., Cambon-Thomsen, A., Archambeau, J., Calot, M. & Fathallah, D. (1985) *Hum. Genet.* **69**, 181
72. Woods, D. E., Edge, M. D. & Colten, M. R. (1984) *J. Clin. Invest.* **14**, 634–638
73. Cross, S. J., Edwards, J. H., Bentley, D. R. & Campbell, R. D. (1985) *Immunogenetics* **21**, 39–48
74. Fathallah, D., Abbal, M., Thomsen, M., Cambon-Thomsen, A. & Campbell, R. D. (1985) *J. Immunogenetics*, in the press
75. Higgs, D. R., Goodbourn, S. E. Y., Wainscoat, J. S., Clegg, J. B. & Weatherall, D. J. (1981) *Nucleic Acids Res.* **9**, 4213–4224
76. Proudfoot, N. J., Gil, A. & Maniatis, T. (1982) *Cell* **31**, 553–563
77. Goodbourn, S. E. Y., Higgs, D. R., Clegg, J. B. & Weatherall, D. J. (1985) *Mol. Biol. Med.* **2**, 223–238
78. Bell, G. I., Karam, J. H. & Rutter, W. J. (1982) *Proc. Natl. Acad. Sci. U.S.A.* **78**, 5759–5763
79. Mauff, G., Adam, R., Wachauf, B., Hitzeroth, H. W. & Hiller, C. (1980) *Immunobiology* **158**, 86–90
80. Mortensen, J. P. & Lamm, L. U. (1981) *Immunology* **42**, 505–511

*Biochem. Soc. Symp.* **51**, 19–28
*Printed in Great Britain*

# Complement Genes of the Major Histocompatibility Complex (Complotypes), Extended Haplotypes and Disease Markers

CHESTER A. ALPER, ZUHEIR AWDEH, DONALD RAUM and EDMOND J. YUNIS

*The Center for Blood Research, 800 Huntington Avenue, Boston, MA 02115, U.S.A.*

## Synopsis

The human major histocompatibility complex (MHC)-linked genes *C2,BF,C4A,C4B* occur in populations and segregate in families as single genetic units or complotypes. Analysis for significant three-point linkage disequilibrium between HLA-B, DR and complotype on normal caucasian chromosomes 6p yields about a dozen haplotypes that account for most of the known HLA-B/HLA-DR linkage disequilibrium pairs previously noted in normal caucasian populations. We refer to the HLA-B/DR/complotype sets with significant linkage disequilibrium as extended haplotypes since they often show limited variation at other MHC-linked loci. From the study of MHC haplotypes in 21-hydroxylase deficiency, C2 deficiency and type 1 diabetes, it is becoming apparent that it is extended haplotypes rather than their individual alleles that are markers for these MHC-associated diseases.

## Introduction

As is true of other genes of the major histocompatibility complex (MHC), alleles of the loci controlling synthesis of C2, factor B and C4 have different frequencies in populations of patients with some diseases compared with control populations. Those alleles with increased frequencies in patients are said to be 'associated' with the disease, those with decreased frequencies are considered 'protective'. Because of sometimes striking differences in frequency of MHC alleles among different ethnic groups and subgroups, it is essential that patient and control genes be ethnically as closely matched as possible to rule out a trivial explanation for association. For example, if one studies patients with $\beta$-thalassaemia in a large metropolitan population in Northern Europe or North America, one might expect to find an increase in patients of MHC alleles increased in individuals of Greek and Italian origin (in whom the disease is most common) compared with the general population. If the proportion of the latter individuals is low in the overall population but high in the patient population, this would lead to an association of little meaning. For this and other reasons, we have chosen to analyse MHC alleles in families of patients as well as in patients. Because people have tended to marry within ethnic subgroups, all the chromosomes within any given family tend to be ethnically matched. Thus, one can define two sets of ethnically matched alleles, one occurring in patients and therefore 'diseased' and one occurring *only* in healthy family members and hence 'control' (Raum *et al.*, 1984). Furthermore, because family members have been analysed, one can assign the alleles in any individual member to one chromosome 6 or the other. In general, in any family, the MHC alleles on each chromosome will be inherited as a bloc or haplotype. In this fashion, ethnically matched sets of disease and control haplotypes (or chromosomes) are distinguishable.

One explanation for association at the population level is linkage disequilibrium at the chromosomal level between the associated MHC allele and a closely linked disease susceptibility gene. Linkage disequilibrium is defined as the occurrence together on the same chromosome of alleles at linked loci at a frequency significantly different from the product of the frequencies of the individual alleles. To analyse linkage disequilibrium for more than two loci, it is essential to correct for lower order two-point linkage disequilibria (Piazza, 1975).

## Linkage Disequilibrium of Major Histocompatibility Complex (MHC) Alleles on Normal Chromosomes

Before analysing some MHC allele–disease associations, it is instructive first to examine linkage disequilibrium between MHC alleles on normal chromosomes from caucasians. Our own experience is with Bostonian caucasians of mixed subethnicity and may therefore differ significantly from that in the more homogeneous caucasian populations of Northern or Southern Europe, although those, too, are changing. Earlier studies have emphasized pairwise linkage disequilibrium, first between HLA-A and B alleles and later between HLA-B and DR alleles (Bodmer & Bodmer, 1978). Prior studies have found weak linkage disequilibria between individual complement alleles and HLA alleles considered separately or even as trios (Albert *et al.*, 1977).

In our view, the four complement genes form a single genetic unit or complotype in families and at the population level (Alper *et al.*, 1983). Evidence for this concept derives from the analysis of many thousands of informative meioses in which no crossovers between the complement loci have been detected and must, therefore, be rare. Support at the molecular level lies in the fact that from the 5′ to 3′ ends of the coding regions of the *C2,BF,C4A,C4B* complex is only about 100 kb (Carroll *et al.*, 1984) or 0.05–0.1 centimorgans (recombination units), predicting random crossovers at less than 1 in 1000–2000 meioses. Family studies involving normal chromosomes from caucasians have established around a dozen common complotypes (or BF, C2, C4A, C4B allelic combinations) at frequencies of about 0.01 or higher (Alper *et al.*, 1983). Table 1 gives a list of these. At a frequency lower than this there are an additional 35 complotypes. The total is thus 47, far fewer than the nearly 1000 possible combinations predicted by the number of BF, C2, C4A and C4B alleles occurring in this population of chromosomes. This situation is precisely analogous to the haplotypic sets comprising the closely linked immunoglobulin heavy chain genes (Pandey *et al.*, 1984), and the multiple linked genes for the Rh and Kell blood group antigens (Race & Sanger, 1975), to mention a few.

Table 1. *Complotypes in normal caucasians*

Complotypes are arbitrarily designated by their *BF*, *C2*, *C4A* and *C4B* alleles, with '0' = null or Q0 (quantity zero).

| Complotypes | Frequency<br>($n = 643$ chromosomes) |
|---|---|
| SC31 | 0.389 |
| SC01 | 0.127 |
| FC31 | 0.112 |
| SC30 | 0.064 |
| SC42 | 0.050 |
| SC61 | 0.031 |
| SC21 | 0.028 |
| FC(3,2)0 | 0.022 |
| SC02 | 0.022 |
| SC33 | 0.020 |
| FC01 | 0.016 |
| SB42 | 0.016 |

Table 2. *Common extended haplotypes in caucasians*

Based on analysis of 643 normal chromosomes.

| Extended haplotype | Frequency |
|---|---|
| [HLA-B8,DR3,SC01] | 0.0762 |
| [HLA-B7,DR2,SC31] | 0.0700 |
| [HLA-B44,DR4,SC30] | 0.0311 |
| [HLA-B44,DR7,FC31] | 0.0295 |
| [HLA-B57,DR7,SC61] | 0.0171 |
| [HLA-B35,DR1,FC(3,2)0] | 0.0112 |
| [HLA-B62,DR4,SC33] | 0.0109 |
| [HLA-B18,DR2,SO42] | 0.0109 |
| [HLA-B14,DR1,SC2(1,2)] | 0.0109 |
| [HLA-B38,DR4,SC21] | 0.0080 |

## Extended MHC Haplotypes

When one calculates three-point linkage disequilibrium between HLA-B, HLA-DR and complotype allelic sets, over a dozen show significant positive linkage disequilibrium (Awdeh *et al.*, 1983*a*). The more common of these sets are given in Table 2; we have designated them as 'extended haplotypes'. The importance of using the whole complotype (and further justification of the complotype concept) is evident in the fact that HLA-B8,DR3,SC01 but not HLA-B8,DR3,FC01 or HLA-B8,DR3,SC31 forms an extended haplotype. FC01 and SC31 differ from SC01 only at the *BF* and *C4A* loci, respectively.

There is another important characteristic of extended haplotypes: they show limited variation at other MHC or MHC-linked loci (Awdeh *et al.*, 1983*a*). Thus, almost all HLA-B8, DR3, SC01 is found with HLA-A1 (75%), A3 (11%) or A2 (7%), with HLA Cw7 (all in which HLA-C is typable) and DP1 (69%) (Matsui *et al.*, 1984). Alleles at the *GLO* locus, some 5–7 centimorgans from HLA-A, distinguish HLA-B8,DR3,SC01 extended haplotypes with male segregation distortion from those without this property (Awdeh *et al.*, 1983*a,b*). Thus, for many extended haplotypes, it appears that alleles at many loci are the same and behave as if they have been 'frozen' at some point during human evolution (Alper *et al.*, 1982). Further evidence that this is true is the fact that lymphocytes from unrelated individuals heterozygous for the same two different extended haplotypes give reactions in mixed lymphocyte culture similar to lymphocytes from MHC-identical sibs (Awdeh *et al.*, 1985). This lack of reaction implies identity on extended haplotypes at loci (other than HLA-DR) contributing to the mixed lymphocyte reaction. Because of the relatively few loci on chromosome 6p available for polymorphic analysis (the number is increasing but mostly within a short chromosomal distance), it is not clear how much of the region is fixed for each extended haplotype. It is probable that this differs from haplotype to haplotype.

**Origins of Extended Haplotypes**

A number of mechanisms have been proposed to explain linkage disequilibrium among MHC alleles. Some are operative over relatively short periods of time over short chromosomal distances. Recent mutation in the MHC region, recent admixture of two different ethnic groups or population stratification, and genetic drift are among these. To explain linkage disequilibrium over longer periods of time and involving chromosomal map distances of more than a few centimorgans, other mechanisms must be invoked. Certainly a highly likely powerful means of maintaining linkage disequilibrium is via selective pressure (Benacerraf & McDevitt, 1972). By analogy with other mammalian species, particularly the mouse, it is likely that immune response genes are in the human HLA-DR, DP and DQ regions, and selection for specific alleles of these loci has been proposed as a determinant of linkage disequilibrium (Bodmer & Bodmer, 1978). More recently, Porter (1983) has postulated selection for specific combinations of complement alleles. This would operate via the known functions of complement in host resistance to infection and perhaps, via solubilization of immune aggregates (Miller & Nussenzweig, 1975) and resistance to immune complex-mediated inflammation. Effects would be both *cis* and *trans*, in both cases influencing the allelic composition and frequencies of complotypes. This view implies that null alleles would be selectively disadvantageous, particularly in the homozygous state, and be particularly high in frequency in autoimmune disorders.

A different mechanism by which extended haplotypes might arise is analogous to *t*-mutant haplotypes in wild mice. The latter are identified in mice by the recessive lethal genes they carry which cause death during embryogenesis at specific stages of development (Bennett, 1975). They show marked suppression of crossing over in meioses with wild chromosomes leading to striking linkage disequilibrium over considerable portions (up to 14 centimorgans) of chromosome 17, including MHC alleles. The crossover suppression appears to be related to DNA inversion on *t*-haplotypes compared with wild chromosomes (Artzt *et al.*, 1982; Silver, 1982; Shin *et al.*, 1983). Most *t*-mutant haplotypes in males heterozygous for wild chromosomes show marked segregation distortion, and the *t*-haplotypes are transmitted to nearly all offspring (Dunn & Gluecksohn-Schoenheimer, 1939). Evidence that the most common caucasian extended haplotype [HLA-B8,DR3,SC01] may have features similar to murine *t*-haplotypes was obtained in analysis of segregation (Cudworth *et al.*, 1979; Awdeh *et al.*, 1983*a*). Remarkably, male segregation distortion is exhibited by [HLA-B8,DR3,SC01] marked by GLO2 but not GLO1 and is most striking (over 90%) in families with type 1 diabetes mellitus or gluten-sensitive enteropathy (Awdeh *et al.*, 1983*b*). We have postulated that human analogues of *t*-mutants would act as genetic 'sinks', accumulating deleterious mutations balanced by the unusually strong chromosomal selective advantage of segregation distortion and 'locked in' by crossover suppression at some stage of human evolution when populations were more isolated than they are now (Alper *et al.*, 1982).

## MHC Allele–Disease Associations

There are a remarkable number of diseases that show MHC allele associations (Bodmer & Bodmer, 1978; Amos & Kostyu, 1980). A few are given in Table 3. The most common markers in caucasians are HLA-B8 and DR3, found to be elevated in a wide variety of disorders including type 1 diabetes, gluten-sensitive enteropathy, myasthaenia gravis, chronic active hepatitis and a host of others. Many other markers may be parts of extended haplotypes (Alper *et al.*, 1982; Dawkins *et al.*, 1983).

There are only two basic ways in which these associations can arise. Either the associated allele itself is directly or indirectly a disease susceptibility gene or the associated alleles are in linkage disequilibrium with a disease susceptibility gene or genes. It has been assumed that alleles with the highest relative risk (supposedly a measure of the distinction between disease and control gene populations) are the most likely candidates for 'primary' or 'causative' genes (Ryder & Svejgaard, 1981). Alleles with lower relative risks are thought to be 'secondarily' increased in frequency because of their positive linkage disequilibrium with primary alleles. In the case of type 1 diabetes, this argument (with others) has been used to identify HLA-DR3 and DR4 as primary susceptibility alleles because of their higher relative risk than HLA-A or B alleles. Although the complement allele *BF*F1* has a higher relative risk than any known HLA-B or DR allele (Table 3), it has not been seriously proposed as a primary susceptibility gene for type 1 diabetes.

A number of features of many of the diseases listed in Table 3 should be noted. For the most part, they are disorders of unknown aetiology but with autoimmune features. Although a genetic component is clearly involved, as judged by high concordance rates for disease in monozygotic twins and a greater risk for disease in family members compared with the general population, most lack a clear-cut mode of inheritance. Because the monozygotic twin rate is not 100% for most of the diseases, it is clear that not all genetically susceptible individuals have disease. The disease susceptibility genes are thus not fully 'penetrant'. For type 1 diabetes where this question has been most thoroughly studied, penetrance is estimated at 25–50%. Since the prevalance of this disease is 2–3/1000, the

Table 3. *Some MHC-allele–disease associations*

$$\text{RR (relative risk)} = \frac{\text{no. of patients with marker}}{\text{no. of controls with marker}} \times \frac{\text{no. of patients without marker}}{\text{no. of controls without marker}}$$

| | HLA-A | | HLA-B | | HLA-D/DR | | BF | |
|---|---|---|---|---|---|---|---|---|
| Disease | Allele | RR | Allele | RR | Allele | RR | Allele | RR |
| Type 1 diabetes | A1 | 1.6 | B8 | 2.5 | DR3 | 4.5 | | |
|   mellitus | | | B18 | 2.5 | | | F1 | 7.5 |
| | | | B15 | 2.5 | DR4 | 4.5 | | |
| | | | B7 | 0.1 | DR2 | 0.1 | | |
| Gluten-sensitive | A1 | 4.0 | B8 | 8.0 | DR3 | 17.0 | | |
|   enteropathy | | | | | | | | |
| Chronic active | | | B8 | 9.0 | DR3 | 9.0 | | |
|   hepatitis | | | | | | | | |
| Multiple sclerosis | A3 | 1.8 | B7 | 2.5 | DR2 | 4.2 | | |
| Ankylosing | | | B27 | 87.0 | | | | |
|   spondylitis | | | | | | | | |

MHC-linked susceptibility gene must be common, no matter what the mode of inheritance. The postulated susceptibility gene would have a frequency between around 0.0025 for pure dominant, 50% penetrance to about 0.1 for pure recessive, 25% penetrance.

## 21-hydroxylase Deficiency Congenital Adrenal Hyperplasia

An exception among MHC-linked, MHC allele-associated diseases is 21-hydroxylase deficiency (21-OH) congenital adrenal hyperplasia. In contrast to the bulk of these diseases of unknown aetiology, pathogenesis and genetic determination, 21-OH is a clear-cut recessive in which MHC linkage was initially found (Dupont *et al.*, 1977) as a surprise – the disease has no immunological features. The susceptibility locus, i.e., the locus for the enzyme deficient in patients, maps between HLA-B and HLA-DR (Klouda *et al.*, 1980; Pucholt *et al.*, 1980; Bias *et al.*, 1981). Although initially no distortion in frequency of MHC alleles was noted, it later became apparent that HLA-Bw47, the rare complotype FCO,31, and HLA-B14 were increased among patients (Pollack *et al.*, 1979, 1981; Klouda *et al.*, 1980; Awdeh *et al.*, 1981*a*). Most recently, using recombinant DNA techniques, two genes for 21-hydroxylase have been identified, each immediately 3′ to a C4 (or C4-like) gene in the mouse (White *et al.*, 1984*a*) and in man (Carroll *et al.*, 1985). It is probable that only one of these is normally expressed.

Because of its MHC linkage, its MHC allele-association and the precise nature of its genetic determination, we have used the MHC haplotype distribution in 21-OH patients as a model for similar studies in type 1 diabetes and related disorders (Fleischnick *et al.*, 1983). Because of the relative rarity of the disease compared with, for example, type 1 diabetes, one would expect that any extended haplotypes carrying 21-OH would be rare in the general population of normal caucasian chromosomes. Moreover, there are known clinical variants of the disease with a severe, salt-wasting form, and milder forms with simple virilization, late onset and only the biochemical abnormalities but no clinical symptoms or signs. If there are individual MHC allele or extended haplotype markers for 21-OH, one would expect that they would mark specific individual mutations, and thus each would be characteristic of only one form of the disease. This is the case, with [HLA-B14,DR1,SC2(1,2)] increased in cryptic and mild cases only (Pollack *et al.*, 1981), whereas [HLA-Bw47,DR7,FC0,31] marks only the severe salt-wasting form of 21-OH (Dupont *et al.*, 1980). One would also expect predominant ethnicity associations for these markers, and this seems to be true with the former's epicentre in the English midlands (Klouda *et al.*, 1980) and the latter's in the Mediterranean area (Pollack *et al.*, 1981). The salt-losing form of 21-OH is rare and so is its marker extended haplotype. On the other hand, [HLA-B14,DR1,SC2(1,2)] is common (0.5%), implying a higher incidence than suspected for cryptic and mild 21-OH. Of interest is the fact that a haplotype, found in high frequency among Australian patients with 21-OH [HLA-Bw22,Cw3,SB45], when it is found in normal individuals unselected for 21-OH is almost always associated with steroid metabolic abnormalities characteristic of heterozygotes for 21-OH (McCluskey *et al.*, 1983). This is consistent with the postulate that most instances of extended haplotypes, whether in patients or not, have similar alleles including disease susceptibility genes. The corollary of this postulate, that if a disease susceptibility gene is not on an extended haplotype, it will be absent or nearly absent from patient chromosomes, is borne out by the absence of [HLA-B8,DR3,SC01] from chromosomes from patients (Fleischnick *et al.*, 1983).

Analysis of [HLA-Bw47,DR7,FC0,31] as it occurs in patients with severe 21-OH is instructive. It accounts for over 20% of chromosomes from salt-losing 21-OH. HLA-Bw47, HLA-Cw7 and FC0,31 are always present, DR7 is almost always present, most GLO alleles are GLO1 and around 75% of HLA-A is A3, suggesting that the region from HLA-A through GLO is relatively fixed – a distance of around 7 centimorgans (Fleischnick *et al.*, 1983). Evidence from DNA studies suggests that there is a deletion of the 21-hydroxylase locus associated with *C4B* on this chromosome (White *et al.*, 1984*b*). The apparent contradiction with the protein allotyping should not be taken too seriously. The assignment of the single expressed C4 variant on this haplotype was based on Chido antigenic reactivity (Fleischnick *et al.*, 1983). Its electrophoretic mobility is intermediate between C4A and C4B zones. In other respects, such as its low haemolytic activity and the apparent molecular size of its $\alpha$ chain, it is like a C4A variant. It is clear that the assignment of some intermediate C4 variants with some properties of C4A and other properties of C4B at the protein level, and probably also the DNA level is, of necessity, arbitrary.


## C2 Deficiency

The gene for C2 deficiency, *C2*Q0*, is part of the extended haplotype [HLA-A25,B18,DR2,S042], although there is most often variability at the HLA-A locus, less often at HLA-B, still less often at HLA-DR and very rarely in the C4A and C4B alleles (Fu *et al.*, 1974, 1975; Awdeh *et al.*, 1981*b*). C2 deficiency has thus far been reported only in caucasians, suggesting that it arose since the separation of the major races of man some 100000 years ago or so. Because there does not seem to be limited variation at HLA-A other than HLA-A25 (10) and no GLO preference, it is probable that the reason for [HLA-A25,B18,DR2,S042] being an extended haplotype is recent occurrence of the mutation leading to *C2*QO*.

The association of homozygous C2 deficiency with systemic lupus erythematosus or lupus-like disorders (Agnello *et al.*, 1974) could have a number of bases, including deficient dissolution of immune aggregates. It is also possible that an aberrant immune response gene is in linkage disequilibrium with *C2*QO*. If heterozygotes for C2 deficiency are, in fact, increased among lupus patients (Glass *et al.*, 1976), the latter hypothesis would be strengthened.


### Type 1 Diabetes Mellitus

As is evident from Table 3, the highest relative risks for individual MHC alleles are those for HLA-DR3, HLA-DR4 and *BF*F1* in caucasian patients with type 1 diabetes mellitus. There are a number of perplexing aspects concerning the genetics of type 1 diabetes than can only be touched on here. For example, in families with two sibs who have type 1 diabetes, the sibs are MHC identical more than half the time but, in a small percentage of cases, they may share no MHC haplotypes. Although earlier interpretations of these observations (Rubinstein *et al.*, 1977) concluded that susceptibility to the disease was inherited as an autosomal recessive trait, there have been a very large number of alternative models proposed since then. One of the more popular of these proposes a mixed model in which both homozygotes and heterozygotes for susceptibility genes can have diabetes but penetrance in the former group is much higher than in the latter (Spielman *et al.*, 1980).

There appear to be an excess of HLA-DR3/DR4 heterozygotes among patients over homozygotes for either HLA-DR3 or DR4 (Rotter *et al.*, 1983), as predicted by Hardy–Weinberg analysis. Nevertheless, when we analysed over 1100 patients' samples for BF F1, the numbers of heterozygotes and homozygotes agreed precisely with those predicted by recessive inheritance (Raum *et al.*, 1981). One can draw the same conclusion from data obtained from the study of Basque diabetics, in whom the *BF*F1* frequency is particularly high (de Mouzon *et al.*, 1979).

The study of randomly selected patients with type 1 diabetes and their families for haplotypes and, in particular, the distribution of extended haplotypes compared with normal caucasian chromosomes, has been highly instructive (Raum *et al.*, 1984). From this study, it could be shown that four extended haplotypes were increased among the patient haplotypes: [HLA-B8,DR3,SC 01,GLO2], (not [HLA-B8,DR3,SC01,GLO1!]), [HLA-B18,DR3,F1C30], [HLA-B15(w62),DR4,SC33] and, only among Ashkenazi Jewish patients, [HLA-B38,DR4,SC21]. HLA-DR3 not on the increased extended haplotypes was reduced in frequency compared with control chromosomes, thus ruling out HLA-DR3 itself as currently defined as a susceptibility gene. The four increased extended haplotypes contain all the previously noted MHC alleles increased among diabetics. There was, in general, a dearth of extended haplotypes not increased in patients but normally found in caucasians such as [HLA-B7,DR2,SC31], [HLA-Bw57,DR7,SC61] and [HLA-Bw44,DR4,SC30].

Because it is likely that extended haplotype are fixed over a considerable portion of chromosome 6p, it is very difficult to localize a susceptibility gene which may be none of those yet uncovered. Just as it is unlikely that HLA-DR3 is a susceptibility gene, so it is unlikely that any complement allele, such as $BF*F1,C4A*Q0$ (Raum *et al.*, 1981), or $C4B*Q0$, all of which are more or less increased in our population of patients, are susceptibility genes. Again, if one removes those alleles which form parts of extended haplotypes from the population of patient chromosomes, there are no increases in frequency in any complement allele.

Principal support for the original work in this review was provided by NIH grants AM 26844, AM 16392, AI 14157, HD 17461, CA 19589, CA 20531 and CA 06516. Mrs. Louise Viehmann and Mrs. Diane Sullivan provided superb secretarial assistance.

# References


Agnello, V., de Bracco, M. M. E. & Kunkel, H. G. (1972) *J. Immunol.* **108**, 837–840

Albert, E. D., Rittner, C., Scholz, S., Kuntz, B. & Mickey, M. R. (1977) *Scand. J. Immunol.* **6**, 459–464

Alper, C. A., Awdeh, Z. L., Raum, D. D. & Yunis, E. J. (1982) *Clin. Immunol. Immunopathol.* **24**, 276–285

Alper, C. A., Raum, D., Karp, S., Awdeh, Z. L. & Yunis, E. J. (1983) *Vox Sang.* **45**, 62–67

Amos, D. B. & Kostyu, D. D. (1980) *Adv. Hum. Genet.* **10**, 137–208

Artzt, K., Shin, H.-S. & Bennett, D. (1982) *Cell* **28**, 471–476

Awdeh, Z. L., Raum, D., Fleischnick, E., Crigler, J. F., Jr., Gerald, P. S. & Alper, C. A. (1981*a*) *Clin. Res.* **29**, 287A

Awdeh, Z. L., Raum, D. D., Glass, D., Agnello, V., Schur, P. H., Johnson, R. B., Jr., Gelfand, E. W., Ballow, M., Yunis, E. J. & Alper, C. A. (1981*b*) *J. Clin. Invest.* **67**, 581–583

Awdeh, Z. L., Raum, D., Yunis, E. J. & Alper, C. A. (1983*a*) *Proc. Natl. Acad. Sci. U.S.A.* **80**, 259–263

Awdeh, Z., Raum, D., Yunis, E. J., Katz. A., Gabbay, K. H. & Alper, C. A. (1983*b*) *Hum. Immunology* **8**, 274 (abstr.)

Awdeh, Z. L., Alper, C. A., Eynon, E., Alosco, S. M., Stein, R. & Yunis, E. J. (1985) *Lancet* **ii**, 853–855

Benacerraf, B. & McDevitt, H. O. (1972) *Science* **175**, 273–279

Bennett, D. (1975) *Cell* **6**, 441–454

Bias, W. B., Urban, M. D., Migeon, C. J., Hsu, S. H. & Lee, P. A. (1981) *Hum. Immunol.* **2**, 139–145

Bodmer, W. F. (1980) *J. Exp. Med.* **152** (*Suppl.*), 353s–357s

Bodmer, W. F. & Bodmer, J. G. (1978) *Br. Med. Bull.* **34**, 309–316

Carroll, M. C., Campbell, R. D., Bentley, D. R. & Porter, R. R. (1984) *Nature* (*London*) **307**, 237–241

Carroll, M. C., Campbell, R. D. & Porter, R. R. (1985) *Proc. Natl. Acad. Sci. U.S.A.* **82**, 521–525

Cudworth, A. G., Wolf, F., Gorsuch, A. N. & Festenstein, H. (1979) *Lancet* **ii**, 389–390

Dawkins, R. L., Christiansen, F. T., Kay, P. H., Garlepp, M., McCluskey, J., Hollingsworth, P. N. & Zilko, P. J. (1983) *Immunol. Rev.* **70**, 5–22

de Mouzon, A., Ohayon, E., Ducos, J. & Hauptmann, G. (1979) *Lancet* **ii**, 1364

Dunn, L. C. & Glueksohn-Schoenheimer, S. (1939) *Genetics* **24**, 587–609

Dupont, B., Oberfeld, S. E., Smithwick, E. M., Lee, T. D. & Levine, L. S. (1977) *Lancet* **ii**, 1309–1311

Dupont, B., Pollack, M. S., Levine, L. S., O'Neill, G. J., Hawkins, B. R. & New, M. I. (1980) in *Histocompatability Testing 1980* (Terasaki, P. I., ed.), pp. 693–706, UCLA Tissue Typing Laboratory, Los Angeles

Fleischnick, E., Awdeh, Z. L., Raum, D., Granados, J., Alosco, S. M., Crigler, J. F., Jr., Gerald, P. S., Giles, C. M., Yunis, E. J. & Alper, C. A. (1983) *Lancet* **i**, 152–156

Fu, S. M., Kunkel, H. G., Brusman, H. P., Allen, F. H., Jr. & Fotino, M. (1974) *J. Exp. Med.* **140**, 1108–1111

Fu, S. M., Stern, R., Kunkel, H. G., Dupon, B., Hansen, J. A., Day, N. K., Good, R. A., Jersild, C. & Fotino, M. (1975) *J. Exp. Med.* **142**, 495–506

Glass, D., Raum, D., Gibson, D., Stillman, J. S. & Schur, P. H. (1976) *J. Clin. Invest.* **58**, 853–861
Klouda, P. T., Harris, R. & Price, D. A. (1980) *J. Med. Genet.* **17**, 337–341
Matsui, Y., Alosco, S. M., Awdeh, Z., Duquesnoy, R. J., Page, P. L., Hartzman, R. J., Alper, C. A. & Yunis, E. J. (1984) *Immunogenetics* **20**, 623–663
McCluskey, J., Kay, P. H., Stuckey, M., Christiansen, F. T., Dawkins, R. L. & Wilson, G. (1983) *Lancet* **i**, 764–765
Miller, L. H. & Nussenzweig, V. (1975) *Proc. Natl. Acad. Sci. U.S.A.* **72**, 418–422
Pandey, J. P., Whitten, H. D. & Fudenberg, H. H. (1984) in *Immunogenetics* (Panayi, G. S. & David, C. S., eds.), pp. 92–109, Butterworths, London
Piazza, A. (1975) in *Histocompatibility Testing 1975* (Kissmeyer-Nielsen, F., ed.), pp. 923–927, Munksgaard, Copenhagen
Pollack, M. S., Levine, L. S., O'Neill, G. J., Pang, S., Lorenzen, F., Kohn, B., Rondanini, G. F., Chuimello, G., New, M. I. & Dupont, B. (1981) *Am. J. Hum. Genet.* **33**, 540–550
Porter, R. R. (1983) *Mol. Biol. Med.* **1**, 161–168
Pucholt, V., Fitzsimmons, J. S., Gelsthorpe, K., Reynolds, M. A. & Milner, R. D. G. (1980) *J. Med. Genet.* **17**, 447–452
Race, R. R. & Sanger, R. (1975) *Blood Groups in Man*, 6th edn., Blackwell, Oxford
Raum, D., Awdeh, Z. & Alper, C. A. (1981) *Immunogenetics* **12**, 59–74
Raum, D., Awdeh, Z., Yunis, E. J., Alper, C. A. & Gabbay, K. H. (1984) *J. Clin. Invest.* **74**, 449–454
Rotter, J. I., Anderson, C. E., Rubin, R., Congleton, J. E., Terasaki, P. I. & Rimon, D. L. (1983) *Diabetes* **32**, 169–174
Rubinstein, P., Suciu-Foca, N. & Nicholson, J. F. (1977) *N. Engl. J. Med.* **297**, 1036–1040
Ryder, L. P. & Svejgaard, A. (1981) *Annu. Rev. Genet.* **15**, 169–187
Shin, H.-S., Flaherty, L., Artzt, K., Bennett, D. & Ravetch, J. (1983) *Nature (London)* **306**, 380–383
Silver, L. M. (1982) *Cell* **29**, 961–968
Spielman, R. S., Baker, L. & Zmijewski, C. M. (1980) *Am. J. Hum. Genet.* **44**, 135–150
White, P. C., Chaplin, D. D., Weis, J. H., Dupont, B., New, M. I. & Seidman, J. G. (1984a) *Nature (London)* **312**, 465–467
White, P. C., New, M. I. & Dupont, B. (1984b) *Proc. Natl. Acad. Sci. U.S.A.* **81**, 7505–7509

*Biochem. Soc. Symp.* **51**, 29–36
*Printed in Great Britain*

# Molecular Genetics of the Fourth Component of Human Complement

MICHAEL C. CARROLL,* ASTRIDUR PALSDOTTIR,† K. TERTIA BELT†
and C.-YUNG YU†

** Department of Pediatrics, Harvard Medical School, Children's Hospital, Boston, MA, U.S.A.
and † Department of Biochemistry, University of Oxford, Oxford, U.K.*

## Synopsis

The fourth component of human complement is encoded by two separate, but closely linked, loci, C4A and C4B, that have been positioned within the class III region of the HLA complex. While the two isotypes vary by only six amino acid residues, they differ significantly in haemolytic activity. Both loci are considerably polymorphic and this may be biologically relevant to ensure interaction with a wide range of pathogens. The number of C4 genes expressed is polymorphic as null alleles, total deficiency and duplication has been shown based on protein studies. Southern analysis of 24 different haplotypes with either C4A or C4B null alleles using the C4 probes showed that three of the null alleles were due to deleted genes but the majority appeared normal. A cosmid library was prepared from DNA of one of the deleted haplotypes and the region of deletion analysed by restriction mapping.

## Introduction

The complement system is a family of 20 or more serum proteins and membrane receptors that serves as the chief effector function of the humoral immune system [1]. This system may be activated specifically by antigen–antibody aggregates through either the classical or alternative pathways or non-specifically by polysaccharide antigen through the alternative pathway. Activation of the fourth component in the early classical pathway is critical for focusing assembly of the latter components at the site of the antigen–antibody aggregate. This is accomplished by formation of a covalent bond between the C4 protein and the immune complex [2].

There are two forms of C4 protein in the serum, i.e. C4A (acidic) and C4B (basic), which are distinguished by electrophoretic [3–5] and antigenic [6] differences. Both forms are considerably polymorphic and more than 35 variants have been detected by electrophoretic studies [7]. A model has been proposed suggesting that variation of C4 in the population is important for ensuring interaction with the other components of the complement system, with antibody and with a wide range of antigenic structures [8].

Recent functional studies have demonstrated that the level of total complement

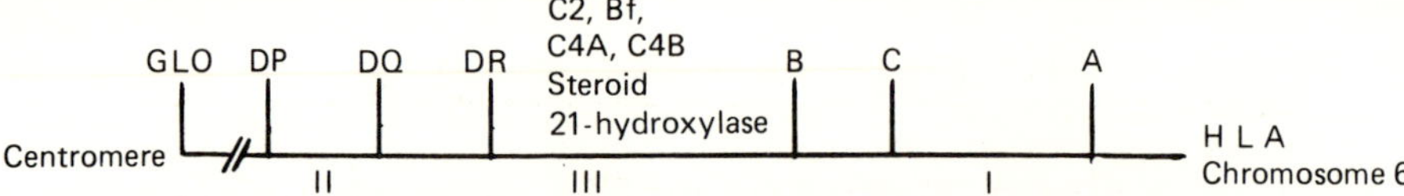


Fig. 1. *Genetic map of the HLA region on the short arm of chromosome 6*

The class III genes, encoding the complement proteins, i.e. C2, Factor B (Bf), C4A and C4B and steroid 21-hydroxylase, lie between the class I genes, i.e. B, C and A, and the class II genes, i.e. DP, DQ and DR. The glyoxylase gene (GlO) is between DP and the centromere.

activated may be dramatically affected by the efficiency of binding of the C4 protein with the antigen–antibody aggregate [9, 10]. The C4A protein was found to bind more efficiently to protein antigen and the C4B protein was found to bind more efficiently to carbohydrate antigen. For example, when C4A was used in the haemolytic assay, 4-10-fold less lysis of the red cell antigen was observed compared to when C4B protein was used in the assay.

The C4 mRNA has been cloned [11] and the complete and near-complete derived amino acid sequences for C4A and C4B, respectively, have been determined [12,13]. Comparison of the two isotypes shows only 12 amino acid residue differences of which 10 are clustered in the C4d region of the $\alpha$ chain near the covalent binding site. Six residues were identified tentatively as isotypic differences and provide a structural basis for the difference in efficiency of binding to antigen–antibody aggregate. The remaining four substitutions in the C4d region probably represent allotypic differences. Analysis of a second C4A cDNA clone and the homologous region of three genomic clones suggested that allotypic differences are due to single amino acid substitutions [13].

The genes encoding C4A and C4B have been mapped to the class III region of the major histocompatibility complex (HLA in man) on chromosome 6 [14] (Fig. 1). This region between HLA B (class I) and D (class II) regions includes the genes encoding complement proteins C2 [15] and factor B (Bf) [16] as well as steroid 21-hydroxylase [17]. Steroid 21-hydroxylase, which is a cytochrome *P*-450 protein, is involved in the synthesis of steroids, principally cortisol and mineral corticoids in the adrenal cortex [18].

The six genes have been positioned within 120 kb of DNA by overlapping cloned genomic fragments isolated from a cosmid library of human genomic DNA [19] (Fig. 2). The genes for C2 and Bf, which are separated by less than 2 kb, are about 30 kb upstream from the two C4 genes, which are separated by 10 kb. A similar organization of complement genes in the mouse has been shown [20]. The C4A and C4B genes were identified by nucleic acid sequencing of restriction fragments isolated from the C4d region which contains the isotypic differences [13, 21]. The two genes are similar by restriction mapping analysis, except for a 6–7 kb intron in the 5′ region that is missing in some alleles of C4B (M. C. Carroll, A. Palsdottir, K. T. Belt and C.-Y. Yu, unpublished work). Comparison of the nucleic acid sequence of the coding region has shown only 14 nucleotide differences out of > 4600 compared [13]. The introns appear to be equally conserved, as less than 1% difference was found on comparison of introns in the C4d region [13]. This degree of homology compared with only about 80% homology between the coding sequences of murine and human C4

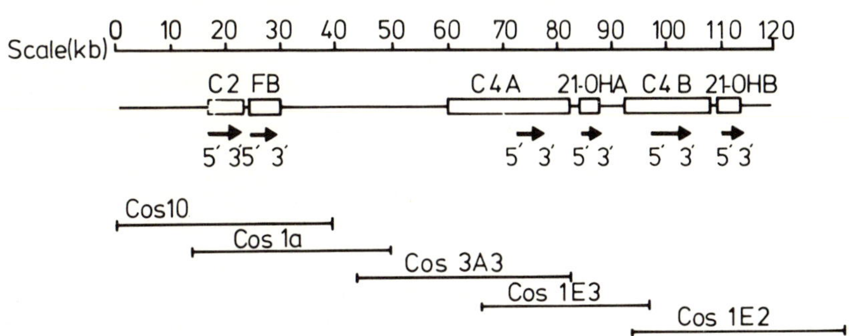


Fig. 2. *Arrangement of the genes encoding complement proteins C2, Bf, C4A and C4B and steroid 21-hydroxylase based on overlapping cloned genomic fragments*

[22] suggests that the genes, i.e. C4 and steroid 21-hydroxylase, duplicated relatively recently.

A gene for steroid 21-hydroxylase (21-OH) has been positioned less than 3 kb from the 3' end of each C4 gene [23]. The genes were localized by hybridization of human adrenal RNA with cloned fragments that were isolated from the C4 cosmid clones. Confirmation that the cloned fragments coded for steroid 21-hydroxylase was shown by homology of the derived amino acid sequence with porcine and bovine steroid 21-hydroxylase. White *et al.* (1984) have mapped the two steroid 21-OH genes in both mouse [24] and man [25] using a cDNA probe specific for bovine steroid 21-hydroxylase.

Deletion of the 21-OHB gene in humans results in congenital adrenal hyperplasia, a relatively common inherited disorder [26]. However, expression studies in the mouse have shown that the 21-OHA gene, not 21-OHB, is expressed primarily in the adrenal gland [27]. It is possible that the 21-OHA gene is expressed either in low levels or in different tissues. A fraction of mRNA in the human liver hybridizes with the 21-OH genomic probe [23] and in the mouse a cDNA clone was isolated from a murine liver library [36].

In addition to polymorphism in the number of alleles at each C4 locus, there is polymorphism in the number of genes expressed (Fig. 3). Null alleles, i.e. the gene product is not detected in the serum, occur at a frequency of 10–15% at each C4 locus [28]. The null alleles have been shown to be associated with susceptibility to autoimmune disease, especially those involving immune complexes [29]. Total deficiency of C4 is relatively rare and it is observed in the population at a frequency less than predicted. Individuals totally deficient almost always suffer from systemic lupus erythrematosus [28]. Duplications of either C4A or C4B occur at 1–2% in the population [30, 31]. Duplication of a C4B gene on one haplotype has been demonstrated at the DNA level [21]. The null alleles have been proposed [31] as the result of unequal crossover between sister chromatids during meiosis, as seen in other multiple gene families such as the α-globin genes [32]. Alternatively, the C4 genes may not have been duplicated on all chromosomes [5].

In order to determine the molecular basis for the null alleles, the C4 cDNA

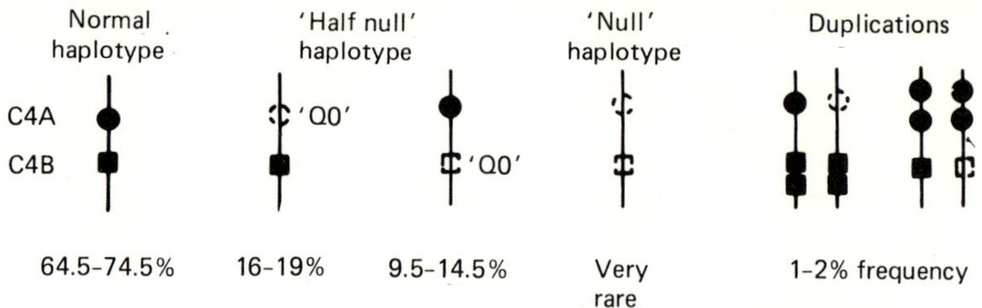


Fig. 3. *Polymorphism in the number of C4 genes expressed*

Null alleles, total deficiency and duplication of C4 genes have been identified in the population based on protein studies. From [28].

probes were used for Southern analysis [33] of genomic DNA extracted from individuals shown to have a null allele at either the C4A or C4B locus [34]. In all, 18 different HLA haplotypes that have a null allele at the C4A locus were examined. Two of the haplotypes had large deletions that included both the C4A and flanking 21-OHA gene, but the majority appeared normal. Of six different HLA haplotypes that have a null allele at the C4B locus, one had a deletion that included both the C4B and 21-OHA genes. Thus, the majority of null alleles did not appear to be due to large deletions.

## Results and Discussion

The two haplotypes with an apparent deletion of the C4A gene were analysed in an attempt to find evidence of a deletion event (Fig. 4). A molecular map of the C4 and 21-OH region was prepared for both haplotypes. Comparison of the maps of the two haplotypes showed that the region of deletion was similar. This suggested that if the genes were deleted, there was a common mechanism. One of the haplotypes, i.e. HLA *A1B*, *B*8, *DR*3, occurs in the population with a frequency of 5–7% [35]. Thus, this single haplotype would account for about one-half of the null alleles at the C4A locus.

The limits of the deletion were estimated by comparison of the null haplotypes with a normal, i.e. two C4 gene, haplotype. The 5′ limit was estimated based on the absence of the large intron that characterizes all C4A and some alleles of C4B. The 3′ limit was estimated based on known restriction site differences between 21-OHA and 21-OHB.

In an attempt to locate the site of recombination, a cosmid library was constructed from genomic DNA of an individual homozygous for the deleted C4A and 21-OHA genes. These individuals did not have congenital adrenal hyperplasia, confirming the observation by White et al. (1984) [26] that the 21-OHA gene was not important for this disease. A cosmid clone KEM-1, which contained a C4B gene, was isolated and characterized by restriction mapping and partial nucleotide sequencing. A restriction fragment was isolated from the 5′ region that included about 2 kb of the 5′ flanking region and subcloned for comparison with a homologous fragment isolated from cosmid clone 3A3 that contained the C4A gene. The restriction maps of the two cloned fragments were identical (results not shown). Because of the homology between the C4A and

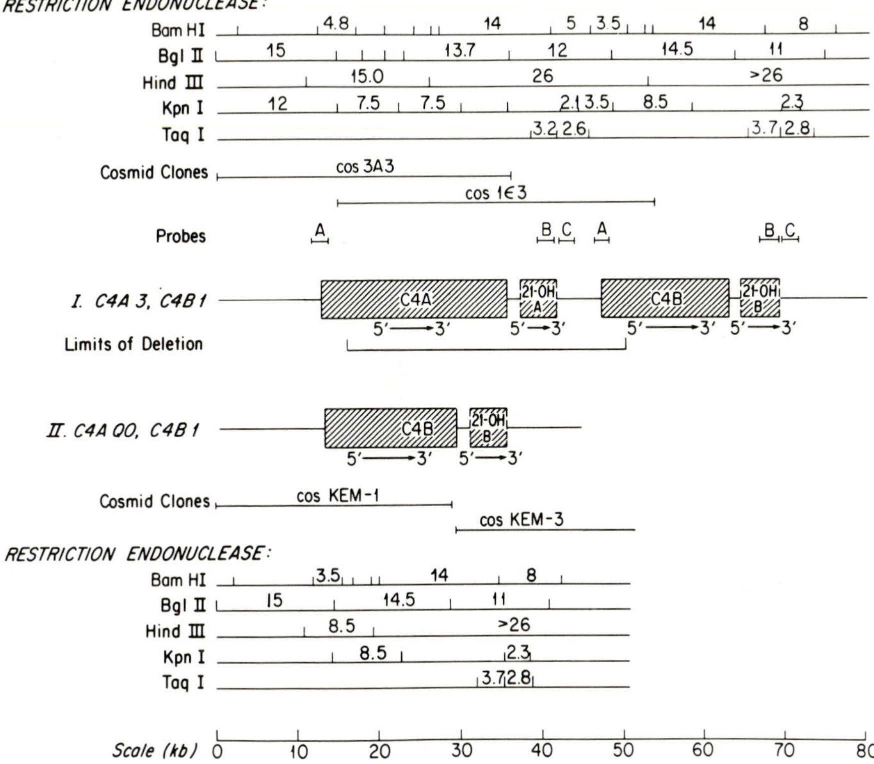



Fig. 4. *Molecular map comparison of C4 haplotypes C4A 3,C4B 1 (I) and C4A Q0,C4B 1 (II)*

Broken lines indicate approximate limits of the steroid 21-hydroxylase genes. Horizontal bar indicates approximate limits of deletion based on current markers. Cosmid clones KEM-1 and KEM-3 were isolated from a cosmid library prepared from genomic DNA of individual K.E.M. (HLA: *A 2,1; B 7,8; DR 4,5; C4A Q0; C4B 1,1; Bf S,S; C2 C,C*). Horizontal arrows indicate direction of transcription. From [34].

C4B genes, a recombination may not be detectable. It is possible that the C4B gene on this haplotype is a hybrid of C4A and C4B. Nucleic acid sequence comparison between the two fragments is under way.

In a similar study of the C4B null haplotypes, only one out of the six examined appeared to have a deletion of the C4B gene (Fig. 5). Comparison of the molecular map with the normal map showed that the deletion of about 28 kb included the 21-OHA genes. The limits of deletion were estimated on the basis of loss of known restriction sites which characterize the region between the C4A and C4B genes. However, as discussed above for the deleted C4 haplotype, the C4A gene may represent a hybrid of C4A and C4B.

The results from restriction mapping alone are not sufficient to determine if the two genes were deleted on these chromosomes. An alternative explanation is that they were not duplicated originally [5]. Analysis of the HLA B47, *DR7* haplotype which is linked to the congenital adrenal hyperplasia disease phenotype and has a deletion of the 21-OHB gene [26], showed that the C4B gene was deleted also (results not shown). Since the 21-OHB gene is critical for normal

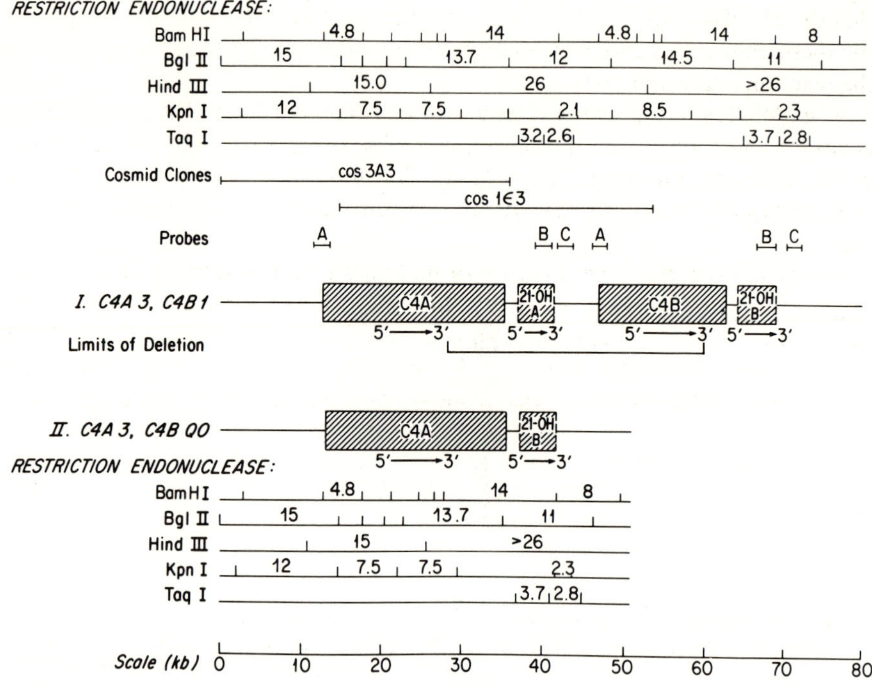



Fig. 5. *Molecular map comparison of C4 haplotypes C4A 3,C4B 1 (I) and C4A 3,C4B Q0 (II)*

The molecular map of the haplotype II was based on restriction mapping of uncloned genomic DNA from an individual homozygous for HLA: *A 2,B 7,DR 2,C4A 3,C4B Q0,Bf S,C2 C*. From [34].

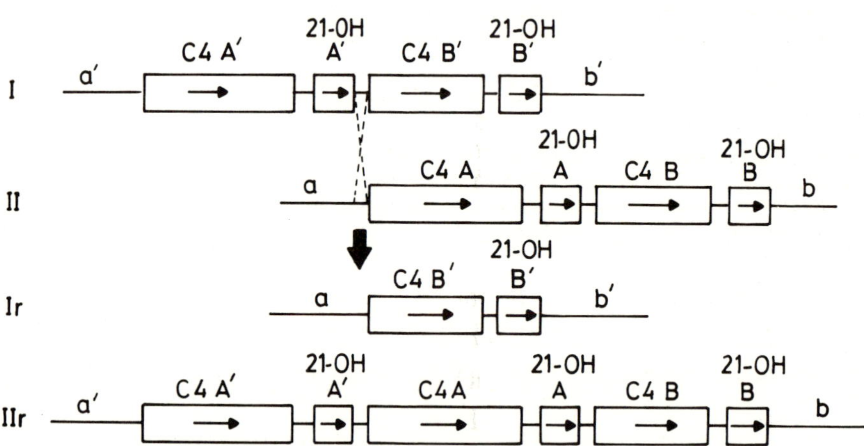



Fig. 6. *Model illustrating unequal crossover in C4 region*

Unequal crossover is one mechanism that would explain deletion of C4 and 21-OH genes on some haplotypes and duplication on others.

development, it is probable that the 21-OHB was the precursor gene. The finding that the 21-OHB and C4B genes were missing on at least one haplotype supports the thesis that both genes were duplicated on all chromosomes in man and that the deletions occurred later.

A model illustrating unequal crossover shows loss of both C4A and 21-OHA genes on one chromosome (Ir) and duplication of the C4A and 21-OHA genes on the other chromosome (IIr) (Fig. 6). This model would explain the finding of haplotypes with a single C4 and 21-OH genes as well as those with three C4 loci.

As discussed earlier, the majority of null alleles do not seem to be the result of gene deletion. An alternative explanation for the relatively high frequency of null alleles is that they represent loci that have been converted so that both C4 loci express the same isotype. A possible example is the haplotype *C4A 2,C4A 3,C4B Q*0 which was proposed as a duplication of C4A [30, 31].

M.C.C. is an Investigator of the American Arthritis Foundation; A.P. is a recipient of an E. P. Abraham Graduate Studentship Award and K.T.B. holds an MRC studentship.


## References

1. Reid, K. B. M. & Porter, R. R. (1981) *Annu. Rev. Biochem.* **50**, 433–464
2. Law, S. K., Lichtenberg, N. A., Holcombe, F. H. & Levine, R. P. (1980) *J. Immunol.* **125**, 634–639
3. O'Neill, G. J., Yang, S. Y. & Dupont, B. (1978) *Proc. Natl. Acad. Sci, U.S.A.* **75**, 5165–5169
4. Awdeh, Z. L. & Alper, C. A. (1980) *Proc. Natl. Acad. Sci. U.S.A.* **77**, 3576–3580
5. Olaisen, B., Teisberg, R., Nordhagen, R., Michaelson, T. & Geede-Dahl, T. (1979) *Nature (London)* **279**, 736–737
6. O'Neill, G. J., Yang, S. Y., Tegoli, J., Berger, R. & Dupont, B. (1978) *Nature (London)* **273**, 668–670
7. Mauff, G., Alper, C. A., Awdeh, Z., Batchelor, J. O. R., Bertrams, T., Bruun-Peterson, G., Dawkins, R. L., Demant, P., Edwards, J., Grosse-Wilde, H., Hauptmann, G., Klonda, P., Lamm, L., Mullenhauer, E., Nerl, C., Olaisen, B., O'Neill, G. O., Rittner, C., Roos, M. H., Skanes, V., Teisberg, P. & Wells, L. (1983) *Immunobiology* **164**, 184–191
8. Porter, R. R. (1983) *Mol. Biol. Med.* **1**, 161–168
9. Law, S. K. A., Dodds, A. W. & Porter, R. R. (1984) *EMBO J.* **3**, 1819–1823
10. Isenman, D. & Young, J. R. (1984) *J. Immunol.* **132**, 3019–3027
11. Carroll, M. C. & Porter, R. R. (1983) *Proc. Natl. Acad. Sci. U.S.A.* **80**, 264–267
12. Belt, K. T., Carroll, M. C. & Porter, R. R. (1984) *Cell* **36**, 907–914
13. Belt, K. T., Yu, Y., Carroll, M. C. & Porter, R. R. (1985) *Immunogenetics* **21**, 173–180
14. Rittner, C., Hauptmann, G., Grosswilde, H., Grosshaus, E., Tongio, M. M. & Mayer, S. (1975) in *Histocompatibility Testing*, pp. 945–953, Munksgaard, Copenhagen
15. Fu, S. F., Kunkel, H. G., Brusman, H. P., Allen, F. H. & Fotino, M. (1974) *J. Exp. Med.* **140**, 1108–1110
16. Allen, F. H. Jr. (1974) *Vox Sang.* **27**, 382–384
17. Dupont, B., Oberfield, S. E., Smithwick, E. M., Lee, T. D. & Levine, L. S. (1979) *Lancet* **ii**, 1309–1311
18. New, M. I., Grumbach, K. & Levine, L. S. (1983) *Principles Practice Med. Genet.* **61**, 36–38
19. Carroll, M. C., Campbell, R. D., Bentley, D. R. & Porter, R. R. (1984) *Nature (London)* **307**, 237–241
20. Chaplin, D. D., Woods, D. E., Whitehead, A. S., Goldberger, G., Colten, H. R. & Seidman, J. G. (1983) *Proc. Natl. Acad. Sci. U.S.A.* **80**, 6947–6951
21. Carroll, M. C., Belt, K. T., Palsdottir, A. & Porter, R. R. (1984) *Philos. Trans. R. Soc. London Ser. B* **306**, 379–388
22. Sepich, D. S., Noonan, D. J. & Ogata, R. T. (1985) *Proc. Natl. Acad. Sci. U.S.A.* **82**, 5895–5899
23. Carroll, M. C., Campbell, R. D. & Porter, R. R. (1985) *Proc. Natl. Acad. Sci. U.S.A.* **82**, 521–525
24. White, P. C., Chaplin, D. D., Weiss, J. H., Dupont, B., New, M. I. & Seidman, J. (1984) *Nature (London)* **312**, 465–470

25. White, P. C., Grossberger, D., Onufer, B. J., New, M. I., Dupont, B. & Strominger, J. L. (1985) *Proc. Natl. Acad. Sci. U.S.A.* **82**, 1089–1093
26. White, P. C., New, M. I. & Dupont, B. (1984) *Proc. Natl. Acad. Sci. U.S.A.* **81**, 7505–7509
27. Parker, K., Chaplin, D. D., Wong, M., Seidman, J., Smith, J. & Schimmer, B. (1985) *Proc. Natl. Acad. Sci. U.S.A.* **82**, 7860–7864
28. Hauptmann, G., Goetz, I., Uring-Lambert, B. & Grossehaus, E. (1986) *Prog. Allergy*, in the press
29. Fielder, A. H. L., Walport, M. J., Batchelor, J. R., Rynes, R. I., Black, C. M., Dodi, I. A. & Hughes, G. R. V. (1983) *Br. Med. J.* **186**, 425–428
30. Uring-Lambert, B., Goetz, J., Tongio, M. M., Mayer, S. & Hauptmann, G. (1985) in *Histocompatibility Testing 1984*, Springer-Verlag, Berlin, Heidelberg, New York
31. Raum, D., Awdeh, Z., Anderson, J., Strong, L., Granados, J., Pevan, L., Giblett, E., Yunis, E. J. & Alper, C. A. (1984) *Am. J. Human Genet.* **36**, 72–79
32. Dozy, A. M., Kan, Y. W., Embury, S. H., Mentzer, W. C., Wang, W. C., Lubin, B., Davis, J. R. & Koenig, H. M. (1979) *Nature (London)* **280**, 605–607
33. Southern, E. M. (1975) *J. Mol. Biol.* **98**, 503–517
34. Carroll, M. C., Palsdottir, A., Belt, K. T. & Porter, R. R. (1985) *EMBO J.* **4**, 2547–2552
35. Schendel, D. J., O'Neill, J. G. & Wank, R. (1984) *Immunogenetics* **20**, 23–31
36. Amor, M., Tosi, M., Duponchel, C., Steinmetz, M. & Meo, T. (1985) *Proc. Natl. Acad. Sci. U.S.A.* **82**, 4453–4458

*Biochem. Soc. Symp.* **51**, 37–46
*Printed in Great Britain*

# Regulation of Complement Gene Expression

HARVEY R. COLTEN and S. BRUCE DOWTON

*Department of Pediatrics, Harvard Medical School, Boston MA 02115, U.S.A.*

## Synopsis

The availability of molecular probes for approximately half of the 20 known complement genes now permits a detailed examination of the regulation of complement expression in liver and at extrahepatic sites in tissue macrophages. Primary cell culture, cell lines and cells transfected with DNA bearing complement genes have been used in this analysis. Pretranslational regulation of complement production has been induced by well defined cytokines such as interleukin-1 and $\gamma$-interferon, as well as directly by endotoxin. The effect of those agents on complement genes is tissue and species specific and is developmentally regulated. These data form the basis for the elucidation of the genomic structural requirements for regulation of inflammation and, by extension, specific immune responsiveness.

## Introduction

Complementary DNA (cDNA) clones corresponding to 12 of the 20 complement proteins have been isolated since 1982 when the reports of the first of these clones appeared (Wiebauer *et al.*, 1982; Woods *et al.*, 1982). For some, the cDNA probes have already been used for the elucidation of structure, organization and chromosomal localization of the corresponding genes (reviewed by Reid, 1985). Recently, the availability of these cDNA clones has allowed more detailed investigation of the molecular regulation of complement gene expression. Emphasis has been focused primarily on the complement genes within the major histocompatibility complex (the class III MHC genes) but others such as the third (C3) and fifth (C5) components have also received some attention because of their key role as mediators of inflammation.

## Gene Mapping

The genes encoding C2, factor B and C4 are on the short arm of human chromosome 6 and mouse chromosome 17 within the HLA and H-2 gene complex respectively and are designated major histocompatibility class III genes. The human (Carroll *et al.*, 1984*a,b*) and murine (Chaplin *et al.*, 1983, 1984; Perlmutter *et al.*, 1985) MHC class III genes have been mapped with a series of overlapping cosmid clones. These studies reveal a striking similarity between the species (Fig. 1). The 3′ terminus of the C2 gene is upstream of and close to

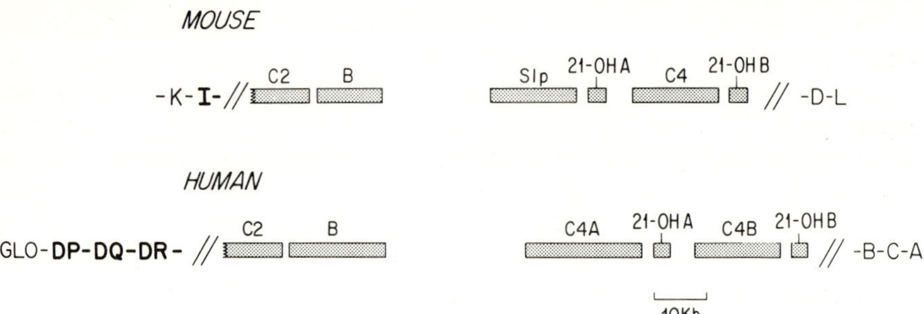


Fig. 1. *Organization of the murine and human major histocompatibility complex*

(for human $< 1$ kb) the 5′ end of the factor B gene (Campbell *et al.*, 1984). Downstream and in the same orientation are two genes encoding C4 or C4-like proteins. For the human, these are designated C4A and C4B; for the mouse, Slp (an haemolytically inactive C4 gene product under sex hormone control) and C4. Immediately adjacent to the C4 genes are two genes (21-OH A and 2-OH B) with sequences corresponding to the cytochrome *P*-450 enzyme, 21-hydroxylase (Carroll *et al.*, 1985; White *et al.*, 1985). The C4 loci are separated by about 10 kb. Whereas the C2/factor B region of this complex is relatively stable, considerable genetic variation (deletions, duplications and possibly inversions) have been detected in the area of the C4 loci. The entire MHC class III gene compex is between HLA-DR and HLA-B in the human genome (Whitehead *et al.*, 1985) and between H-2I and H-2D in the mouse but the orientation of the complex relative to the flanking MHC class I and II genes is unknown. The third (C3) and fifth (C5) components are encoded by genes homologous to the C4 gene and are thought to be products of gene duplication events. However the C3 and C5 genes are located outside of the MHC in both species. In the mouse, the C3 gene is on chromosome 17 distant from the H-2 region (Natsuume-Sakai *et al.*, 1978) and C5 is on chromosome 2 (Itakura *et al.*, 1972). The human C3 gene has been mapped to chromosome 19 (Whitehead *et al.*, 1982). The location of the human C5 gene is not yet known but work on this subject is in progress.


## Complement Proteins

*Second (C2) component of complement*

The C2 protein isolated from plasma is a single chain glycoprotein ($M_r$ 100 000) which serves as substrate for the C1s enzyme of the classical activation pathway (Cooper, 1975). Cleavage of C2 by C1s generates a *C*-terminal fragment (C2a) with serine protease activity. This polypeptide, in complex with a fragment of C4 (C4b), serves as the C3 cleaving enzyme of the classical pathway. There is a single copy of the C2 gene per haploid (Carroll *et al.*, 1984*b*), but the fine details of C2 genomic structure have not yet been published. Preliminary data suggests many similarities between the C2 and factor B genes. The latter has been more extensively studied (see below). Human liver mRNA directs synthesis of three C2 primary products when translated *in vitro* (Perlmutter *et al.*, 1984) or in *Xenopus* oocytes (D. E. Woods, G. Goldberger & H. R. Colten, unpublished work). It is likely that post-transcriptional modification or different primary transcripts account for the three forms of C2 found in hepatocyte-derived hepatoma cell lines and in human monocyte/macrophage cultures. Translation of the most abundant C2 mRNA (2.8 kb) results in an 84-kDa primary product which is glycosylated and secreted within 1–2 h. Two forms of C2 of lower molecular mass (79 kD and 70 kDa) remain cell-associated for at least 12 h after synthesis. These multiple forms of C2 have been detected in murine L cells transfected with a genomic fragment bearing the human C2 and factor B genes (Perlmutter *et al.*, 1985). In these cells, specific C2 mRNA and C2 protein (detected immunochemically and functionally) are found, but the lower molecular mass C2 mRNA is present in greater amounts than in the primary cell cultures or cell lines. Initially, we believed that these would be useful for further analysis of the cell associated C2 proteins. The studies of regulation of C2 and factor B expression in cells transfected with cosmid DNA raise questions about this expectation.

In extrahepatic mononuclear phagocytes the proportion of C2-producing cells and the rates of synthesis per cell vary as a function of cellular maturation and tissue of origin (Alpert *et al.*, 1983). For example, the proportion of C2-producing cells increases from none detectable in marrow, 10% in blood monocytes, up to 45% in spleen and peritoneal cavity. Only 2.5% of lung (bronchoalveolar lavage) macrophages produce C2, but the higher rate of synthesis of C2 per bronchoalveolar cell compensates for the lower proportion of C2-producing cells (Cole *et al.*, 1980). These tissue-specific differences in C2 gene expression are observed in human and guinea pig cells and, for both, the mechanism of regulation is pretranslational (Cole *et al.*, 1985). Further modulation of C2 synthesis occurs during an inflammatory response. That is, the net tissue concentration of C2 is increased not only by the increase in cell number but also by increased C2 mRNA and rate of synthesis of C2 per cell (Cole *et al.*, 1983, 1985). Finally, preliminary data (F. S. Cole & H. R. Colten, unpublished work) indicate that postsynthetic modification of C2 in activated macrophages increases the specific biological activity of locally synthesized C2, possibly by oxidation-dependent stabilization of the classical pathway C3 convertase. These data and evidence for independent regulation of the factor B gene prompted the studies of C2 and B gene expression cited below.


*Factor B*

Factor B is an approx. 95 kDa single-chain glycoprotein component of the alternative pathway of complement activation (Christie *et al.*, 1980). Cleavage of factor B by the serine protease factor D generates an *N*-terminal fragment Ba (30 kDa) and a C-terminal fragment Bb which bears the active enzymic site. The Bb fragment, in complex with a cleavage product of the third component C3b, functions as the alternative pathway C3-cleaving enzyme. Additional functions of activated factor B include enhancement of macrophage-mediated cytoxicity (Hall *et al.*, 1982), induction of macrophage spreading (Sundsmo & Goetze, 1981), and activation of plasminogen (Sundsmo & Wood, 1981).

Synthesis of the factor B primary translation product (83 kDa) (Woods *et al.*, 1982) is directed by a 2.6 kb polyadenylated mRNA. Single loci for the human (Carroll *et al.*, 1984*b*; Campbell *et al.*, 1984) (6 kb) and murine (Chaplin *et al.*, 1983) factor B genes have been identified. The human factor B gene is divided into 18 exons (Campbell *et al.*, 1984), three of which, coding for regions within the Ba fragment, show significant sequence homology to one another and may therefore have evolved by tandem duplication. Active site residues in the Bb fragment are encoded on separate exons. With the exception of an unique exon found in Bb, the organization of the 3′ two-thirds of the factor B gene is similar to that observed for other serine protease genes.

A comparison of human and murine factor B reveals a high degree of conservation of structure (Sackstein *et al.*, 1983). Overall, for the 477 amino acids determined in the Bb fragment of murine factor B, 83% of the residues are identical with the corresponding region of the human protein. When conservative substitutions are considered, the homology is 90%. Even in the 3′ untranslated region there is 59% nucleotide sequence homology, and an unusual polyadenylation signal is found in both species. This unusual signal is also found in the human C4 cDNA sequence.

The synthesis and secretion of factor B is similar to that described for C2 but important differences, including the response to cytokines that regulate B and C2 gene expression, have been observed (see below). Human factor B is synthesized in liver and at extrahepatic sites in mononuclear phagocytes. Murine macrophages and hepatocytes also synthesize factor B as well as C2. Control of mouse factor B synthesis is rather complex and almost certainly is exerted as multiple levels. Recent data (A. Falus & H. R. Colten, unpublished work) indicate independent regulation of B and C2 in murine hepatocytes and macrophages, similar to the tissue-specific control of murine C4 and S1p noted in earlier studies (Sackstein & Colten, 1984).

*The third component (C3)*

The lynchpin of the complement cascade, the third (C3) component, is the complement protein present in plasma at the highest concentration. In its native form, C3 is a 185 kDa two-chain disulphide-linked glycoprotein which is synthesized as a single chain precursor (Brade *et al.*, 1977). The synthesis of prepro-C3 is programmed by a 5 kb mRNA (Wiebauer *et al.*, 1982). Postsynthetic cleavage by signal peptidase and a plasmin-like enzyme generates native C3 (Goldberger *et al.*, 1981). As is the case for the precursors of C4 (pro-C4) (Goldberger *et al.*, 1980) and C5 (pro-C5) (Lundwall *et al.*, 1985), $\beta$ chain is the *N*-terminal segment in pro-C3. The $\beta$ chain is separated in pro-C3 from $\alpha$-chain by an arginine-rich intersubunit linking peptide.

The murine C3 gene is approximately 24 kb (Fey *et al.*, 1983) and although genomic clones that include a region 1.2 kb 5′ to the gene have been isolated, only limited information about the sequence of the flanking region is available (Lundwall *et al.*, 1984). Nucleotide sequence data for this 5′ flanking region are relevant to regulation of C3 expression by endotoxin, interleukin 1 and $\gamma$-interferon (see below), especially since inter-species differences have already been recognized.

*The fourth component (C4)*

The fourth component of complement (C4) is a serum glycoprotein composed of three disulphide-linked polypeptides (Schreiber & Muller-Eberhard, 1974). Synthesis of the 185 kDa single-chain precursor of C4, pro-C4 (Hall & Colten, 1977; Roos *et al.*, 1978) is directed by a polyadenylated mRNA of approx. 5 kb (Ogata *et al.*, 1983; Whitehead *et al.*, 1983). The $\beta$ chain (about 78 kDa) is the *N*-terminal subunit, the $\alpha$ chain (about 95 kDa) the central subunit and the $\gamma$ chain (about 31 kDa) the *C*-terminal segment of pro-C4 (Goldberger *et al.*, 1980; Parker *et al.*, 1980). These chains are separated in the precursors protein by arginine-rich intersubunit linking peptides (Ogata *et al.*, 1983; Whitehead *et al.*, 1983). Postsynthetic processing of pro-C4 involves proteolytic excision of the linking peptides, sulphation (Karp, 1983*a*) modification of residues within the region of the thiolester site (Karp, 1983*b*) and glycosylation of $\alpha$ and $\beta$ chains (Roos *et al.*, 1980; Matthews *et al.*, 1982). Finally, extracellular modification of $\alpha$ chain results in the apparent loss of a *C*-terminal peptide of uncertain length (Chan *et al.*, 1983).

In the murine and human genome, there are two C4-like genes per haploid but a relatively high frequency of duplications, deletions and rearrangements leads to variations in this region. Some species such as guinea pigs (Whitehead *et al.*, 1983) and hamster (Levi-Strauss *et al.*, 1985) appear to have only a single C4 gene. The human C4 genes differ in size (C4A, 22 kb; C4B, 16 kb). This size difference results primarily from the presence of a larger intron in the 5′ region of the C4A gene, but complete details of genomic structure, particularly in the 5′ flanking region have not yet been published (Carroll *et al.*, 1984*a,b*). Nonaka *et al.* (1985) have isolated and characterized genomic clones corresponding to the analogous murine genes. The S1p and C4 genes are each 16 kb and, in contrast to human C4A and C4B, are similar in exon–intron structure and nucleotide sequence. Of particular interest is the observation that C4 and S1p differ in their 5′ flanking regions. This is of importance in view of the extensive sequence homology within coding regions (96% nucleotide: 94% amino acid) and the differences in regulation of C4 and S1p. The divergence in the flanking region between the two is accounted for by many -CACA- repeats 5′ to the C4 gene. A possible role for these sequences in regulation of expression of other genes has been proposed (Rich *et al.*, 1984). The possible role of the -CACA- repeats must be accepted with caution, however, because C4 and S1p from only a single mouse strain have been studied in sufficient detail. Analysis of the C4 and S1p genes and 5′ flanking sequences in strains of the H-2$^{w7}$, W16, and W19 haplotypes will be of particular interest since in these strains the 'C4 gene complex' has been duplicated (Levi-Strauss *et al.*, 1985) and S1p is not under sex hormone control (i.e., is constitutively expressed in the female) (Karp *et al.*, 1982).

*The fifth component (C5)*

C5 is synthesized in hepatocytes and at extrahepatic sites in macrophages as a single-chain precursor protein which is then processed to the two chain native protein (Ooi & Colten, 1979). The amino acid sequence derived from nucleotide sequence analysis of human (Lundwall *et al.*, 1985) and murine (B. F. Tack, unpublished work) C5 cDNA clones indicates that $\beta$ chain is the *N*-terminal subunit of pro-C5 and it is separated from $\alpha$ chain by an arginine-rich linking peptide. The structure of the C5 gene and the factors that govern its expression are not yet known.

## Regulation of Expression

The complement effector proteins of the alternative and classical pathways (up to C5) and several of the complement regulatory proteins are synthesized in hepatocytes and at extrahepatic sites in mononuclear phagocytes. Examples of tissue specific regulation and the effects of cellular maturation, species specificity and other variables have been uncovered. Together with the emerging nucleotide sequence data outlined in the earlier sections, the information on regulation of complement gene expression now permits speculation about the structural basis for these differences. Differences in hepatic and extrahepatic regulation provide independent control of the size of the complement protein reservoir in plasma, on the one hand, and the complement proteins produced at local sites of tissue injury or inflammation on the other. The effect of this differential control is illuminated in recent studies demonstrating that endotoxin, interleukin-1 and $\gamma$-interferon are modulators of complement gene expression.

### Endotoxin

The direct effect of lipopolysaccharide (LPS) on complement gene expression has been studied in adult and newborn human mononuclear phagocytes and in mice. In addition, since LPS also stimulates interleukin 1 (IL-1) production by macrophages (reviewed by Dinarello, 1984), an indirect effect of this agent on complement gene expression has also been revealed (see below). Strunk *et al.* (1985*a*) have recently shown that the lipid A component of LPS increases net C3 protein synthesis 5–30-fold in adult human mononuclear phagocytes and that the mechanism is largely pretranslational, since the concentration of C3-specific mRNA is increased at least 5-fold. Smaller but significant increases in C2 and factor B mRNA were induced by LPS. A corresponding increase in factor B synthesis (2–3-fold) was noted but net C2 protein synthesis was not affected by LPS. It is possible that a second signal is required to initiate a change in C2 translation rates, or that the catabolism of C2 protein is increased by LPS at the same time, so that no net change in C2 synthesis is observed. It is this aspect of the study that resulted in a search for signals other than LPS that regulate complement expression in macrophages.

The C3 response to LPS stimulation is developmentally regulated, as indicated by a comparison of human adult and neonatal monocytes (M. B. St. John-Sutton, R. C. Strunk & F. S. Cole, unpublished work). C3 and factor B gene expression are not affected by lipid A in neonatal monocytes, though other monocyte functions (e.g. superoxide generation) are increased. Although the mechanisms are not clear, preliminary work suggests that C3 and B expression is enhanced in the neonatal cells when exposed to precursors of lipid A, to proteins associated with endotoxin and when the neonatal cells are coincubated with adult monocytes exposed to LPS. The response of the neonatal monocyte to LPS is in some respects analogous to the response in mice of the C3H/HeJ strain. Macrophages from these animals fail to respond to lipid A but are responsive to lipid A precursors (e.g. as judged by IL-1 generation). This strain has been useful for studies of factor B and other acute phase gene expression (Ramadori et al., 1985; G. Ramadori, J. D. Sipe, C. A. Dinarello, S. D. Mizel & H. R. Colten, unpublished work). For instance, endotoxin injection into an LPS-responsive (C3H/FeJ) strain, but not the LPS-unresponsive strain (C3H/HeJ), induces a dose-dependent increase in factor B gene expression in liver as well as in lung, kidney, intestine, spleen, heart and peritoneal macrophages. In experiments with IL-1 the C3H/HeJ strain served as an important control for the specificity of the IL-1 effect. In that study, direct evidence was obtained that, in hepatocytes, the increased expression of factor B, C3 and other acute phase proteins induced by endotoxin or sterile inflammation could be accounted for by IL-1, a product of macrophages. Whether regulation of extrahepatic factor B gene expression (especially in peritoneal macrophages) is regulated directly by IL-1 (autocrine or paracrine regulation) or via another cytokine (e.g. $\gamma$-interferon) is uncertain.

*Interleukin 1*

Recently the purification of human IL-1 to homogeneity (Dinarello, 1984) and isolation of cDNA clones corresponding to murine interleukin 1 (LoMedico et al., 1984) have allowed generation of well-defined reagents for examination of the effect of these polypeptides on complement gene expression. For these and the studies of $\gamma$-interferon regulation we have initially focused our attention on the C2/factor B complex, but the importance of these mediators in the control of other genes has been recognized.

In primary murine hepatocyte cultures both recombinant generated murine IL-1 (rIL-1) and highly purified human IL-1 induce a dose- and time-dependent, reversible increase in expression of the factor B gene and a decrease in albumin gene expression (Ramadori et al., 1985). This regulation is pretranslational since the kinetics and direction of change in specific mRNA for factor B and albumin correspond to the change in synthesis of the respective proteins.

A similar study showed species-independent IL-1 regulation of factor B and C3 gene expression (increase) and albumin expression (decrease) in a human hepatoma derived cell line (D. H. Perlmutter, G. Goldberger, C. A. Dinarello, S. B. Mizel & H. R. Colten, unpublished work). The expression of the C2 gene is unaffected by IL-1, a point of some significance in comparison to the regulation of C2 and B by $\gamma$-interferon. Evidence suggests that the effect of IL-1 on human C3, factor B and albumin expression is mediated by a change in transcription, though the precise mechanism and structural requirements are yet to be determined. We have undertaken a preliminary comparison of nucleotide sequences 5' to several acute phase responsive genes [human $\alpha_1$-proteinase inhibitor, rat fibrinogen, mouse C3, human C-reactive protein (CRP), rat $\alpha_1$ acid glycoprotein and human haptoglobin] for which sequence data are available in the GenBank. The sequence for the mouse albumin gene, a negative acute phase reactant, was also included in the analysis. Thus far, no regions of sequence homology 5′ to the coding regions are common to all the positive acute phase reactants and no sequences 5′ to the mouse albumin gene suggest a basis for the decrease in expression induced by IL-1. Nevertheless, the small size of the factor B 5′ flanking region (limited by the 3′ terminus of the C2 gene) provides a well-defined region to probe the structural basis for the differential effect of IL-1 on B and the closely related IL-1-unresponsive, gene C2. A cosmid bearing both genes on a 30 kb segment of human DNA has been introduced into mouse L cells where each is constitutively expressed (Perlmutter *et al.*, 1985). These class III transfected cells synthesize and secrete C2 (functionally active) and factor B proteins indistinguishable from the products produced in primary human cell cultures. In the transfectant, human factor B gene expression is increased when the cells are incubated in murine rIL-1 or human IL-1 (D. H. Perlmutter, G. Goldberger, C. A. Dinarello, S. B. Mizel & H. R. Colten, unpublished work). The response is dose-dependent and reversible. Expression of the endogenous murine C3 gene (constitutively expressed in murine fibroblasts) is increased by IL-1 but the human C2 gene, as in primary cells, is unaffected by IL-1.

### *γ-Interferon*

The expression of the human C2 and factor B gene is regulated by γ-IFN in primary cell culture and in the class III transfectants that constitutively express the C2 and Bf genes (Strunk *et al.*, 1985b). γ-IFN induces a dose-dependent increase in synthesis of C2 and factor B in both cell types; an effect detected at concentrations of γ-IFN of less than 50 pg/ml. The regulation is mediated at a pretranslational level, but the kinetics and magnitude of the response to γ-IFN for the two proteins is different. In both primary human monocytes and in the class III transfectants, the γ-IFN-induced increase in factor B gene expression always exceeds the increase in C2 gene expression. In primary cell cultures, the γ-IFN effect on factor B is detected earlier than the effect on C2, but in the class III transfectant the kinetics of γ-IFN induction are similar for both genes. These data suggest differences in transcriptional and/or post-transcriptional events for the C2 and Bf genes in the two cell types. Alternatively, since the response is species restricted and probably specified by the receptor (human γ-IFN on human mononuclear phagocytes and murine γ-IFN for L-cells), the effect could be due to a difference in species-specific signal transduction. Transfection of the human C2-Bf gene complex into human fibroblasts might clarify this point.

A similar explanation may account for the difference in response of the human and murine C3 genes to γ-IFN in the mononuclear phagocyte and L-cell; i.e., γ-IFN has no effect on or decreases expression of human monocyte C3, whereas γ-IFN induces an increase in expression of the endogenous murine C3 in the L-cell. On the other hand, the differences in C3 regulation by γ-IFN may reflect differences in regulatory sequences within or flanking the corresponding human and murine genes. Once the human C3 5′ flanking sequences are determined, a direct comparison and site specific mutations will determine which of the alternatives is correct.

Previous work indicated that the single copy human C2 gene generates multiple C2 primary translation products, only one of which is secreted following postsynthetic processing (Perlmutter *et al.*, 1984). $\gamma$-IFN induces an increase in both the secreted and cell-associated forms of C2 in the mononuclear phagocyte. In the L-cell the C2 response to $\gamma$-IFN is more difficult to interpret; i.e. $\gamma$-IFN induced an increase in expression of the 2.8 kb C2 mRNA without an effect on steady state concentration of the C2 mRNA of the lower molecular mass (2.0 kb). The latter may be mostly C2 mRNA derived from incomplete transcripts or products of abnormal post-transcriptional processing of a conventional transcript. In either case it is not likely that the majority of the 2.0 kb C2 mRNA corresponds to that which programs synthesis of the cell-associated forms since, as in the mononuclear phagocyte, $\gamma$-IFN induction of cell-associated C2 protein was observed in the L-cell.

## Conclusion

The primary structure of several genes encoding acute phase proteins, C-reactive protein (Woo *et al.*, 1985) haptoglobin (Maeda, 1985) $\alpha_1$-acid glycoprotein (Reinke & Feigelson, 1985) as well as IL-1-responsive complement genes have been published. These data plus the detailed analysis of the C2 factor B gene complex should provide guidelines for initiating direct studies that will reveal the structural basis for molecular regulation of inflammation and, by extension, genes important in specific immunity (e.g. IL-2 class I and II MHC proteins, etc.).

## References

Alpert, S. E., Auerbach, H. S., Cole, F. S. & Colten, H. R. (1983) *J. Immunol.* **130**, 102–107
Brade, V., Hall, R. E. & Colten, H. R. (1977) *J. Exp. Med.* **146**, 759–765
Campbell, R. D., Bentley, D. R. & Morley, B. J. (1984) *Philos. Trans. R. Soc. London Ser. B.* **306**, 367–378
Carroll, M. C., Belt, T., Palsdottir, A. & Porter, R. R. (1984*a*) *Philos. Trans. R. Soc. London Ser B.* **306**, 379–388
Carroll, M. C., Campbell, R. D., Bentley, D. R. & Porter, R. R. (1984*b*) *Nature London* **307**, 237–241
Carroll, M. C., Campbell, R. D. & Porter, R. R. (1985) *Proc. Natl. Acad. Sci. U.S.A.* **82**, 521–525
Chan, A. C., Mitchell, K. R., Munns, T. W. & Atkinson, J. P. (1983) *Proc. Natl. Acad. Sci. U.S.A.* **80**, 268–272
Chaplin, D. D., Woods, D. E., Whitehead, A. S., Goldberger, G., Colten, H. R. & Seidman, J. G. (1983) *Proc. Natl. Acad. Sci. U.S.A.* **80**, 6947–6951
Chaplin, D. D., Sackstein, R., Perlmutter, D. H., Weis, J. H., Coligan, J., Colten, H. R. & Seidman, J. G. (1984) *Cell* **37**, 569–576
Christie, D. L., Gagnon, J. & Porter, R. R. (1980) *Proc. Natl. Acad. Sci. U.S.A.* **77**, 4923–4927
Cole, F. S., Matthews, W. J., Marino, J. T., Gash, D. J. & Colten, H. R. (1980) *J. Immunol.* **125**, 1120–1124
Cole, F. S., Mathews, W. J., Rossing, T. H., Gash, D. J., Lichtenberg, N. A. & Pennington, J. E. (1983) *Clin. Immunol. Immunopathol.* **27**, 153–159
Cole, F. S., Auerbach, H. S., Goldberger, G. & Colten, H. R. (1985) *J. Immunol.* **134**, 2610–2616
Cooper, N. R. (1975) *Biochemistry* **14**, 4245–4249
Dinarello, C. A. (1984) *Rev. Infect. Dis.* **6**, 51–94
Fey, G., Domdey, H., Wiebauer, K., Whitehead, A. S. & Odink, K. (1983) *Springer Semin. Immunol.* **6**, 119–147

Goldberger, G., Abraham, G. N., Williams, J. & Colten, H. R. (1980) *J. Biol. Chem.* **255**, 7071–7074

Goldberger, G., Thomas, M. L., Tack, B. F., Williams, J., Colten, H. R. & Abraham, G. N. (1981) *J. Biol. Chem.* **256**, 12617–12619

Hall, R. E. & Colten, H. R. (1977) *Proc. Natl. Acad. Sci. U.S.A.* **74**, 1707–1710

Hall, R. E., Blaese, R. M., Davis, A. E., III, Tack, B. F., Colten, H. R. & Muchmore, A. V. (1982) *J. Exp. Med.* **156**, 834–843

Itakura, K., Hutton, J. J., Boyse, E. A. & Old, L. J. (1972) *Transplantation* **13**, 239–243

Karp, D. R. (1983*a*) *J. Biol. Chem.* **258**, 12745–12748

Karp, D. R. (1983*b*) *J. Biol. Chem.* **258**, 14490–14495

Karp, D. R., Atkinson, J. P. & Shreffler, D. C. (1982) *J. Biol. Chem.* **257**, 7330–7335

Levi-Strauss, M., Tosi, M., Steinmetz, M., Klein, J. & Meo, T. (1985) *Proc. Natl. Acad. Sci. U.S.A.* **82**, 1746–1750

LoMedico, P. T., Gubler, U., Hellman, C. P., Dukovich, M., Giri, J. G., Pan, Y. C. E., Collier, K., Semionow, R., Chua, A. O. & Mizel, S. B. (1984) *Nature (London)* **312**, 458–460

Lundwall, A. B., Wetsel, R. A., Domdey, H., Tack, B. F. & Fey, G. H. (1984) *J. Biol. Chem.* **259**, 13851–13856

Lundwall, A. B., Wetsel, R. A., Kristensen, T., Whitehead, A. S., Woods, D. E., Ogden, R. C., Colten, H. R. & Tack, B. F. (1985) *J. Biol. Chem.* **260**, 2108–2112

Maeda, N. (1985) *J. Biol. Chem.* **260**, 6698–6709

Matthews, W. J., Goldberger, G., Marino, J. T., Einsten, L. P., Gash, D. J. & Colten, H. R. (1982) *Biochem. J.* **204**, 839–846

Natsuume-Sakai, S., Hayarawa, J. I. & Takahashi, M. (1978) *J. Immunol.* **121**, 491–498

Nonaka, M., Nakayama, K., Yeul, Y. D., Shimizu, A. & Takahashi, M. (1985) *J. Biol. Chem.* **260**, 10936–10943

Ogata, R. T., Shreffler, D. C., Sepich, D. S. & Lilly, S. P. (1983) *Proc. Natl. Acad. Sci. U.S.A.* **80**, 5061–5065

Ooi, Y. M. & Colten, H. R. (1979) *J. Immunol.* **123**, 2494–2498

Parker, K. L., Capra, J. D. & Shreffler, D. C. (1980) *Proc. Natl. Acad. Sci. U.S.A.* **77**, 4275

Perlmutter, D., Cole, F. S., Goldberger, G. & Colten, H. R. (1984) *J. Biol.Chem.* **259**, 10380–10385

Perlmutter, D. H., Colten, H. R., Grossberger, D., Strominger, J., Seidman, J. G. & Chaplin, D. D. (1985) *J. Clin. Invest.* **76**, 1449–1454

Ramadori, G., Sipe, J. D., Dinarello, C. A., Mizel, S. B. & Colten, H. R. (1985) *J. Exp. Med.* **162**, 930–942

Reid, K. B. M. (1985) *Immunology* **55**, 185–196

Reinke, R. & Feigelson, P. (1985) *J. Biol. Chem.* **260**, 4397–4403

Rich, A., Nordheim, A. & Wang, A. H. J. (1984) *Annu. Rev. Biochem.* **53**, 791–846

Roos, M. H., Atkinson, J. P. & Shreffler, D. C. (1978) *J. Immunol.* **121**, 1106–1115

Roos, M. H., Kornfeld, S. & Shreffler, D. C. (1980) *J. Immunol.* **124**, 2860–2863

Sackstein, R. & Colten, H. R. (1984) *J. Immunol.* **133**, 1618–1626

Sackstein, R., Colten, H. R. & Woods, D. E. (1983) *J. Biol. Chem.* **258**, 14693–14697

Schreiber, R. D. & Muller-Eberhard, H. J. (1974) *J. Exp. Med.* **140**, 1324–1335

Strunk, R. C., Whitehead, A. S. & Cole, F. S. (1985*a*) *J. Clin. Invest.* **76**, 985–990

Strunk, R. C., Cole, F. S., Perlmutter, D. H. & Colten, H. R. (1985*b*) *J. Biol. Chem.* **260**, 15280–15285

Sundsmo, J. S. & Goetze, O. (1981) *J. Exp. Med.* **154**, 763–777

Sundsmo, J. S. & Wood, L. M. (1981) *J. Immunol.* **127**, 877–880

White, P. C., Grossberger, D., Onufer, B. J., Chaplin, D. D., New, M. I., Dupont, B. & Strominger, J. L. (1985) *Proc. Natl. Acad. Sci. U.S.A.* **82**, 1089–1093

Whitehead, A. S., Solomon, E., Chambers, S., Bodmer, W. F., Povey, S. & Fey, G. (1982) *Proc. Natl. Acad. Sci. U.S.A.* **79**, 5021–5025

Whitehead, A. S., Goldberger, G., Woods, D. E., Markham, A. F. & Colten, H. R. (1982) *Proc. Natl. Acad. Sci. U.S.A.* **80**, 5387–5391

Whitehead, A. S., Colten, H. R., Chang, C. C. & DeMars, R. J. (1985) *J. Immunol.* **134**, 641–643

Wiebauer, K., Domdey, H., Diggelman, H. & Fey, G. (1982) *Proc. Natl. Acad. Sci. U.S.A.* **79**, 7077–7081

Woo, P., Korenberg, J. & Whitehead, A. S. (1985) *J. Biol. Chem.* **260**, 13384–13388

Woods, D. W., Markham, A. F., Ricker, A. T., Goldberger, G. & Colten, H. R. (1982) *Proc. Natl. Acad. Sci. U.S.A.* **79**, 5661–5665

*Biochem. Soc. Symp.* **51**, 47–57
*Printed in Great Britain*

# Leukocyte Complement Receptors and Adhesion Proteins in the Inflammatory Response: Insights from an Experiment of Nature

TIMOTHY A. SPRINGER* and DONALD C. ANDERSON†

* *Dana–Farber Cancer Institute and Harvard Medical School, 44 Binney Street, Boston, MA 02115, U.S.A., and †Baylor College of Medicine and Texas Children's Hospital, 6621 Fannin, Houston, TX 77030, U.S.A.*

## Synopsis

The complement receptor type 3 (CR3) mediates phagocytosis and degradation of iC3b-opsonized particles by macrophages and granulocytes. The CR3 is identical to the Mac-1 molecule, which is composed of two non-covalently associated glycoprotein subunits, $\alpha$M of $M_r$ 170000 and $\beta$ of $M_r$ 95000. Patients with recurring, life-threatening bacterial infections have been identified who have moderate (95%) or severe (> 99%) deficiency of Mac-1 and of the related LFA-1 and p150,95 molecules. The primary defect is in the shared $\beta$ subunit of these molecules. Patient leukocytes are not only deficient in CR3 but in a wide variety of adhesion-dependent functions, including granulocyte chemotaxis, adherence to surfaces, and aggregation. Monoclonal antibodies to the Mac-1 $\alpha$ subunit and to the $\beta$ subunit block these functions. The hypothesis will be advanced that Mac-1 functions dually as the CR3 and in 'nonspecific' adherence reactions. Adherence functions are stimulated in normal granulocytes by chemoattractants, which also induce a rapid 5-fold increase in Mac-1 and p150,95 on the cell surface. It is proposed that absence of Mac-1 and p150,95 expression and upregulation by patient granulocytes is causally related to their inability to extravasate and migrate into inflammatory sites.

## Introduction

Cell surface adherence reactions are of central importance in a wide spectrum of granulocyte, monocyte, and lymphocyte functions which contribute to host defense against infection. Granulocyte and monocyte translocation *in vitro* and mobilization *in vivo* are influenced by the nature of cell–substrate adherence interactions [1]. Studies employing time-lapse photography have shown that granulocytes adhere preferentially to vascular endothelium adjacent to a site of inflammation prior to their diapedesis into tissues [2]. This 'directed' adherence is facilitated by byproducts of inflammation such as C5a which bind to specific receptors on granulocytes and monocytes and initiate a sequence of events that enhance cellular adherence [3].

Adhesive interactions are fundamental to other granulocyte and monocyte

Table 1. *The Mac-1, LFA-1 family*

Common features: the $\beta$ subunits appear identical. The $\alpha$ subunits $\alpha$M and $\alpha$L are 35% homologous in sequence. The $\alpha$ and $\beta$ subunits are noncovalently associated in $\alpha_1\beta_2$ complexes. Both $\alpha$ and $\beta$ subunits are glycosylated and exposed on the cell surface. All functions shown require divalent cations. An important property of selected proteins of the Mac-1 family, as well as of other critical leukocyte receptors (e.g. CR1) is the capacity of inflammatory stimuli to 'upregulate' or enhance their surface expression (reviewed in [9, 10]).

| Property | Mac-1 (OKM1, Mo1) | LFA-1 | p150.95 |
|---|---|---|---|
| Subunits | $\alpha$M, $\beta$ | $\alpha$L, $\beta$ | $\alpha$X, $\beta$ |
| ($M_r \times 10^{-3}$) | (170, 95) | (180, 95) | (150, 95) |
| Cell distribution | Monocytes, macrophages, granulocytes, large granular lymphocytes | Lymphocytes, monocytes, granulocytes, large granular lymphocytes | Monocytes, macrophages, granulocytes |
| Stimulation increases surface expression | + | − | + |
| Functions inhibited by MAbs | CR3; granulocyte adherence; stimulated adherence; spreading, aggregation and chemotaxis | Cytolytic T lymphocyte-mediated killing and T helper cell responses; natural killing; antibody-dependent cellular cytotoxicity; phorbol ester-stimulated lymphocyte aggregation | ? |

functions. Specific recognition of opsonized mircro-organisms is facilitated by membrane receptors for IgG and for the third component of complement (C3), which mediate cell–microbe adhesion prior to triggering of endocytosis. Adhesion mediated by IgG (Fc) receptors is also required for antibody-dependent cytotoxicity of target cells. In the absence of opsonins, some micro-organisms may adhere to granulocytes and monocyte–macrophages without undergoing ingestion or may be phagocytosed non-specifically, depending on the physical properties of the micro-organism.

Many different cell surface proteins are important in these events. Increasing evidence indicates that the Mac-1, LFA-1 family of glycoproteins (Table 1) are of ubiquitous importance in the aforementioned granulocyte and monocyte adhesion reactions, and additionally in lymphoid adhesive interactions such as those required for T lymphocyte or natural killer cell cytotoxicity. These molecules appear to synergize with other receptors or function independently to regulate or mediate a panoply of functional interactions. The wide variety of these functions, and their common dependence on cell adhesion, suggests that the Mac-1, LFA-1 glycoproteins are of general importance in leukocyte adhesion phenomena, and thus, may be analogous to the adhesion molecules of other tissues, such as the nervous system (N-CAM) or liver (L-CAM).

Importantly, chemotactically relevant concentrations of the chemoattractants fMet-Leu-Phe and C5a, and the secretagogues phorbol myristate acetate and calcium ionophore, stimulate a 5–10-fold increase in the amount of Mac-1 and

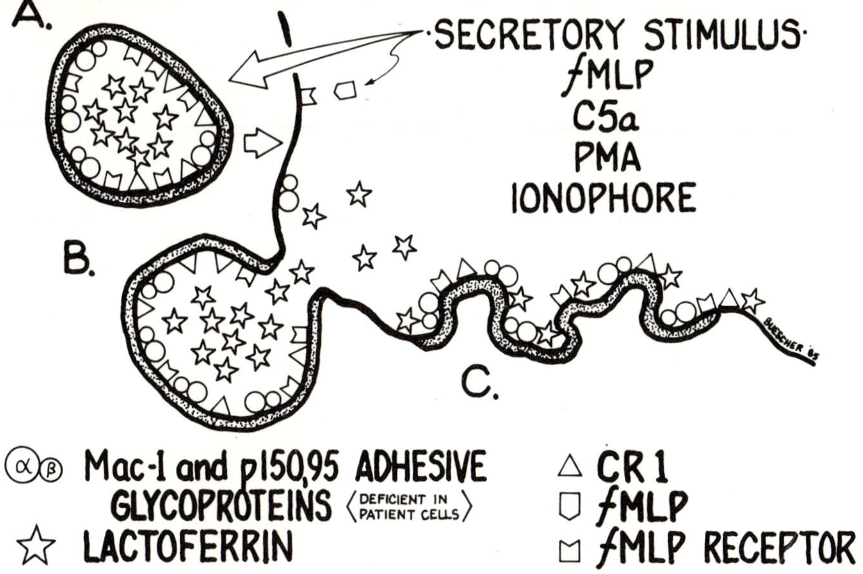


Fig. 1. *Secretory vesicle mobilization in granulocytes*

The components shown are all mobilized to the cell surface or secreted in response to the indicated stimuli. Deficient patient cells lack an intracellular pool of Mac-1 and p150,95 and thus fail to mobilize them to the cell surface; the secretory response and mobilization of other surface components such as the CR1 and fMet-Leu-Phe (fMLP) receptor is otherwise completely normal in patient cells. Mac-1 and p150,95, the CR1, and fMet-Leu-Phe receptor may be in storage sites distinct from one another and from that of lactoferrin in the secondary granule. They are shown in the same secretory vesicle only for ease of representation. (Drawing by Dr. S. Buescher.)


p150,95 expressed on the surface of monocytes and granulocytes [4]. In contrast, the related LFA-1 glycoprotein is not increased. 'Upregulation' is maximal after approx. 8 min at 37 °C, and is not impeded by inhibitors of protein synthesis (T. Springer, L. Miller & D. Anderson, unpublished work). Thus, Mac-1 and p150,95 are stored in a latent pool in granulocytes and monocytes, and can be rapidly mobilized to the cell surface by inflammatory stimuli (Fig. 1). In granulocytes, the intracellular Mac-1 pool co-sediments in sucrose gradients with secondary granules [5,6, T. Springer & N. Borregaard, unpublished work]. Furthermore, permeabilization of granulocytes with saponin greatly increases the amount of fluorescent staining with anti-Mac-1 MAb (T. Springer, unpublished work). Further experiments are needed before it can be definitively established whether Mac-1 and p150,95 are stored in the membrane-enclosing secondary granules or some other secretory vesicle. By a number of different measurement techniques, the amount of Mac-1 stored in the latent pool is 5–10-fold higher than on the resting cell surface. Since a 5–10-fold increase in surface Mac-1 is accompanied by only about 25% increase in granulocyte surface area [7], the density of Mac-1 and p150,95 in the membrane bilayer of the storage vesicle must be approx. 16–36-fold higher than on the cell surface.

A major objective of the present paper is to provide evidence that the Mac-1 glycoproteins regulate monocyte and granulocyte adherence, chemotaxis and other adherence-dependent functions *in vitro*, and are required for leukocyte recruitment into inflammatory sites *in vivo*. Two types of studies are presented. The first utilizes leukocytes from patients with a recently discovered heritable deficiency of the entire Mac-1 glycoprotein family [8]. The other type utilizes MAbs to these glycoproteins [9] together with normal leukocytes *in vitro*. A number of MAbs have been obtained which are specific for the $\alpha M$, $\alpha L$, or $\alpha X$ subunits, and thus react only with Mac-1, LFA-1, or p150,95, respectively. Another type of MAb reacts with the common $\beta$ subunit, and hence with all three of these glycoproteins (Table 1).

## Mac-1, LFA-1 Deficiency Disease

Recently, a disease has been recognized in which the Mac-1, LFA-1, and p150,95 glycoproteins are deficient [4,8]. Patients have recurrent, life-threatening bacterial infections, a lack of pus formation, and persistent granulocytosis. The deficiency affects all cell lineages which normally express the Mac-1, LFA-1 glycoprotein family, i.e., monocytes, granulocytes, and lymphocytes, and cell lines established from patients. Deficiency is inherited as an autosomal, recessive mutation. Each of the three $\alpha$ subunits, and the common $\beta$ subunit, is deficient from the surface of all patients' cells, as shown by immunofluorescent flow cytometry and immunoprecipitation with MAb specific for each subunit. Two phenotypes have been defined, severe deficiency and moderate deficiency, with surface expression of $< 0.2\%$ and $5–10\%$, respectively, of normal amounts of Mac-1, LFA-1, and p150,95 [8] (Fig. 2). Patient granulocytes show normal surface expression of the Fc receptor, the complement receptor type 1 (CR1), and many other markers surveyed in an inernational monoclonal antibody workshop [10].

In both phenotypes, the underlying defect is in the common $\beta$ subunit [4], as summarized in Fig. 3. In normal cells, $\alpha$ and $\beta$ subunit precursors ($\alpha'$ and $\beta'$) are synthesized and become noncovalently associated, probably in the endoplasmic reticulum, and transported to the Golgi, where carbohydrate processing and a slight increase in molecular weight occurs. The mature molecules are then transported to the cell surface or to intracellular storage sites. Patient cells, however, appear to lack $\beta$ subunit synthesis, or to make it only in small amounts. Normal $\alpha'$ precursors are made, but do not undergo carbohydrate processing, suggesting biosynthesis is blocked prior to the Golgi. $\beta$ association appears to be required for processing and transport to the surface. The $\alpha$ chains are not expressed on the surface (severe deficiency) or in amounts which appear stoichiometrically limited by the small quantity of $\beta$ produced (moderate deficiency). In addition to surface expression, patient granulocytes and monocytes lack the intracellular pool of Mac-1 and p150,95. After stimulation with fMet-Leu-Phe or phorbol ester, patient cells show little if any increase in Mac-1 and p150,95 surface expression [4,8].

## Functional Consequences of Deficiency

The effects of this deficiency disease have taught us much about the importance of the Mac-1, LFA-1 glycoprotein family in leukocyte adhesion and

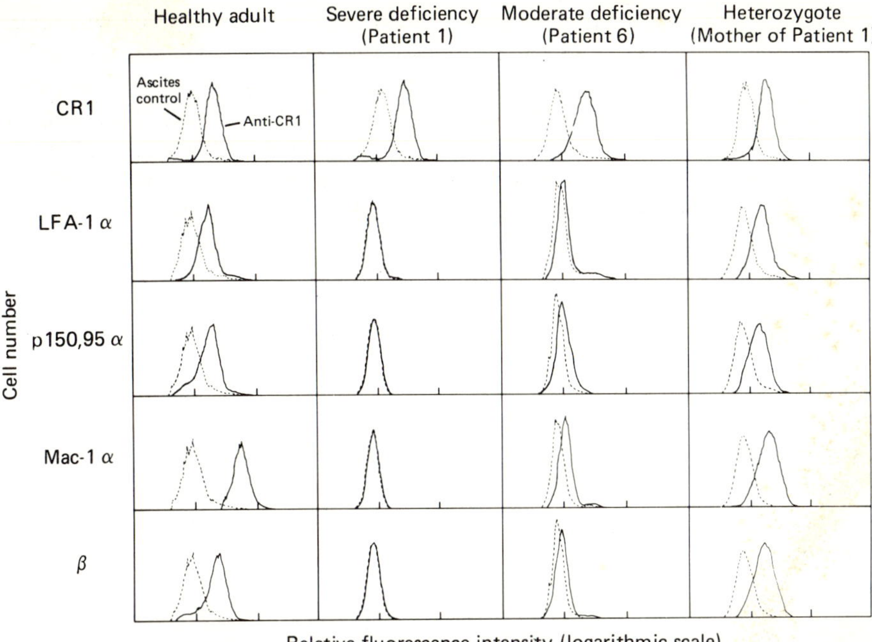


Fig. 2. *Immunofluorescence flow cytometry of granulocytes of representative severe and moderate deficiency patients, a heterozygote, and a healthy adult*

Unstimulated granulocytes were indirectly stained with antibodies to the CR1 or the indicated $\alpha$ or $\beta$ subunits (solid lines) or control MAb (broken lines). A similar degree of deficiency was found if patient granulocytes were stained after fMet-Leu-Phe stimulation. [From Anderson *et al.* (1985) with permission.]

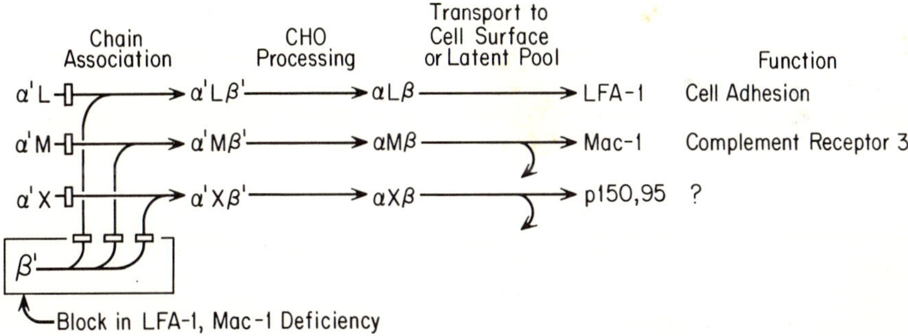


Fig. 3. *Biosynthesis of the Mac-1 , LFA-1 glycoprotein family* [4, 9]

Patient cells have a primary block in $\beta$ subunit synthesis, and a secondary block in $\alpha$ subunit processing.

migration [8,11]. The first known function of Mac-1 was as the complement receptor type 3, which mediates binding and phagocytosis of particles opsonized with the iC3b ligand [11,12]. Indeed, despite some initial controversy [13], it is clear that patients are deficient in the CR3 [11,14]. The increased surface expression of the CR3 induced by chemoattractants may be an important mechanism whereby granulocytes increase their ability to phagocytose opsonized

Table 2. *Assessments of adherence-dependent granulocyte functions*

Data presented with respect to each functional assay is represented by mean $\pm 1$ s.d. value for each patient category derived from individual patient mean values of two to six separate experiments. Summarized from [8].

| Functional assay | Severe* deficiency | Moderate† deficiency | Healthy adults |
|---|---|---|---|
| Chemotaxis | | | |
|   fMet-Leu-Phe ($10^{-8}$M) | $43 \pm 6\ddagger$ | $66 \pm 6$ | $105 \pm 4$ |
|   C5a | $42 \pm 5$ | $68 \pm 14$ | $108 \pm 7$ |
| Adherence | | | |
|   Baseline (saline) | $12 \pm 2\S$ | $16 \pm 9$ | $38 \pm 6$ |
|   fMet-Leu-Phe ($10^{-8}$ M) | $12 \pm 3$ | $28 \pm 12$ | $63 \pm 6$ |
|   PMA (5 $\mu$g/ml) | $16 \pm 4$ | $31 \pm 12$ | $67 \pm 9$ |
| Aggregation | | | |
|   C5a | $16 \pm 11\|$ | $15 \pm 12$ | $100 \pm 0$ |
|   fMet-Leu-Phe ($10^{-7}$ M) | $16 \pm 4$ | $14 \pm 13$ | $40 \pm 6$ |
|   PMA (10 $\mu$g/ml) | $15 \pm 9$ | $22 \pm 3$ | $105 \pm 7$ |
| Phagocytosis | | | |
|   Oil Red O (IgG) | $1.4 \pm 0.6\P$ | $1.4 \pm 0.5$ | $1.7 \pm 0.4$ |
|   Oil Red O (iC3b) | $1.9 \pm 1.2\P$ | $2.4 \pm 1.2$ | $7.0 \pm 3.1$ |
|   C3-opsonized zymosan | $4.6 \pm 0.7$** | $7.9 \pm 3.2$ | $17.4 \pm 4.0$ |

  * Includes assessments on severe deficiency patients 1, 2 and 3.

  † Includes assessments on moderate deficiency patients 4, 6, 7 and 8.

  ‡ Boyden assay values (mean $\pm 1$ s.d.) for fMet-Leu-Phe or C5a (10% zymosan-activated serum) expressed as $\mu$m migration/40 min incubation.

  § Percent of granulocytes adhering to serum (6%) coated glass under baseline or stimulated conditions at 21 °C.

  ‖ Granulocyte aggregation responses to C5a (10% zymosan-activated plasma), fMet-Leu-Phe or phorbol myristate acetate (PMA), at 37 °C, measured by the increase in light transmittance and expressed as the % of the response to C5a.

  ¶ Dionylphthalate uptake ($\mu$g/$10^6$ granulocytes in 15 min).

  ** Slope of chemiluminescence evolution ($10^{-5} \times$ c.p.m.$^2$)

micro-organisms as they migrate into inflammatory sites. However, the functional consequences of Mac-1, LFA-1 deficiency are much broader than for a complement receptor deficiency.

The recurrent soft tissue infections in patients appear due to an inability of granulocytes and monocytes to migrate into inflammatory sites [8,11]. There is an absence of pus formation and consequently, the common occurrence of necrotic, ulcerative, skin or mucous membrane infections. This has been confirmed by biopsies of infected tissues (gingiva and skin) and by Rebuck skin windows, which show profoundly impaired leukocyte mobilization. Following Rebuck skin abrasions, healthy control individuals show immigration of neutrophils at 2 and 4 h followed by monocytes at 6 h. Severely deficient patients show no mobilization of neutrophils or monocytes to the site even at 24 h, and leukocyte mobilization in moderately deficient patients is strikingly diminished and delayed. Thus, Mac-1, LFA-1 deficiency results in a profound defect in diapedesis, i.e., in the ability of leukocytes to leave the circulation by migrating between endothelial cells and through the basement membrane in order to accumulate at inflammatory sites.

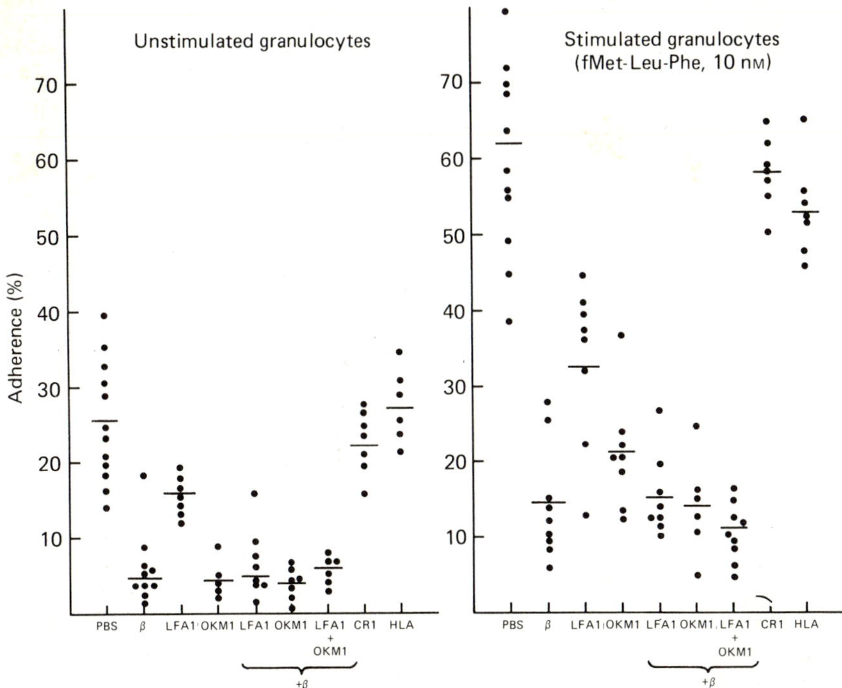



Fig. 4. *Effects of MAbs to the Mac-1, LFA-1 glycoprotein family on granulocyte adherence*

Granulocytes were preincubated with MAbs, washed and then incorporated into Smith Hollers adherence chambers in which they were allowed to adhere to serum (6%) coated glass substrates under unstimulated (phosphate-buffered saline) or stimulated (fMet-Leu-Phe, 10 nM) conditions at 21 °C. MAb preparations used for the studies shown included: the OKM1 MAb to Mac-1 α (5 μg/ml), F(ab')$_2$ fragments of the TS1/22 MAb to LFA-1 α (5 μg/ml), and F(ab')$_2$ fragments of the TS1/18 MAb to the common β subunit of Mac-1 and LFA-1 (5 μg/ml). Control MAbs included saturating concentrations of a F(ab')$_2$ fragment of rabbit IgG directed against the human C3b receptor (anti-CR1) and a MAb against the HLA framework antigen (W6/32).

This dysfunction correlates with defects *in vitro* in adherence and adhesion-dependent functions, including chemotaxis, aggregation, orientation and phago-cytosis of iC3b coated particles (Table 2) [8,11]. In additional patient studies, defects of granulocyte antibody-dependent cellular cytotoxicity, mononuclear cytotoxicity and cytolytic T-lymphocyte responses have been consistently demonstrated; in each assay, disturbed cytotoxic functions are related to diminished target cell binding. Observed abnormalities of lymphocyte prolifer-ative responses to mitogens or allogeneic stimulation are presumably also related to impaired cell–cell interactions. With respect to many of these assays, the relative severity of functional abnormalities documented is directly related to the degree of molecular deficiency and severity of clinical infections or other manifestations in selected patients or kindreds.

In contrast to these characteristic abnormalities of adherence-dependent functions, adherence-independent functions of patient granulocytes or monocytes are uniformly normal. These include specific and saturable binding of the chemotactic factor, fMet-Leu-Phe, normal expression and function of receptors for C5a, IgG Fc and CR1, cellular activation and shape change in suspension, oxidative metabolic and secretory (primary and secondary granule markers) responses to soluble stimulants, and normal assembly or function of cytoskeletal proteins [8].

## Inhibition of Adherence-Dependent Functions by MAb

Binding of MAb to normal granulocytes largely reproduces the defects found in Mac-1, LFA-1 deficient patients. Both baseline and fMet-Leu-Phe-stimulated adherence are strikingly inhibited (Fig. 4). Spreading, chemotaxis, and aggregation are also inhibited. In most assays, the incorporation of MAbs during functional assessments are required to demonstrate effects or to promote the most effective inhibition of cell function. These observations suggest that a continual temporal expression of new Mac-1 binding sites contributes to these functions *in vitro* and presumably *in vivo*. The general order of potency of inhibition is anti-$\beta$ > anti-(Mac-1 $\alpha$) > anti-(p150,95 $\alpha$) > anti-(LFA-1 $\alpha$). This suggests that all members of the glycoprotein family may contribute to these reactions; the order of potency reflects their relative amounts on the granulocyte surface [10]. These effects are quite specific. They are obtained with F(ab')$_2$ fragments [anti-$\beta$ and anti-(LFA-1 $\alpha$)] and are not given by IgG MAb bound to other surface molecules (the CR1 and HLA-A,B). Functions not deficient in patient cells are not inhibited by these MAbs, i.e., shape change in suspension, fMet-Leu-Phe binding, superoxide generation, secondary granule secretion, and phagocytosis of Oil Red O (IgG).

## A Dynamic Model of Chemotaxis and Diapedesis

The clinical and histopathological manifestations observed in Mac-1 deficient individuals as well as the specific molecular and functional deficits documented in patient cells suggest that many adherence-dependent inflammatory reactions of normal cells are mediated by the Mac-1, LFA-1 glycoprotein family, and specifically, that chemoattractant-stimulated hyperadherence and aggregation are mediated by an increased surface expression of Mac-1 and p150,95. MAb-blocking experiments confirm these findings. We believe chemotaxis is profoundly deficient in patients because of the underlying inability to adhere properly to substrates. We propose that, *in vivo*, chemoattractants diffusing from sites of inflammation into the circulation induce Mac-1 and p150,95 'upregulation', leading to increased adherence of monocytes and granulocytes to endothelial substrates. We further propose that, in analogy with the importance of the Mac-1, LFA-1, and p150,95 glycoproteins in chemotaxis *in vitro*, these glycoproteins mediate essential adherence functions during diapedesis and migration into the inflammatory sites (Fig. 5). Thus, we propose that the most important clinical manifestation of Mac-1 deficiency (i.e., the inability of granulocytes to migrate into inflammatory sites and form pus) is due to a lack of 'upregulation' of adhesiveness, which is normally regulated by the increased surface expression of Mac-1 and p150,95.

The molecular mechanisms by which these glycoproteins mediate or regulate

Table 3. *Effects of subunit-specific monoclonal antibodies to Mac-1 proteins on adherence-dependent and adherence-independent granulocyte functions*

For most experiments granulocytes were preincubated in saturating concentrations of MAbs or their $F(ab')_2$ fragments, washed and then assayed. +, Consistently and significantly blocks function; ±, inconsistent or minimal blockage; −, no blockage.

| | Monoclonal antibody specificity | | | | | |
| --- | --- | --- | --- | --- | --- | --- |
| | Mac-1 $\alpha$ | | LFA-1 $\alpha$ | p150,95 $\alpha$ | $\beta$ | HLA-A,B |
| Granulocyte function | (OKM1) | (LM 2/1.6) | [TS1/22 F(ab')$_2$] | (SHCL-3) | [TS1/18 F(ab')$_2$] | (W6/32) |
| Adherence (6% serum-coated glass) | + | + | ± | + | + | − |
| Hyperadherence (6% serum-coated glass + 10 nм-fMet-Leu-Phe) | + | + | ± | + | + | − |
| Spreading (glass) | +* | +* | − | ±* | +* | − |
| Aggregation (C5a) | + | + | ± | + | + | − |
| Chemotaxis (C5a) | +* | | − | ±* | +* | − |
| Phagocytosis (iC3b-Oil Red O) | + | | − | − | + | − |
| Shape change (suspension) | − | − | − | − | − | − |
| fMet-Leu[³H]Phe binding | − | | − | − | − | − |
| Superoxide generation (PMA) | − | − | − | − | − | − |
| Secretion of glucuronidase, vitamin B₁₂ transport protein (PMA) | − | − | − | − | − | − |
| Phagocytosis (IgG-Oil Red O) | − | − | − | − | − | − |

* Requires presence of monoclonal antibody during assay for inhibitory effect.

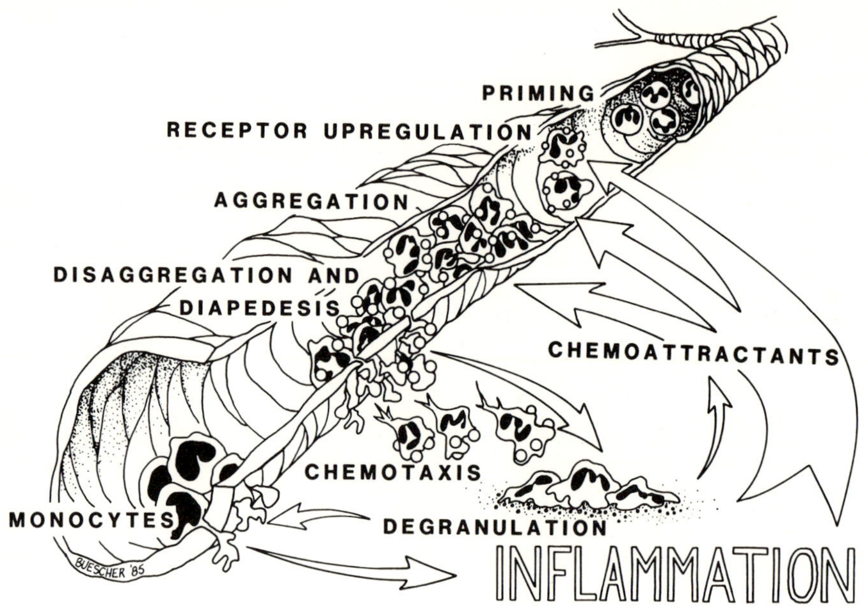


Fig. 5. *Chemoattractant-mediated Mac-1 and p150,95 upregulation, changes in leukocyte adherence, and diapedesis at inflammatory sites*

(Drawing by Dr. S. Buescher.)

adhesivity are not known, and further studies promise to provide important insights into the biology of inflammation. The Mac-1 $\alpha$ subunit gene has recently been cloned in this laboratory, and its sequence may soon yield important clues about structure and function. Whatever the mechanisms, the delivery of adhesive proteins to the cell surface in discrete packages, by fusion of specific granules or other secretory vesicles with the plasma membrane, allows an interesting speculation about the basis of chemotaxis. We propose that during granulocyte chemotaxis, secretory vesicles containing Mac-1 and p150,95 fuse with the plasma membrane at the leading edge of the cell. The site of fusion is hypothesized to be directed by the same chemoattractant-sensing machinery that guides the ruffling lamellipodia toward the chemoattractant. A focal point of high Mac-1 and p150,95 concentration would thus be formed in the plasma membrane at the site of fusion, and adhesion would be initiated or strengthened at this focal point. After further cell translocation, this focal point of attachment would near the uropod, and by this time, diffusion of the adhesion proteins in the plane of the membrane would have lowered their concentration and secondarily, the strength of adhesion. Because each granulocyte contains more than 1000 specific granules, this cycle would be repeated many times. This process could be superimposed on endocytic recycling of membrane, with endocytosis over the entire surface of the cell and readdition at the leading edge of the cell, thus allowing bulk membrane flow from the leading edge to the uropod, as has been hypothesized to effect cell locomotion [15].

## References


1. Smith, C. W. & Hollers, J. C. (1980) *J Clin. Invest.* **65**, 804–812
2. Atherton, A. & Born, G. V. R. (1972) *J. Physiol. (London)* **222**, 447–474
3. Tonnesen, M. G., Smedly, L. A. & Henson, P. M. (1984) *J. Clin. Invest.* **74**, 1581–1592
4. Springer, T. A., Thompson, W. S., Miller, L. J., Schmalstieg, F. C. & Anderson, D. C. (1984) *J. Exp. Med.* **160**, 1901–1918
5. Todd, R. F., III, Arnaout, M. A., Rosin, R. E., Crowley, C. A., Peters, W. A. & Babior, B. M. (1984) *J. Clin. Invest.* **74**, 1280–1290
6. O'Shea, J. J., Brown, E. J., Seligmann, B. E., Metcalf, J. A., Frank, M. M. & Gallin, J. I. (1985) *J. Immunol.* **134**, 2580–2587
7. Hoffstein, S. T., Friedman, R. S. & Weissmann, G. (1982) *J. Cell Biol.* **95**, 234–241
8. Anderson, D. C., Schmalstieg, F. C., Finegold, M. J., Hughes, B. J., Rothlein, R., Miller, L. J., Kohl, S., Tosi, M. F., Jacobs, R. L., Waldrop, T. C., Goldman, A. S., Shearer, W. T. & Springer, T. A. (1985) *J. Inf. Dis.* **152**, 668–689
9. Sanchez-Madrid, F., Nagy, J., Robbins, E., Simon, P. & Springer, T. A. (1983) *J. Exp. Med.* **158**, 1785–1803
10. Springer, T. A. & Anderson, D. C. (1985) in *Leukocyte Typing II: Vol. 3, Human Myeloid and Hematopoietic Cells* (Reinherz, E. L., Haynes, B. F., Nadler, L. M. & Bernstein, I. D., eds.), Springer-Verlag, New York, in the press
11. Anderson, D. C., Schmalstieg, F. C., Arnaout, M. A., Kohl, S., Tosi, M. F., Dana, N., Buffone, G. J., Hughes, B. J., Brinkley, B. R., Dickey, W. D., Abramson, J. S., Springer, T., Boxer, L. A., Hollers, J. M. & Smith, C. W. (1984) *J. Clin. Invest.* **74**, 536–551
12. Beller, D. I., Springer, T. A. & Schreiber, R. D. (1982) *J. Exp. Med.* **156**, 1000–1009
13. Dana, N., Todd, R., Pitt, J., Colten, H. R. & Arnaout, M. A. (1983) *Immunobiology* **164**, 205–206
14. Dana, N., Todd, R. F., III, Pitt, J., Springer, T. A. & Arnaout, M. A. (1984) *J. Clin. Invest.* **73**, 153–159
15. Abercrombie, M., Heaysman, J. E. M. & Pegrum, S. M. (1970) *Exp. Cell Res.* **62**, 389–398

*Biochem. Soc. Symp.* **51**, 59–67
*Printed in Great Britain*

# The Receptor for Immunoglobulin E as a Membrane Protein

HENRY METZGER, GISELE ALCARAZ, JEAN-PIERRE KINET and RODOLFO QUARTO

*Section on Chemical Immunology, Arthritis and Rheumatism Branch, National Institute of Arthritis, Diabetes, and Digestive and Kidney Diseases, National Institutes of Health, Bethesda, MD 20892, U.S.A.*

## Synopsis

Mast cells and related cells have a surface glycoprotein that avidly binds monomeric immunoglobulin E (IgE). This protein is more complex than originally thought, its analysis having been complicated by its lability in mild detergents. The properties of this receptor, especially with respect to its interaction with lipids and detergents, is reviewed and the implications for the study of other membrane protein systems are discussed.

## Introduction

Monomeric immunoglobulin E (IgE) binds with high affinity to a plasma membrane protein found only on mast cells, basophils and related tumour cells. Aggregation of this membrane protein initiates the release of pre-formed and newly synthesized mediators from these cells. In the initial structural analysis of this protein it was determined that the binding site for IgE was on a surface-iodinatable glycopeptide having an $M_r$ of about 50000 [1,2]. More-complete analysis revealed that the glycopeptide ($\alpha$) is closely associated with two other types of polypeptide, $\beta$ and $\gamma$, the latter being present as a disulphide-linked dimer ($\gamma_2$) [3–5].

Initial evidence that the receptor might consist of more than a single $\alpha$ chain came from studies of the $M_r$ of the IgE-binding component. Using an assay by which the binding of IgE could be assessed on solubilized receptors [6], analysis of the activity in crude cell extracts by sedimentation and diffusion revealed an $M_r$ substantially higher than the 50000 determined on polyacrylamide gels in sodium dodecyl sulphate [7]. Nevertheless, amino acid analysis suggested that there was only a single $\alpha$ chain per IgE bound [8]. Studies using the technique of radiation inactivation led to a similar conclusion [9]. Of course, as long as the detection of receptor protein on gels was limited to examining surface-iodinatable components, one could always surmise that there were additional components which simply were not exposed on the surface of the cells. However, even when the receptors from cells that had biosynthetically incorporated radioactive leucine were examined, the $\alpha$ chain was the only polypeptide observed in substantial amounts [3]. Our own studies involved a technique for purification that deliberately avoids the use of any reagents ordinarily considered as denaturants of proteins [10]. Nevertheless, our results from studies using chemical crosslinking reagents [3,4] or a reagent that labels proteins within the membrane bilayer [11], or both, indicated the presence of additional components that were being lost during purification. Ultimately we recognized that prolonged exposure of the receptors to the 'mild' detergents used to solubilize the cells caused dissociation of these additional peptides [12].

In the present paper we review this unusual property and discuss how this dissociation can be taken advantage of for purifying the subunits. We also show why, despite this property, it is appropriate to consider the $\alpha\beta\gamma_2$ tetramer as the unit receptor. Finally, we consider the implications of our results for the study of this and other membrane protein systems.

## Reaction of Receptor with Detergents and Lipids

### *Dissociation of the receptor in detergents*

As already mentioned, we had determined that detergents dissociate $\alpha$ from the other chains. We had also found that addition of phospholipids prevented dissociation [12]. Nevertheless, it was difficult for us to study the dissociation phenomenon for two technical reasons. First, we had no way of labelling each of the polypeptide components efficiently and inexpensively. Second the only solvent by which intact receptors could be purified, micellar detergent plus phospholipid, was a complicated one and therefore served as a poor baseline. After considerable experimentation we observed that (*a*) the intact receptor could be isolated without added lipids if the purification was performed at very low (submicellar) concentrations of detergent [13], and (*b*) the $\alpha$, $\beta$ and $\gamma$ chains could each be extrinsically iodinated if the receptor was briefly exposed to micellar detergent prior to iodination in submicellar detergent [13]. We were then in a position to study the dissociation phenomenon.

Our principal findings have been as follows [14].

(1) Regardless of the extent of dissociation, and of the combination of perturbants (pH, temperature, time etc.) used to produce it, $\beta$ and $\gamma_2$ always dissociate from $\alpha$ in parallel.

(2) It is the concentration of micellar detergent relative to the concentration of the receptors that is the critical variable. For example, if all other experimental conditions are held constant, the amount of dissociation is directly proportional to the concentration of micellar detergent (Fig. 1*a*). Because of the rough correlation between the critical micelle concentration and aggregation number of the detergents we studied, we cannot determine whether the true correlation is between the concentration of the receptor versus. the concentration of micellar detergent or of detergent micelles (compare Figs. 1*a* and 1*b*). Although under otherwise mild conditions the correlation of dissociation with the concentration of micellar detergent is equivalent for certain detergents having critical micellar concentrations that vary over a 50-fold range (Fig. 1*a*), harsher conditions reveal differences in the extent of dissociation induced by this same group of detergents at equivalent concentrations of micellar detergent (see below).

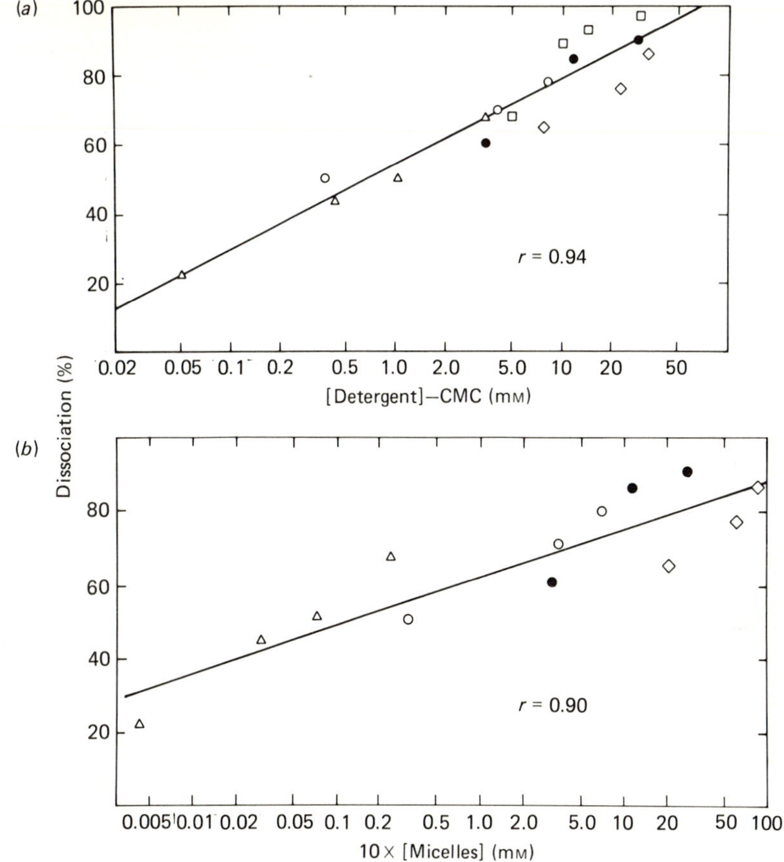

Fig. 1. *Dissociation of receptor of IgE with detergents*

(*a*) Dissociation as a function of the concentration of micellar detergent △, Triton X-114; ○, sodium deoxycholate; ●, Chaps; ◇, sodium cholate; □, octylglucoside. The solvent contained 0.15 M-NaCl and 0.02 M-Tris/HCl, pH 7.6. Dissociation was conducted at 4 °C over 4 h. CMC, critical micellar concentration. (*b*) Same data as in (*a*) except the abcissa shows the concentration of detergent micelles. The latter was calculated from the aggregation number (molecules/micelle). Since this value is not known for octylglucoside the corresponding data were excluded. (From [14] with permission.)

(3) In a fixed volume the molar ratio of (micellar) detergent: receptor must be $10^5$ in order to see substantial dissociation under otherwise mild conditions. This is much higher than the ratio of (micellar) detergent:lipid required to dissolve membranes and disperse suspensions of lipids. Here a ratio of 2–10 is sufficient [15,16].

(4) Those detergents that most efficiently dissolve biomembranes are also those that most efficiently dissociate receptors. Umbreit & Strominger [17] were the first to recognize the usefulness of the hydrophilic/lypophilic balance (HLB number), an empirically determined value which is based on the effectiveness of a detergent to emulsify lipids in an aqueous solvent, for comparing the effectiveness of mild detergents to dissolve biomembranes. They observed that regardless of the specific chemical character of the detergent, those that had

an HLB number between 12 and 14 were most efficient. We found that detergents with HLB numbers in the same range were also maximally effective in inducing the dissociation of the receptor.

(5) Under mild conditions (physiological pH and ionic strength and low temperatures) dissociation proceeds slowly and may take hours to achieve the final value. This final value of dissociation may leave a substantial fraction of the receptors intact. Although we cannot exclude the possibility that heterogeneity of the receptors contributes to incomplete dissociation under such conditions, the combined data suggest rather that an equilibrium is achieved. The data furthermore suggest that this equilibrium involves a third component; that is, a component other than the receptor and the detergent micelles. The likeliest candidate is bound lipid (see below).

(6) Reagents that promote protein denaturation promote chain dissociation also. Conditions which destabilize electrostatic interactions (high salt, extremes of pH, high temperature) as well as those that interfere with hydrophobic interactions (chaotropic ions) all promote dissociation. Because of the complexity of the system (involving a multichain polypeptide complex, lipid and detergent micelles), it is not possible without much more extensive studies to disentangle the possible mechanisms.

(7) Dissociation can be prevented or diminished by addition of phospholipids to the mixture of receptor plus micellar detergent [12,15]. However, so far we have not found simple lipid mixtures to be effective and only complex mixtures of animal cell phospholipids are fully effective (Table 1). The lipids do not simply act by 'neutralizing' the detergent micelles, since the specificity of the protective effect of lipids is not observed if instead the solubilization of membranes by the detergents is analysed. That is, lipids that can prevent the solubilization of membranes are ineffective in preventing dissociation of the receptor [18].

We visualize the action of the detergent as follows. We think it likely that firmly bound lipid is required for the stable interaction of the $\alpha$, $\beta$ and $\gamma$ chains. Whether this is directly through participation of lipid in interchain bonding or indirectly through stabilization of the polypeptide structure by binding of lipid at sites removed from the interface cannot be determined at present. Presumably the micellar detergent can gradually remove this lipid and eventually deplete the receptor to an extent that causes destabilization, providing an exogenous source of lipids is not provided. Investigations are currently under way to quantify the amounts of tightly bound lipid and to identify its nature. (These studies are being performed by Dr. Benjamin Rivnay, Weizmann Institute of Science, Rehovot, Israel.)


*The special case of Triton X-114*

Bordier was the first to recognize the potential utility of Triton X-114 for the characterization and isolation of membrane proteins [19]. This is one of the group of phenoxypolyethoxyethanol detergents that have been found useful for solubilizing membrane proteins. It contains on the average two less ethoxy groups than the more commonly used Triton X-100. As a consequence the cloud point, the temperature at which supramicellar aggregates form, is at 25–30 °C

Table 1. *Effect of lipids in preventing the dissociation of $\alpha\beta\gamma_2$ by detergent*

Extrinsically iodinated IgE-receptor complexes were purified and maintained in 2 mM of the detergent Chaps {3-[(3-cholaminopropyl)dimethylammonio]-1-propanesulphonate}. A small aliquot was then diluted into a solution containing the desired detergent with or without 2 mM lipids. After incubation, the IgE–receptor complexes were precipitated with anti-IgE and the precipitates dissolved in sodium dodecyl sulphate and electrophoresed on polyacrylamide gels. Autoradiographs of the latter were densitometrically scanned and the ratio of $\beta$ and $\gamma_2$ to $\alpha$ was assessed. This ratio was compared to the ratio observed with the unincubated control samples in order to calculate the percentage dissociation. Modified from [18].

| Detergent (mM)* | Lipid | Temperature (°C) | Time (h) | Dissociation (%) |
|---|---|---|---|---|
| Chaps (10) | None | 25 | 12 | 57 |
| Chaps (10) | Tumour phospholipid | 25 | 12 | 0 |
| Triton X-100 (5.4) | None | 25 | 12 | 96 |
| Triton X-100 (5.4) | Tumour phospholipid | 25 | 12 | 45 |
| Octylglucoside (24.5) | None | 25 | 12 | 96 |
| Octylglucoside (24.5) | Tumour phospholipid | 25 | 12 | 74 |
| Chaps (10) | Individual phospholipids or various mixtures thereof | 25 | 6 | 50–60 |

* The molarities shown yield a concentration of micellar detergent of about 4 mM in each case.

for Triton X-114 instead of at 60–100 °C. Such aggregates form a separate phase and after centrifugation an upper phase is observed which contains 95% of the volume but only a very small percentage of the detergent. Bordier observed that integral membrane proteins were preferentially distributed in the lower, detergent-rich phase, whereas ordinary 'water-soluble' proteins remained in the supernatant.

We have examined the distribution of the IgE-$\alpha\beta\gamma_2$ complex as well as the distribution of the individual components of the complex in the two phases, and recently reported our results [13]. These are summarized in Table 2. The striking observation was that components which were likely to bind equivalent amounts of detergent, such as IgE-$\alpha\beta\gamma_2$ and $\alpha\beta\gamma_2$ or IgE-$\alpha$ and $\alpha$, distributed quite differently in the two phases (Table 2). It appears as if by being complexed with a large hydrophilic component, a hydrophobic component will distribute more in the detergent-depleted phase than would otherwise be the case. We proposed that the distribution of a component was therefore not simply related to whether it bound detergent or not, as had originally been surmised, but that the overall balance of surface hydrophobicity/hydrophilicity determined the partitioning. Subsequently Maher & Singer [20] reported on the distribution in Triton X-114 of the acetylcholine receptor from the electric organ of *Torpedo californica*. Their basic observation was that this pentamer of transmembrane peptides distributes in the upper phase of phase-separated Triton X-114 instead of in the lower phase, as Bordier's hypothesis would have predicted. They propose that, unlike other integral membrane proteins, those that form hydrophilic central channels may have an unusual transmembrane configuration resulting in a non-smooth outer surface. This, therefore, would prevent intercalation into the detergent micelles. However, they could not exclude the possibility that the large hydrophilic membrane-protruding domains of the receptor might have been responsible for

Table 2. *Distribution of IgE, IgE–receptor complex and its component parts in phase-separated Triton X-114*

This is a summary of results from Alcaraz *et al.* [13]. The values have been rounded off.

| | Amount (%) in: | |
| --- | --- | --- |
| Species | Upper (detergent-depleted) phase | Lower (detergent-enriched) phase |
| IgE | 100 | 0 |
| IgE–$\alpha\beta\gamma_2$ | 80 | 20 |
| $\alpha\beta\gamma_2$ | 20 | 80 |
| IgE–$\alpha$ | 80 | 20 |
| $\alpha$ | 20 | 80 |
| $\beta\alpha_2$ | 0 | 100 |

their results. The hypothesis of Maher & Singer may be directly testable if X-ray-diffraction data of higher resolution can be obtained. The initial diffraction data do not appear to be adequate to resolve this issue [20]. Clearly our own results are difficult to explain on the basis of Maher & Singer's hypothesis and indeed the unexpected distribution of the acetylcholine receptor and the IgE–receptor complex in the upper phase of Triton X-114 may have quite different explanations.

Regardless of the mechanism, the phase separation technique can be highly useful for the isolation of the different subunits of the receptor for IgE, as is illustrated in Fig. 2. In this experiment we utilized the Triton X-114 first to dissociate IgE-$\alpha$ from $\beta\gamma_2$, and then to separate them physically by phase separation. This procedure has turned out to be extremely helpful as an initial step in the preparative isolation of the subunits for analysis of their covalent structure.

*Interaction of receptors with lipid*

We have already described the evidence that the $\alpha\beta\gamma_2$ complex contains firmly bound lipid. We have data that indicate that these chains interact with lipid in other ways also [18]. In addition to the firmly bound lipid, there is evidence for more loosely attached lipid. The observations are these. The $\alpha$ chain, which contains the binding site for IgE, cannot be oxidatively iodinated efficiently in the presence of IgE. This is so regardless of whether the iodination is carried out on intact cells or with solubilized receptors in micellar detergent [22]. Dissociating the $\beta$ and $\gamma$ chains does not improve the iodination. Performing the iodination in submicellar concentrations of detergent also yields little modification of the $\alpha$ chain. If, however, the IgE–$\alpha\beta\gamma_2$ complex is exposed, even briefly, to lipid-free micellar detergent (under conditions where there is no dissociation of $\beta$ and $\gamma$ from $\alpha$) and then iodinated in submicellar detergent, the iodination of the $\alpha$ is highly effective [13,18]. Adding back micellar detergent with or without lipid prior to iodination makes the $\alpha$ chain inaccessible again.

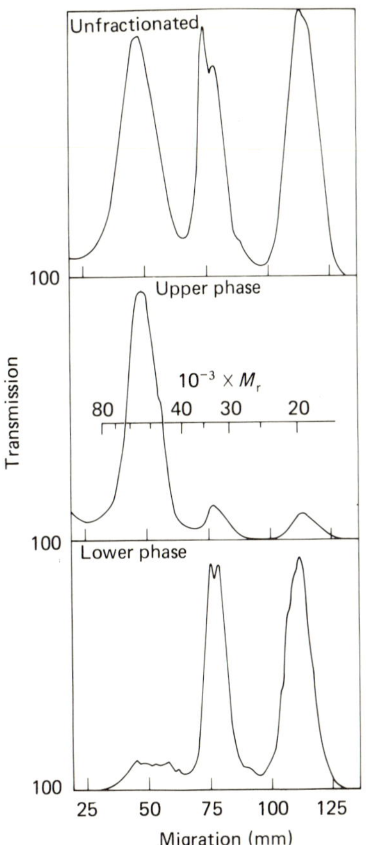



Fig. 2. *Distribution of α, β and γ chains of the receptor of IgE in the supernatant and pellet of phase-separated Triton X-114*

The panels show the results of densitometric scans of autoradiographs from polyacrylamide gels on which the samples had been electrophoresed. The iodinated IgE (complexed to α) which was present in the starting material and in the material from the upper phase is not shown. Top panel: starting material containing (left to right) the iodinated α, β and dimer of γ chains; middle panel: upper, detergent-depleted, phase; lower panel: lower, detergent-enriched phase. (From Alcaraz *et al.* [13] with permission.)

The simplest explanation of these findings is that the α chains contain iodinatable tyrosines in a hydrophobic surface area which presumably interacts with lipid *in situ*. This lipid is easily replaced by micellar detergent and either the lipid or the detergent prevents access of the iodide to the phenolic ring.

In addition to the non-covalently bound lipid, studies using radioactive fatty acids indicate that these can be biosynthetically esterified to the α, β and γ chains. The incorporation is poor, however, so that it is not possible at this time to quantify the molar ratio of covalently bound lipid to receptor peptide. Therefore, we cannot determine whether a significant fraction of the α, β and γ chains are modified in this way. Thus, whether this covalently bound lipid contributes significantly to the hydrophobicity of any of the chains, e.g. as manifested by their interaction with detergent (Table 2) or with hydrophobic matrices [13], is unclear. Published data on the amino acid composition of the α chain [8] and our initial results on $\beta$ and $\gamma$ (G. Alcaraz, J. P. Kinet, R. Quarto & H. Metzger, unpublished work) suggest that the peptide structure itself could account for the hydrophobicity.

### Is the $\alpha\beta\gamma_2$ Complex a Single Protein?

The unusual instability of the $\alpha\beta\gamma_2$ complex in detergents (we are unaware of any multisubunit protein that has analogous properties) raised the question of whether this complex should be considered a single protein or not. This question has no final answer and in one sense is meaningless because it is too imprecise. What we can answer are more specific questions. Thus, it is clear that $\alpha$ does not appear to exist on the membrane (or in the cell's interior) uncomplexed to $\beta$ and $\gamma_2$. (The opposite question cannot yet be answered. That is, it is not known whether $\beta$ and $\gamma_2$ can exist independently or possibly in association with other polypeptides. This could most readily be determined by the use of anti-$\beta$ and anti-$\gamma$ antibodies, but so far generation of such antibodies has not been reported.) It can also be definitely stated that there is no evidence for complexes other than $\alpha(\beta\gamma_2)_1$, e.g. such as $\alpha(\beta\gamma_2)_2$ etc. The most persuasive evidence that the $\alpha\beta\gamma_2$ is a unit entity comes from biosynthetic studies [5,23]. From these one can say that the $\alpha$, $\beta$ and $\gamma$ chains are co-ordinately synthesized and degraded and that, once inserted into the membrane, no independent shuffling or exchange of the chains appears to occur [23].

### Lessons Learned

The results of studies on the receptor for IgE have implications for investigations on other membrane proteins. Although as stated the results are so far unprecedented, it seems unlikely that this receptor is unique. At least there are other examples where one 'mild' detergent or another has been found to induce dissociation of complexes of several discrete proteins. Two recent examples are noteworthy. Arad *et al.* [24] have provided new evidence that in the plasma membrane of turkey erythrocytes the GTP binding regulatory protein ($N_s$) is stably complexed with adenylate cyclase and that these proteins are not independently diffusing units as had been proposed. They found that, by using an especially mild detergent, $CH_3[CH_2]_7\text{-}[OCH_2\text{-}CH_2]_5OH$ [25], the two proteins could be purified as a unit. Interestingly, adding exogenous lipid (sonicated asolectin) and a variety of 'carrier' proteins such as bovine serum albumin or bovine globin enhanced the stabilization of the complex. The possibility that specific lipids may be involved has not yet been explored.

A further example concerns proteins of the mitochondrial respiratory chain from beef heart. Shinomura *et al.* [26] showed that the iron–sulphur protein could be dissociated from its complex with cytochrome $b$ and cytochrome $c$, by thorough depletion of bound lipid. Once separated, an active molecular complex could be demonstrated by readdition of soybean phospholipids to the isolated components and chromatography of the mixture on a gel filtration column.

In both of these examples the role of the lipid remains to be defined. As with the receptor for IgE, it is unclear whether the lipid is acting indirectly to stabilize the individual components or directly at the 'domain of bonding' [27] between them.

In the context of this discussion, it is reasonable to ask whether interactions between the $\alpha\beta\gamma_2$ complex and still other membrane proteins or cytoplasmic constituents could so far have been missed. We believe that this is a real possibility. It can be shown that a variety of factors tend to promote protein interactions in membranes. These include the high concentration of the individual proteins themselves, the preorientation of these proteins due to their intercalation in the bilayer, and the presence of high concentrations of 'bystander' proteins (B. Grasberger, A. Minton, C. Delisi & H. Metzger, unpublished work). These factors, as well as the disruption of the hydrophobic milieu that inevitably results from the solubilization of the membrane, can contribute to the dissociation of multiprotein interactions. Gentle procedures that minimize the disruption of lipid–protein interactions, as well as very different approaches such as the use of crosslinking reagents and analysis of intact systems by radiation inactivation, need to be explored.

G.A. is Attaché de Recherche, Institut National de la Santé, et de la Recherche Médicale, on leave from INSERM, Unit 174, Marseille, France, and partially supported by a grant from the Ministère de l'Industrie et de la Recherche, France and from DNAX, U.S.A.

## References

1. Conrad, D. H. & Forese, A. (1976). *J. Immunol.* **116**, 319–326
2. Kulczycki, A., Jr., McNearney, T. A. & Parker, C. W. (1976) *J. Immunol.* **117**, 661–665
3. Holowka, D., Hartmann, H., Kanellopoulos, J. & Metzger, H. (1980) *J. Receptor Res.* **1**, 41–68
4. Holowka, D. & Metzger, H. (1982) *Mol. Immunol.* **19**, 219–227
5. Perez-Montfort, R., Kinet, J.-P. & Metzger, H. (1983*a*) *Biochemistry* **22**, 5722–5728
6. Rossi, G., Newman, S. A. & Metzger, H. (1977) *J. Biol. Chem.* **252**, 704–711
7. Newman, S. A., Rossi, G. & Metzger, H. (1977) *Proc. Natl. Acad. Sci. U.S.A.* **74**, 869–872
8. Kanellopoulos, J. M., Liu, T. Y., Poy, G. & Metzger, H. (1980) *J. Biol. Chem.* **255**, 9060–9066
9. Fewtrell, C., Kempner, E., Poy, G. & Metzger, H. (1981) *Biochemistry* **20**, 6589–6594
10. Kanellopoulos, J., Rossi, G. & Metzger, H. (1979) *J. Biol. Chem.* **254**, 7691–7697
11. Holowka, D., Gitler, C., Bercovici, T. & Metzger, H. (1981) *Biochemistry* **20**, 6589–6594
12. Rivnay, B., Wank, S. A. & Metzger, H. (1982) *Biochemistry* **21**, 6922–6927
13. Alcaraz, G., Kinet, J.-P., Kumar, N., Wank, S. A. & Metzger, H. (1984) *J. Biol. Chem.* **259**, 14922–14927
14. Kinet, J.-P., Alcaraz, G., Leonard, A., Wank, S. & Metzger, H. (1985) *Biochemistry* **24**, 4117–4124
15. Rivnay, B. & Metzger, H. (1982) *J. Biol. Chem.* **257**, 12800–12818
16. Mimms, L. T., Zampighi, G., Nozaki, Y., Tanford, C. & Reynolds, J. A. (1981) *Biochemistry* **20**, 833–840
17. Umbreit, J. N. & Strominger, J. L. (1973) *Proc. Natl. Acad. Sci. U.S.A.* **70**, 2997-3001
18. Kinet, J.-P., Quarto, R., Perez-Montfort, R. & Metzger, H. (1985) *Biochemistry* **24**, 7342–7348
19. Bordier, C. (1981) *J. Biol. Chem.* **256**, 1604–1607
20. Maher, P. A. & Singer, S. J. (1985) *Proc. Natl. Acad. Sci. U.S.A.* **82**, 958–962
21. Brisson, A. & Unwin, P. N. T. (1985) *Nature (London)* **315**, 474–477
22. Pecoud, A. & Conrad, D. (1981) *J. Immunol.* **127**, 2208–2214
23. Quarto, R., Kinet, J.-P. & Metzger, H. (1985) *Mol. Immunol.* **22**, 1045–1052
24. Arad, H., Rosenbusch, J.-P. & Levitski, A. (1984) *Proc. Natl. Acad. Sci. U.S.A.* **81**, 6579–6583
25. Rosenbusch, J. P., Garavito, R. M., Dorset, C. L. & Engel, A. (1982). *Protides Biol. Fluids* **30**, 171–174
26. Shimomura, Y., Nishikima, M. & Ozawa, T. (1984) *J. Biol. Chem.* **259**, 14059–14063
27. Monod, J., Wyman, J. & Changeaux, J. P. (1965) *J. Mol. Biol.* **12**, 88–118

*Biochem. Soc. Symp.* **51**, 69–81
*Printed in Great Britain*

# Cellular Receptors to the Anaphylatoxins C3a and C5a

RUTH HUEY, YOSHIHIRO FUKUOKA, PAUL D. HOEPRICH, JR.
and TONY E. HUGLI

*Department of Immunology, IMM18, Scripps Clinical and Research Foundation,
10666 North Torrey Pines Road, La Jolla, CA 92037, U.S.A.*

## Synopsis

A single component in the plasma membrane of human polymorphonuclear leukocytes (hPMN) has been identified as the binding site for C5a. Labelled human C5a can be cross-linked to a 48000-$M_r$ membrane component on the hPMN surface by using the bifunctional reagent ethylene glycol bis(succinimidyl succinate). The membrane component is believed to be the C5a receptor or the binding subunit of a C5a receptor complex. Our ligand-uptake data indicate that the hPMN has high-affinity binding sites for C5a with a $K_d$ of the order of 1–2 nM and an estimated 50000–113000 binding sites/cell. Preliminary binding studies of C3a and C5a to rat peritoneal mast cells indicate that non-specific uptake by these cells is so great that it obscures characterization of specific receptor interactions. Data recently reported [Gervasoni, Conrad, Hugli, Schwartz & Ruddy (1986) *J. Immunol.* **136**, 285–292] suggest that non-specific binding of C3a to mast cells is caused by electrostatic interactions between the cationic ligand and anionic heparin-proteoglycan on the cell surface, with an additional complication of the bound ligand undergoing proteolytic degradation. It is therefore proposed that synthetic analogue peptides designed to minimize non-specific interactions with the cell will be useful tools for demonstrating anaphylatoxin receptors on mast cells and may prove essential for receptor isolation.

## Introduction

Complement activation fragments C3a and C5a (e.g. anaphylatoxins) are potent spasmogens and stimulators of a variety of circulating and tissue cell types. These include neutrophils, monocytes, eosinophils, basophils and mast cells (Hugli, 1981). More recent evidence indicates a role for the anaphylatoxins in immunoregulation either by direct or indirect action on T-lymphocyte subpopulations (Hugli & Morgan, 1984). Indirect interactions also exist with vascular endothelial cells that respond to mediators released from neighbouring cells stimulated by the anaphylatoxins (Marceau & Hugli, 1984; Hugli & Marceau, 1985). The varied biological responses elicited by the anaphylatoxins are mediated via cell surface receptors that, on neutrophils, undergo dynamic flux and redistribution when they bind the ligand (Chenoweth & Goodman, 1983). In fact, modulation of C5a receptors on neutrophils by the ligand appears to play a very important physiological function that results in down-regulation of the inflammatory cellular response during excessive or repeated complement activation. Furthermore, the role of C5a in initiating inflammatory responses by recruitment of granulocytes via chemotactic migration and by activation of various intracellular biochemical pathways points to the biological importance of this potent humoral mediator in host defence. We have recently demonstrated a component in the membrane of human polymorphonuclear leukocytes (hPMN) that specifically binds human C5a and to which the C5a can be covalently cross-linked using bifunctional reagents (Huey & Hugli, 1984). This component of the hPMN membrane is believed to be the binding site for the putative C5a receptor. Whether or not the C5a receptor is composed of a multi-unit complex, as are certain other receptors (Pilch & Czech, 1984; Blecher & Bar, 1981), has not been established. What has been determined is that the binding element or putative subunit of the C5a receptor is not unlike that identified for the formylated chemotactic peptides (Niedel *et al.*, 1984) or for the chemotactic leukotriene $LTB_4$ (Goldman & Goetz, 1984) in terms of size.

In addition to our identification and characterization of a putative receptor for C5a on hPMN, we examined the binding characteristics of anaphylatoxins in mast cells and cultured mast cell lines. Recent evidence indicates that tissue contractile and relaxant responses elicited by the anaphylatoxins are mediated by a combination of vasoamines, prostanoids and leukotrienes (Stimler *et al.*, 1983; Regal, 1982; Hugli & Marceau, 1985). Many, if not most, of the cellular mediators released from stimulated tissue originate with the mast cell. Thus it can be postulated that the spasmogenic action of anaphylatoxins result primarily from effects of products released by mast cells after C3a or C5a activation. Direct binding and ligand uptake measurements of C3a and C5a on mast cells is seriously complicated by interference from polyanions such as heparin- or chondroitin-proteoglycans on the cell surface (Gervasoni *et al.*, 1986), as well as proteases such as chymase and tryptase (Benditt & Arase, 1959; Schwartz *et al.*, 1983). These complications suggest to us advantages in using synthetic analogues of the anaphylatoxins designed specifically to circumvent the problem of non-specific binding of the natural ligands to this cell type. A brief discussion is given of the potential value that these synthetic reagents may have in characterizing the specific receptors or binding sites for anaphylatoxins on the mast cell surface.


**Results and Discussion**

The ligand used in these studies was purified human C5a radiolabelled with [125]I by the lactoperoxidase–glucose oxidase method (Hugli & Chenoweth, 1981). Uptake of the labelled ligand was saturable only at 0 °C (Fig. 1); however total binding was greater at higher temperatures, presumably because increased membrane fluidity at elevated temperatures may increase receptor expression. As noted in previous studies, binding capacity of C5a to hPMN decreases with time at 37 °C due to active internalization of the ligand–receptor complex (Chenoweth & Hugli, 1980). This phenomenon led us to perform cross-linking experiments at 0 °C in order to avoid loss of available receptor on the cell

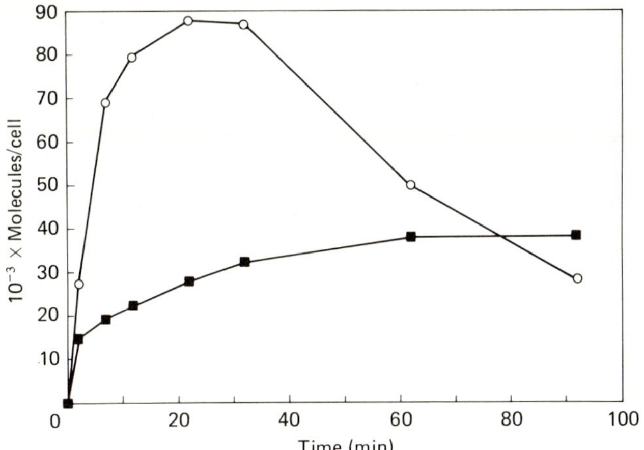

Fig. 1. *Time course of uptake of* $^{125}I$-C5a *by human PMNs*

PMNs ($5 \times 10^6$ cells/ml) were incubated with $1.24 \times 10^{-9}$ M-$^{125}$I-C5a in RPMI/bovine serum albumin buffer at pH 7.4 for various periods of times. Incubations were carried out at 0 °C (■) and at 37 °C (○). Uptake at 37 °C reaches an apparent maximum and then falls with time as a result of internalization of the receptor–ligand complex. Uptake at 0 °C reaches equilibrium by 60 min; however, fewer receptor sites are available for ligand binding.

surface, a process which involves digestion of the ligand by intracellular proteases following internalization (Chenoweth & Goodman, 1983).

Labelled C5a at 0–60 nM is incubated with $5 \times 10^6$ hPMN for 60 min at 0 °C in order to saturate the ligand receptor sites. When the same conditions are employed for $^{125}$I-C5a binding in the presence of a 50-fold excess of unlabelled C5a most, but not all, of the uptake is competed for by the unlabelled factor. The difference in $^{125}$I-C5a binding with and without unlabelled C5a is concluded to be specific uptake and defines a C5a receptor or membrane binding site (Fig. 2). Scatchard analysis of the specific uptake data for C5a on hPMN indicates a $K_d$ of approx. 2 nM and an estimated receptor population of 78 500 sites/cell for this particular human donor (Fig. 3). A more general analysis suggests that the average number of receptors is 80 000 sites/cell, based on data obtained from five different individuals, but that the range is from 50 000 to 113 000 (see Table 1). This estimate of C5a receptors per cell is similar to the estimated number of receptors for the formylated peptides and for leukotriene B$_4$ on the hPMN surface.

Demonstration of a singular component on the PMN surface that is capable of binding C5a in a specific manner was accomplished by using a cross-linking technique (Huey & Hugli, 1985). The ligand $^{125}$I-C5a was covalently attached to the hPMN surface by the cross-linking reagent ethylene glycol bis(succinimidyl succinate) (EGS). When 16 nM-$^{125}$I-C5a was exposed to $4 \times 10^8$ hPMN for 60 min at 0 °C and then treated with 1 mM-EGS for 15 min, some of the label was cross-linked to elements on the membrane surface. Evidence that $^{125}$I-C5a was covalently associated with a component(s) on the plasma membrane surface was provided by results obtained from a sucrose gradient of the material recovered from PMNs disrupted by using nitrogen cavitation. Fig. 4 illustrates

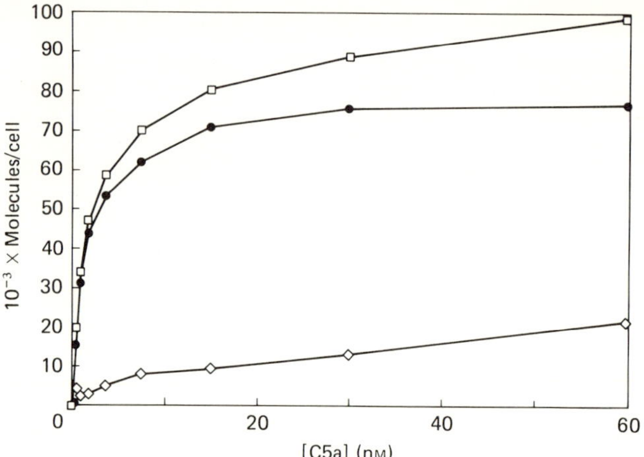

Fig. 2. *Dose–response curve for binding of* $^{125}I$*-C5a by human PMNs*

PMNs ($5 \times 10^6$ cells/ml) were incubated with various concentrations of $^{125}$I-C5a in RPMI/bovine serum albumin buffer for 1 h at 0 °C (□). Non-specific binding was estimated by adding a 50-fold excess of unlabelled C5a to the reaction mixture (◇). The specific binding data were obtained by subtracting the non-specific binding (e.g. uptake measured in the presence of a 50-fold excess of unlabelled C5a) from the total uptake (●). Putative C5a receptors on human PMNs are saturated by 20–40 nM-C5a.

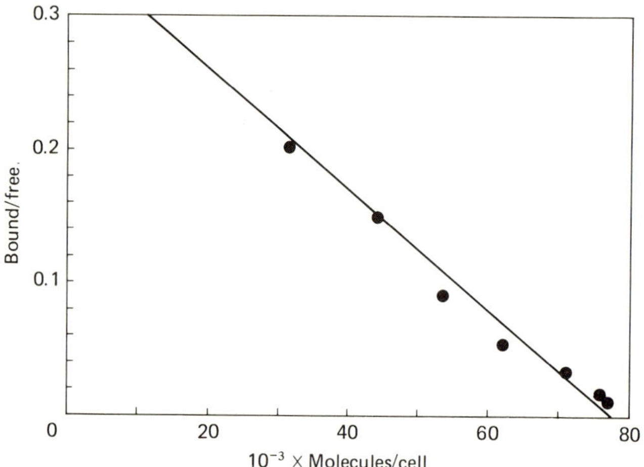

Fig. 3. *Scatchard analysis of the specific uptake of* $^{125}I$*-C5a by human PMNs*

Adapted from data shown in Fig. 2. The number of molecules specifically bound per cell at varying $^{125}$I-C5a concentrations is plotted versus the ratio of bound $^{125}$I-C5a molecules to those remaining free in solution. The correlation coefficient ($r$) was 0.992. From the $x$-intercept of the linear least-squares fit of the data, the number of C5a molecules bound/cell for this particular experiment was 78 500. From the inverse of the slope a $K_d$ value of $1.99 \times 10^{-9}$ M was obtained.

the profile of the $^{125}$I-C5a-containing band which coincides with the plasma membrane fraction on these sucrose gradients. Association of the ligand with the plasma membrane fraction is as expected for 0 °C conditions, but had these cells been incubated at 37 °C the ligand would have been degraded and distributed more widely among other cellular fractions and not so uniformly associated with just plasma membrane.

Table 1. *Receptors to three major chemotactic factors on human PMNs*

C5a receptors were estimated at an average of 80000 sites/cell from a mean of 19 separate determinations. The data ranges from 50000 to 113000 sites/cell for determinations obtained on cells isolated from five different individuals and from the same individuals on different days.

| Factor | Estimated number/cell | Estimated $M_r$ | Reference |
|---|---|---|---|
| Human C5a | 80000 (50000–113000) | 40000–48000 | Huey & Hugli (1985) |
| fMet-Leu-Phe | 88000 (55000–120000) | 50000–70000 | Niedel et al. (1980) and Schmitt et al. (1983) |
| Leukotriene B4 | 26000–40000 | 60000 | Goldman & Goetzl (1984) and Goldman et al. (1985) |
| 20-Hydroxyleukotriene $B_4$ | 42000 | Not determined | Clancy et al. (1984) |

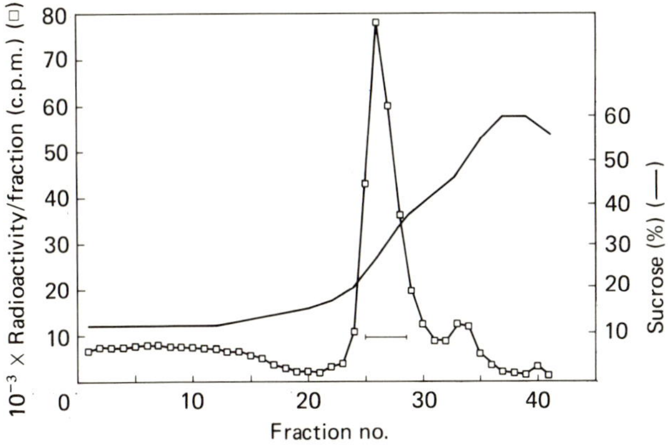

Fig. 4. *Covalent association of $^{125}I$-C5a with a plasma membrane component*

$^{125}$I-C5a was crosslinked to intact PMNs and then PMN membrane was isolated after $N_2$ cavitation to disrupt cellular integrity. The Figure shows radioactivity/fraction plotted versus fraction number, as well as the sucrose concentration of each fraction as determined by refractometry. Peak radioactivity occurs at 25–35% sucrose which coincides with the position of the plasma membrane marker band.

Analysis of components in this plasma membrane fraction to which the labelled C5a is cross-linked was performed by SDS/polyacrylamide-gel electrophoresis and visualization accomplished by autoradiography. These results indicate a band of label at $M_r$ approx. 60000 (Fig. 5). We interpret these results to indicate that a specific component exists that binds C5a and which may represent a putative receptor. The complex formed between C5a and this membrane component is of $M_r$ 60000 whether the electrophoretic analysis is carried out with or without reducing agent present. These data indicate that the membrane component that binds C5a is not covalently bound to any other subunit or structural element in the membrane. Two other groups of investigators have identified a C5a-binding component in hPMN membrane. Rollins & Springer (1985) estimated the complex between C5a and the membrane component to be of $M_r$ 52000 by using the non-specific cross-linking reagent

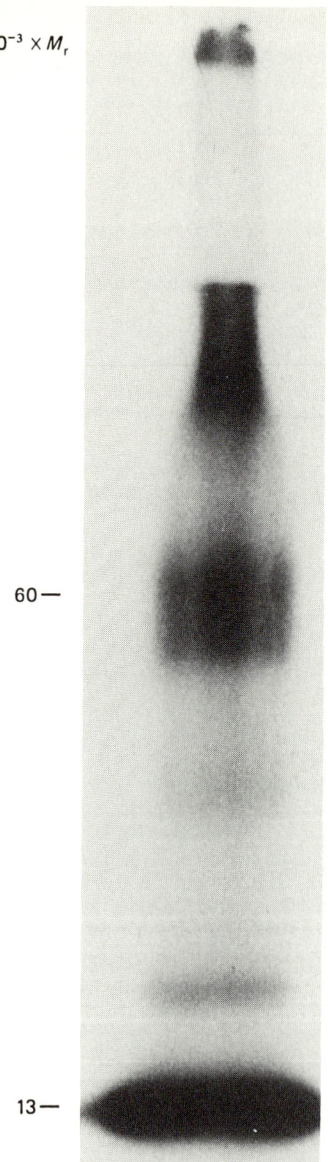



Fig. 5. *Crosslinking of* $^{125}I$-*C5a to intact PMNs by EGS*

The PMN membrane was isolated using $N_2$ cavitation and discontinuous sucrose gradient centrifugation. The membrane material was dissolved in SDS-containing buffer and analysed by SDS/polyacrylamide-gel electrophoresis. Bands were visualized by autoradiography. The 13000-$M_r$ band is tentatively assigned as C5a and the 60000-$M_r$ bands as a complex between C5a and a putative receptor molecule of $M_r$ approx. 48000.

disuccinimidyl suberate. Johnson & Chenoweth (1985) also estimated a complex of C5a and membrane component to be of $M_r$ 52000 by SDS/polyacrylamide-gel electrophoresis analysis. They used a heterobifunctional photoreactive reagent ($p$-azidobenzoyl-2-mercapto-$N$-ethylamide-2-thiopyridine disulphide) to deriva-

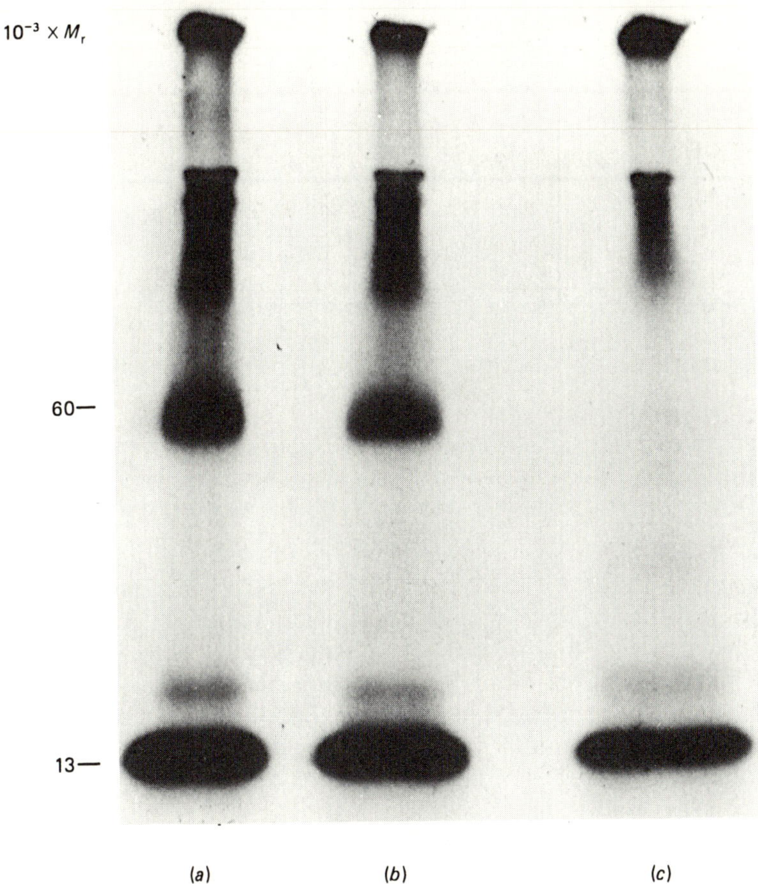



Fig. 6. *Competition by C3a and C5a for crosslinking of* $^{125}I$-*C5a to PMN plasma membrane*

Isolated PMN plasma membrane was incubated with $3.2 \times 10^{-8}$ M-$^{125}$I-C5a alone (lane *a*) or in combination with either $1.4 \times 10^{-5}$ M-C3a (lane *b*) or $1.5 \times 10^{-5}$ M-C5a (lane *c*) for 30 min at 0 °C followed by crosslinking with 1 mM-EGS. Samples were analysed by SDS/10% (w/v) polyacrylamide-gel electrophoresis and autoradiography. Note that the 60000-$M_r$ band is absent only when unlabelled C5a is introduced as a competing ligand. The 13000-$M_r$ band is $^{125}$I-C5a.

tize C5a at Cys-27 before offering the ligand to the neutrophil surface. Photoactivation of the derivatized C5a leads to a specific reaction that links C5a with a component estimated at $M_r$ 40000. Therefore, three independent studies have demonstrated a complex between C5a and a single membrane component estimated to be $M_r$ 40000–48000 after C5a is subtracted from the total weight.

We were also able to demonstrate C5a binding to isolated PMN membrane. When EGS was added to membrane fractions previously saturated with $^{125}$I-C5a, a complex of exactly the same size as that observed with intact PMNs was visualized by autoradiography after SDS/polyacrylamide-gel electrophoresis analysis. Addition of a 500-fold excess of unlabelled C5a eliminated the complex from the gel (see Fig. 6) and confirmed the specificity of C5a binding to this particular membrane component. Since the linear dimension of the EGS reagent

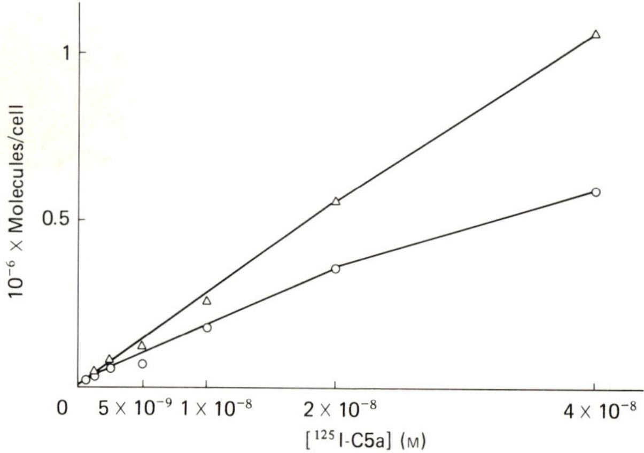

Fig. 7. *Uptake of rat (△) and human (○) C5a by rat peritoneal mast cells*

Uptake of the labelled rat ligand was greater than that of human C5a, possibly indicating a significant difference in affinity for the binding of homologous versus heterologous C5a.

is 1.5 nm (15Å) between its two reactive groups, potentially it could reach beyond the receptor and couple C5a to an adjacent proteins. Therefore, evidence for specificity given in Fig. 6 still could be misinterpreted if a neighbouring component were cross-linked with a high degree of efficiency. However, since cross-linking by two independent means (Rollins & Springer, 1985; Johnson & Chenoweth, 1985) also identified a membrane component of approximately the same size as that linked by EGS, a misassignment of the $M_r$ 40000 component as a receptor is less likely to have occurred.

In a separate set of experiments we have examined the uptake and binding characteristics of C3a and C5a with isolated rat peritoneal mast cells. We find that rat C5a is taken up more efficiently than is human C5a (Fig. 7). Labelled human C3a is taken up by the mast cell in a concentration-dependent manner, but saturation is not reached even at C3a concentrations of $8 \times 10^{-7}$ M. When the labelled ligand is competed by unlabelled des-Arg-C3a or C3a, a portion of the bound C3a is competed away but a majority of the binding is non-specific. Moreover as seen in Fig. 8, the difference between [125]I-C3a uptake, with and without unlabelled C3a present, represents specific binding which is not saturable even up to $8 \times 10^{-7}$ M-C3a (results not shown). In addition, the estimated number of binding sites taken from the uptake data is extremely large. A partial explanation for these binding kinetics can be found in data presented by Gervasoni *et al.* (1986) who observed that C3a exposed to rat mast cells was degraded. They further characterized uptake of [125]I-C3a by mast cells and identified a prominent non-specific binding related to a charge–charge interaction with heparin-proteoglycan which is expressed on the mast cell surface. This anionic material apparently has a high affinity for the cationic anaphylatoxins, which it binds by electrostatic interactions. Binding of the anaphylatoxin to the proteoglycan renders the anaphylatoxin susceptible to proteolysis by chymase which is present on the peritoneal mast cell surface. This degradation step is minimized at 0 °C but not eliminated completely. Therefore, our data charac-

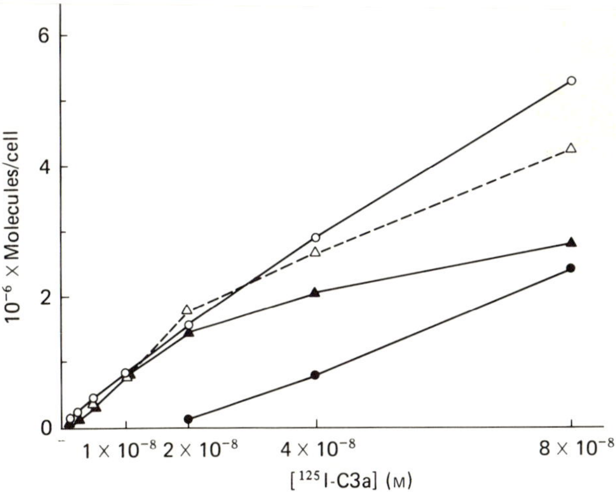



Fig. 8. *Uptake of $^{125}$I-C3a by isolated rat peritoneal mast cells*

The $^{125}$I-labelled ligand (○) was competed by unlabelled C3ai (des-Arg-C3a) (△) or C3a (▲) at 100-fold excess. The lower curve (●) represents the difference between the binding of labelled C3a in the presence and absence of unlabelled ligand (○–▲). At these levels of C3a the binding sites on mast cells are not saturated. Note that these data were obtained for a heterologous system (e.g. human C3a and rat cells).

terizing concentration-dependent uptake must be interpreted in light of these complexities. The fact that binding is not saturable even at elevated concentrations may be explained if the ligand is being slowly digested by the chymase. We are presently repeating these experiments using protease inhibitors in an attempt to circumvent this problem.

Preliminary studies with cells other than the rat peritoneal mast cell indicate that major differences exist between mast cell populations in terms of binding the anaphylatoxin. For example, uptake of $^{125}$I-C3a by bone marrow mast cells from the mouse is concentration-dependent and addition of 100-fold excess of unlabelled ligand markedly decreases uptake of the labelled ligand, which indicates that specific binding exists (Fig. 9). Conversely, a cloned mast cell line PT18 derived from the mouse bone marrow cell shows very little uptake of labelled ligand. Moreover, this small amount of binding appears to be non-specific in nature according to competition experiments. Other cell lines, such as U-937, a human monocyte line or RBL-1, a rat leukaemic basophil cell line, show intermediate binding properties and estimates of C3a receptor number by Scatchard analysis indicate fewer than 10000 sites/cell type.

Consequently, it appears that two distinct problems exist for the characterization of anaphylatoxin receptors on mast cells. One aspect is the surface protease (e.g. chymase or tryptase) that may degrade the ligand when it comes in contact with the cell surface. It is anticipated that this problem can be remedied by protease inhibitors. The second problem of non-specific binding is related to the ligand itself. These molecules are cationic and the cell surface is anionic. In addition, the active site *C*-terminal portion of the molecule is an amphipathic helix having an exposed surface that is hydrophobic (Lu *et al.*, 1984)

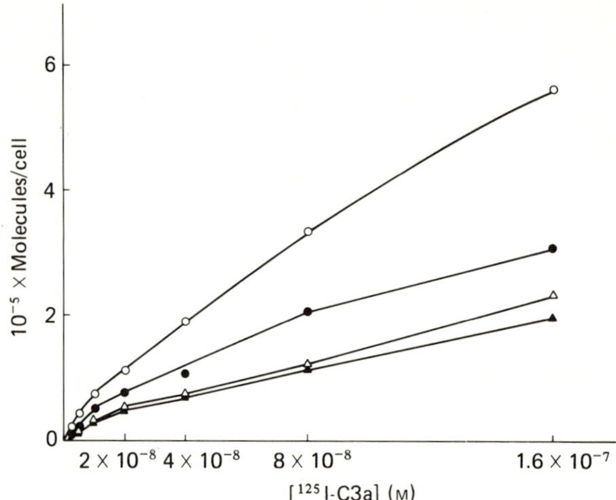

Fig. 9. *Uptake of human* $^{125}I$-C3a *by bone marrow mast cells (BMMC)* ($\bigcirc$,$\bullet$) *and by a marrow cell line (PT18)* ($\triangle$, $\blacktriangle$)

Apparent binding of the ligand is much greater for isolated bone marrow cells than for cultured cells. When unlabelled C3a is present at 100-fold excess ($\bullet$,$\blacktriangle$) a significant proportion of the total uptake by the BMMC is inhibited but none of the uptake by the PT18 cell line is competed by unlabelled ligand. The BMMC and PT18 cell line were kindly provided by Dr. Steven Wasserman (University of California-San Diego, La Jolla, CA, U.S.A.).

and capable of binding to similar hydrophobic sites on the mast cell surface. Therefore, non-specific interactions between C3a or C5a and the mast cell surface are prevalent, especially since polyanionic heparin-proteoglycan and hydrophobic membrane sites are relatively abundant. For this reason we anticipate using synthetic ligands as receptor probes, ligands that retain full functional activity but are designed to minimize cationic charge character or hydrophobicity.

A series of C3a analogue peptides have been synthesized with a goal of maximizing the helical content of these fragments (see Table 2). Peptides II and III assume a helical conformation more readily than C3a 21R and exhibit a greater intrinsic activity than does the corresponding peptide (e.g. C3a, 21R) that is based on the sequence of the natural factor (see Table 3). Therefore, it appears feasible to design C3a analogue peptides that maintain the ability to assume a helical conformation and express a high level of activity but are less cationic than C3a and have fewer hydrophobic side chains on the surface of the helix. Such analogues should serve as effective probes of the specific C3a receptor without binding to the non-specific anionic or hydrophobic sites that attract C3a.

Since the spasmogenic activity of C3a or C3a peptides is dependent on histamine release from activated tissue mast cells, it is probable that the ability of these synthetic fragments to induce ileal contraction relates directly to their affinity for C3a receptors on mast cells. If this premise proves true then such peptides as C3a 21R substituted with 2-aminobutyric acid or 2-aminoisobutyric acid promise to bind to C3a receptors on mast cells with greater affinity than does intact C3a. As predicted, substitution of helix-inducing residues such as 2-aminobutyric acid or 2-aminoisobutyric acid enhance activity, whereas

Table 2. *Synthetic analogue peptides of the anaphylatoxins*

Residue replacements that differ from the sequence of natural human C3a 57–77 are underlined in the analogue peptides. C3a peptide II contains six positions substituted by alanine; peptide III contains six positions substituted by alanine and four by 2-aminobutyric acid; peptide IV contains six alanine, four 2-aminobutyric acid and two proline substitutions; peptide V contains six alanine and four-aminoisobutyric acid substitution sites. C5a 21YR contains an *N*-terminal tyrosyl residue and 20 residues identical to the C5a 55–74 sequence. Peptide C5a 21YR was synthesized in Dr. Bruce Erickson's laboratory at Rockefeller University. Abbreviations: Ab, 2-aminobutyric acid; Aib, 2-aminoisobutyric acid.

**C3a peptides**

|  |  | 57                      77 |  |
|---|---|---|---|
| I | C3a 21R | C–N–Y–I–T–E–L–R–R–Q–H–A–R–A–S–H–L–G–L–A–R | native factor sequence |
| II | C3a 21R,Ala | A–N–A–A–A–E–A–R–R–Q–H–A–R–A–A–H–L–G–L–A–R | helix enhanced sequence |
| III | C3a 21R,Ala,Ab | A–N–A–Ab–A–E–A–Ab–R–Q–A–Ab–R–A–A–Ab–L–G–L–A–R | helix enhanced sequence |
| IV | C3a 21R,Ala,Ab,Pro | A–N–A–Ab–A–P–A–Ab–R–Q–A–Ab–R–P–A–Ab–L–G–L–A–R | helix disrupted sequence |
| V | C3a 21R,Ala,Aib | A–N–A–Aib–A–E–A–Aib–R–Q–A–Aib–R–A–A–Aib–L–G–L–A–R | helix enhanced sequence |

**C5a peptides**

|  |  | 70    74 |  |
|---|---|---|---|
| I | C5a 5R | M–Q–L–G–R | native factor sequence |
| II | C5a 2IYR | 55            74 <br> Y–C–V–V–A–S–Q–L–R–A–N–I–S–H–K–D–M–Q–L–G–R | native factor sequence |

Table 3. *Biological activity of C3a 21R and peptide analogues*

| | Ileal assay* | | |
| Peptide | Concentration (M) | Relative activity (%) | Vascular permeability assay activity† |
| --- | --- | --- | --- |
| C3a | $3 \times 10^{-9}$ | 100 | + + + |
| C3a 21R | $2.9 \times 10^{-9}$ | 100 | + + |
| C3a 21R,Ala,Ab | $1.2 \times 10^{-9}$ | 250 | + + + |
| C3a 21R,Ala,Aib | $1.2 \times 10^{-9}$ | 250 | + + + |
| C3a 21R,Ala,Ab,Pro$_2$ | $9 \times 10^{-7}$ | 0.33 | − |
| Leu-Gly-Leu-Ala-Arg | $5 \times 10^{-6}$‡ | 0.20 | Not done |

* Guinea pig ileal strips (1.0 cm) were suspended in a 2.0 ml tissue bath containing Tyrode's buffer at pH 7.4 and 37 °C with continuous oxygen purge. Concentrations indicate dose required to induce muscle contraction.
† Anesthetized guinea pigs were given 0.5 ml of a 1% (w/v) solution of Evan's blue via cardiac puncture; 50 $\mu$l aliquots of peptide solutions containing $5 \times 10^{-10}$ nmol of peptide were injected subcutaneously. After 30 min the animals were killed and dye infiltration qualitatively scored.
‡ Assay done in 5.0 ml tissue bath.

introduction of helix-breaking residues such as proline in the 21-residue C3a analogue virtually eliminated activity (Table 3). Therefore, C3a peptides designed to maintain helical structure, but with fewer cationic and hydrophobic side chains, should be ideal ligands. These synthetic peptides would have an affinity for the C3a receptor equal to or stronger than that of the natural factor and therefore become valuable tools for receptor identification and isolation. The same may be true for C5a analogue peptides, since C3a 21YR (see Table 2), based on the sequence of human C5a, exhibited weak spasmogenic activity in preliminary experiments (Khan *et al.*, 1985). Synthetic ligands offer an attractive alternative to isolated products such as C5a which are difficult to obtain in quantity from plasma.

The authors acknowledge the efforts of Ellye Lukaschewsky in preparing this manuscript. Figs. 1–6 were adapted from Huey & Hugli (1985). The research was supported by U.S. Public Health Service Grants HL25658 and HL16411. R.H. is supported by a New Investigator Research Grant R23 HL30719.

# References

Benditt, E. P. & Arase, M. (1959) *J. Exp. Med.* **110**, 451–460
Blecher, M. & Bar, R. S. (1981) *Receptors and Human Disease*, pp. 2–23, Williams and Wilkins, Baltimore
Chenoweth, D. E. & Goodman, M. G. (1983) in *Leukocyte Locomotion and Chemotaxis* (Keller, H. & Till, G. O., eds.), pp. 252–321, Birkhauser-Verlag, Basle and Boston
Chenoweth, D. E. & Hugli, T. E. (1980) *J. Immunol.* **124**, 1517 (abstr.)
Clancy, R. M., Dahinden, C. A. & Hugli, T. E. (1984) *Proc. Natl. Acad. Sci. U.S.A.* **81**, 5729–5733
Gervasoni, J. E., Jr., Conrad, D. H., Hugli, T. E., Schwartz, L. B. & Ruddy, S. (1986) *J. Immunol.*, **136**, 282–292
Goldman, D. W. & Goetzl, Y. Y. (1984) *J. Exp. Med.* **159**, 1027
Goldman, D. W., Gifford, L. A., Young, R. N. & Goetzl, E. J. (1985) *Fed. Proc. Fed. Am. Soc. Exp. Biol.* **44**, 781 (abstr.)

Hoeprich, P. D. & Hugli, T. E. (1984) *Fed. Proc. Fed. Am. Soc. Exp. Biol.* **44**, 1185 (abstr.)

Huey, R. & Hugli, T. E. (1985) *J. Immunol.* **135**, 2063–2068

Huey, R. & Hugli, T. E. (1984) *Complement* **1**, 166 (abstr.)

Hugli, T. E. (1981) *Crit. Rev. Immunol.* *I* **4**, 321–366

Hugli, T. E. & Chenoweth, D. E. (1981) in *Immunoassays: Clinical Laboratory Techniques for the 1980's* (Nakamura, R. M., Dito, W. R. & Tucker, E. S., III, eds.), vol. 4, pp. 443–460, Alan R. Liss, New York

Hugli, T. E. & Morgan, E. L. (1984) *Contemp. Topics Immunobiol.* **14**, 109–153

Hugli, T. E. & Marceau, F. M. (1985) *Br. J. Pharmacol.* **84**, 725–733

Johnson, R. J. & Chenoweth, D. E. (1985) *J. Biol. Chem.* **260**, 7161–7164

Khan, S. A., Erickson, B. W., Kawahara, M. S. & Hugli, T. E. (1985) *Complement* **2**, 118 (abstr.)

Lu, Z-X., Fok, K. F.; Erickson, B. W. & Hugli, T. E. (1984) *J. Biol. Chem.* **259**, 7367–7370

Marceau, F. M. & Hugli, T. E. (1984) *J. Pharmacol. Exp. Ther.* **230**, 749–754

Niedel, J., Davis, J. & Cuatrecasas, P. (1980) *J. Biol. Chem.* **255**, 7063–7066

Pilch, P. F. & Czech, M. P. (1984) in *Membranes, Detergents, and Receptor Solubilization* (Venter, J. C. & Harrison, L. D., eds.), pp. 161–175, Alan R. Liss, New York

Regal, J. F. (1982) *J. Pharmacol. Exp. Ther.* **220**, 102–107

Rollins, T. E & Springer, M. S. (1985) *J. Biol. Chem.* **260**, 7157–7160

Schmitt, M., Painter, R. G., Jesaitis, A. J., Preissner, K., Sklar, L. A. & Cochrane, C. G. (1983) *J. Biol. Chem.* **258**, 649–654

Schwartz, L. B., Kawahara, M. S., Hugli, T. E., Vik, D., Fearon, D. T. & Austen, K. F. (1983) *J. Immunol.* **130**, 1891–1895

Stimler, N. P., Bloor, C. M. & Hugli, T. E. (1983) *Immunopharmacology* **5**, 251–257

*Biochem. Soc. Symp.* **51**, 83–96
*Printed in Great Britain*

# Complement Receptors and Related Complement Control Proteins

ROBERT B. SIM, VIVEK MALHOTRA, JEAN RIPOCHE, ANTHONY J. DAY,
KINGSLEY J. MICKLEM* and EDITH SIM†

*MRC Immunochemistry Unit, Department of Biochemistry, University of Oxford,
South Parks Road, Oxford OX1 3QU, U.K.*

## The Complement System

The complement system in human plasma is activated in response to the formation of antibody–antigen complexes, or the entry of various types of foreign materials into the circulation or tissue spaces (for review see [1–4]). Activation occurs via two routes, named the alternative and classical pathways (Fig. 1). A key step in either pathway is the formation, on the surface of the complement activator, of complex, unstable proteolytic enzymes termed 'C3 convertases' (Fig. 1) which cleave and activate the most abundant complement protein C3. Once C3 is cleaved into its two fragments, C3a and C3b, the C3b fragment binds covalently on to the surface of the complement activator, and this reaction leads on to three of the important biological functions of the complement system. These are:

(1) *Lysis of complement activators with a lipid bilayer membrane.* C3b molecules deposited on to the surface of the complement activator serve as a binding site for C5, which is then cleaved and activated by the convertase enzymes (Fig. 1). Activation of C5 leads to the assembly of the membrane attack complex, which becomes inserted into the target membrane and causes membrane disruption (for review see [6]).

(2) *Phagocytosis of particulate material.* Surface-bound C3b and its degradation products, as discussed below, assist in promoting phagocytosis of particulate material. This is mediated via receptors for C3b and other C3 fragments on phagocytic cells (for review see [7]).

(3) *Regulation of immune complex size.* Deposition of C3b and other, earlier complement components on immune complexes regulates their size and has a role in preventing formation of large insoluble aggregates which, in pathological

Abbreviations: The nomenclature of complement components and C3 and its fragments is as recommended by the World Health Organization [103,104]. Further C3 fragments are as defined by Davis *et al.* [35]. Complement receptors type 1, 2 and 3 are abbreviated as CR1, CR2 and CR3 [7]. C4b-binding protein and decay accelerating factor are abbreviated as C4bp and DAF respectively.

*Present address: Nuffield Department of Pathology, Level 1, John Radcliffe Hospital 1, Headington, Oxford OX1 9DU, U.K.

† Present address: Wellcome Toxicology Unit, Department of Pharmacology, University of Oxford, South Parks Road, Oxford OX1 3QT, U.K.

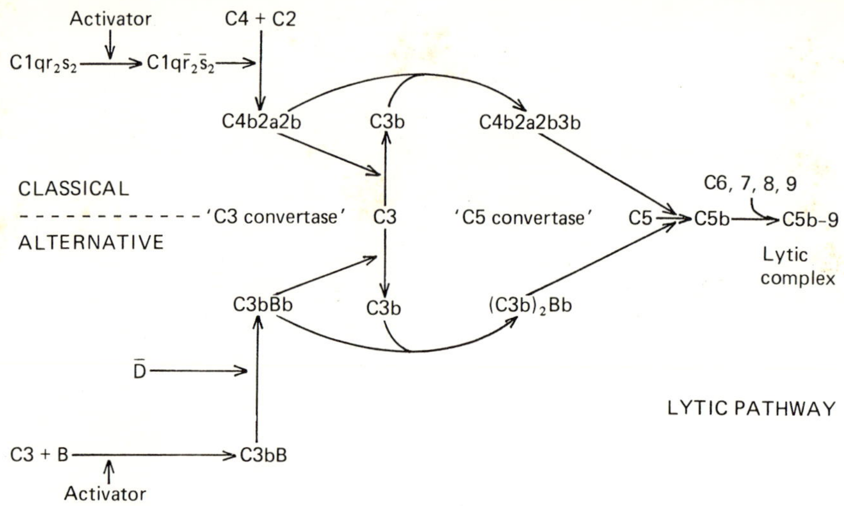



Fig 1. *The reaction pathways of the complement system*

Activation of the classical pathway proceeds by interaction of the C1q subcomponent of C1 with a complement activator. Binding of C1q in turn converts C1r in the C1 complex into its proteolytically active form C1r̄. C1r̄ activates the serine proteinase proenzyme C1s, to form C1s̄, which in turn cleaves and activates C4 and C2 which form the complex proteinase C4b2a2b, the 'C3 convertase' of this pathway. Since activated C4, like C3, can undergo covalent binding reactions, the convertase enzyme becomes localized on the surface of the complement activator. C3 is cleaved by the proteolytically active subunit, C2a, of the convertase. C3b, which is deposited covalently on the complement activator surface close to the site of attachment of the convertase, serves as a binding site for C5, which is then activated by the convertase. Activation of C5 is followed by non-enzymic assembly of the C5b6789 lytic complex. Initial events in the activation of the alternative pathway are less well understood, but early events include the assembly of a C3bB complex on the surface of the activator. This is in turn activated by factor D to form the C3bBb complex, which is homologous to C4b2a2b and is the 'C3 convertase' of the alternative pathway.

conditions, may become deposited in the tissues (for review and discussion see [8–10]). C3b deposition also causes immune complexes to bind to erythrocytes via complement receptor type 1 (CR1), and erythrocyte-bound immune complexes appear to be cleared via the liver [11–13].

From the considerations above, it is clear that very important biological functions of complement are mediated by C3 receptors.

## The Covalent Binding Reaction of C3b, and Subsequent Degradation of C3b

The covalent binding of C3 is mediated by an acyl transfer from an activated internal thiolester, a mechanism which is now partially understood [14–19]. Complement component C4, a homologue of C3 [20], possesses the same covalent binding mechanism. When C3 is cleaved by a convertase, the activated internal thiolester becomes exposed (Fig. 2). It is very reactive, and will bind to any nucleophilic group in the vicinity. Since C3 is generally cleaved by a convertase enzyme which is localized on the surface of a complement activator, there is a possibility of activated C3 reacting with a nucleophile, e.g. an OH or

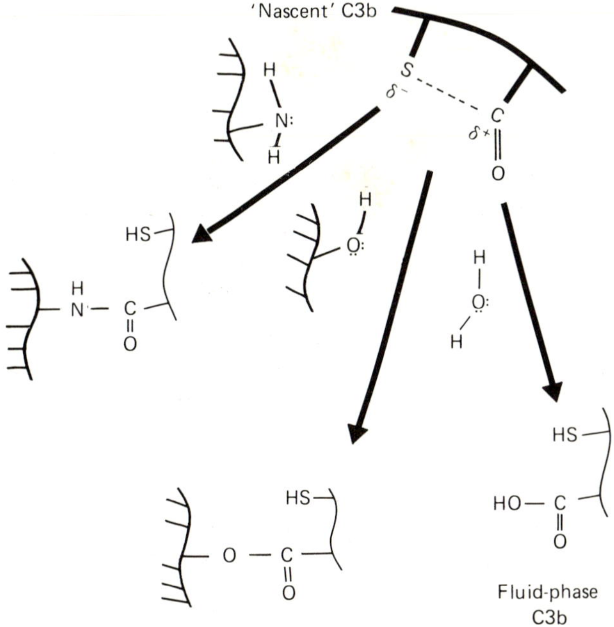


Fig. 2. *The Covalent Binding Reaction of C3*

On cleavage of C3 by a convertase enzyme, an activated internal thiolester, which in native C3 is protected from the solvent, becomes exposed and will react with any available nucleophile. Activated C3 ('nascent C3b') is shown reacting with OH or $NH_2$ groups on the surface of complement activators to form ester or amide bonds respectively, or with water, to form 'fluid-phase' C3b.

$NH_2$ group, on the complement activator surface. Alternatively, activated C3 may merely react with water, the most abundant nucleophile in biological systems. In most circumstances, only about 5–10% of the C3 which is activated binds to an activator surface or macromolecule, while 90% or more reacts with water. This creates two populations of C3b, i.e. surface-bound or soluble, which are antigenically different, [21] and may be recognized differently by proteins responsible for C3b degradation, or by cellular receptors for C3 fragments. This will be discussed briefly in later sections. In addition, since the covalent binding reaction of C3 is so non-specific, activated C3 can form other minor populations, by, e.g. binding to itself to form C3 (C3b) dimers, or binding to bystander proteins [22–25]. This has been shown *in vitro*, but it is not known whether these have any special functional significance *in vivo*.

Once C3b is formed, it undergoes a well-characterized series of degradation reactions which produce fragments which interact with various cell-surface proteins or receptors. The route of degradation of C3 is summarized in Fig. 3. C4 undergoes a similar series of breakdown steps (Fig. 4), and certain C4 fragments also interact with cell-surface proteins. Cell-surface proteins which interact with many of these C3 and C4 fragments have been identified. Receptors for C3a and C4a are discussed elsewhere [42]. Cell surface proteins which interact with other C3 or C4 fragments are summarized in Table 1, and structural features of these proteins, where known, are shown in Table 2. The cellular

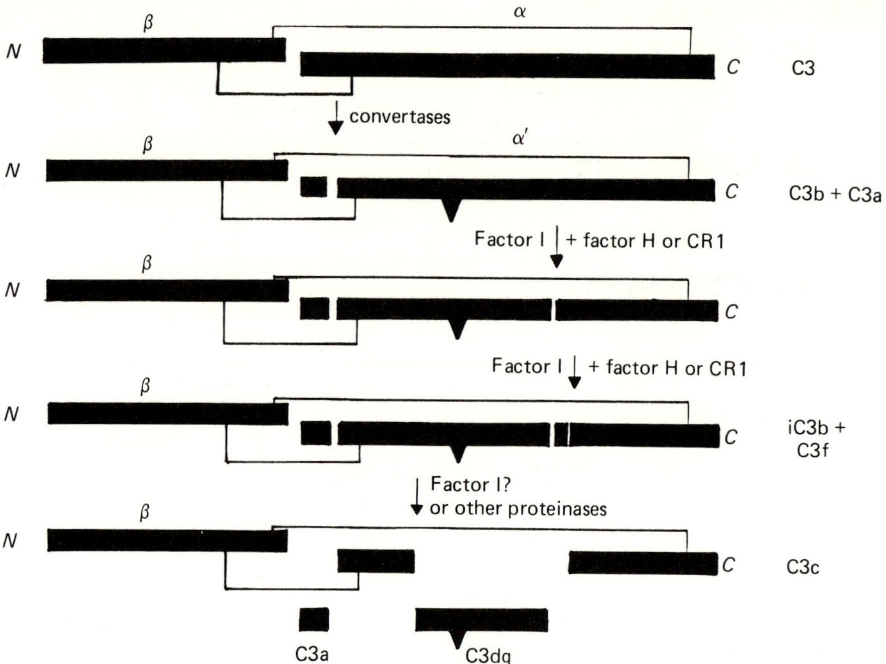



Fig. 3. *Degradation of C3*

Cleavage of the α chain of C3 by convertase enzymes splits C3 into C3a (9000 $M_r$) and C3b (184000 $M_r$). The thiolester site of C3b, through which C3b may bind covalently to surfaces, is indicated by a spike. Once C3b is formed it is rapidly cleaved in two places by the serine proteinase, factor I, to form iC3b. A 2000-$M_r$ fragment, C3f, is liberated [26,27]. For soluble C3b, this reaction has an absolute requirement for a cofactor protein, which binds to C3b. Only C3b in the C3b–cofactor complex is susceptible to factor I. For surface-bound C3b, the cofactor protein accelerates iC3b formation, but is not absolutely required. Factor H and the cell-surface protein CR1 are the physiological cofactors for this reaction [28–31]. iC3b is then slowly cleaved at a single site to form C3c (143000 $M_r$) and C3dg (39000 $M_r$). The identity of the proteinase responsible for this reaction is uncertain. Many authors have presented evidence that factor I, with CR1 (only) as cofactor, is responsible for this step [32–34]. However, other evidence indicates that CR1 is not involved [31] or, at least, is not involved in the breakdown of soluble, as opposed to surface-bound, iC3b to C3c plus C3dg [35,36]. In mouse iC3b, there is no potential factor I cleavage site in the appropriate region of the polypeptide chain [37], although mouse iC3b is also degraded in the circulation to C3c and C3dg-like fragments [38]. This casts doubt on the involvement of factor I at this stage of degradation. iC3b can also be cleaved by other enzymes, e.g. leukocyte elastase, trypsin, to form C3d, a fragment smaller than C3dg, plus C3c (for summary see [37]). Degradation of surface-bound C3b results in retention of iC3b and C3dg or C3d on the surface, while C3c dissociates.

distribution of these proteins, and brief indications of their main roles or functions, are indicated in Table 3.

The better-characterized of the proteins listed in Table 3, i.e. CR1, CR2 and CR3, have been described principally in terms of their ability to mediate adherence of particles coated with surface-bound C3 fragments. CR1, and probably also CR2, will also recognise soluble monomeric C3 fragments. Another group of these proteins however, are not known to mediate adherence of C3 fragment-coated particles. These include factor H, p150,95, and gp 45-70. It is possible that the function of these proteins is to recognize not surface-bound

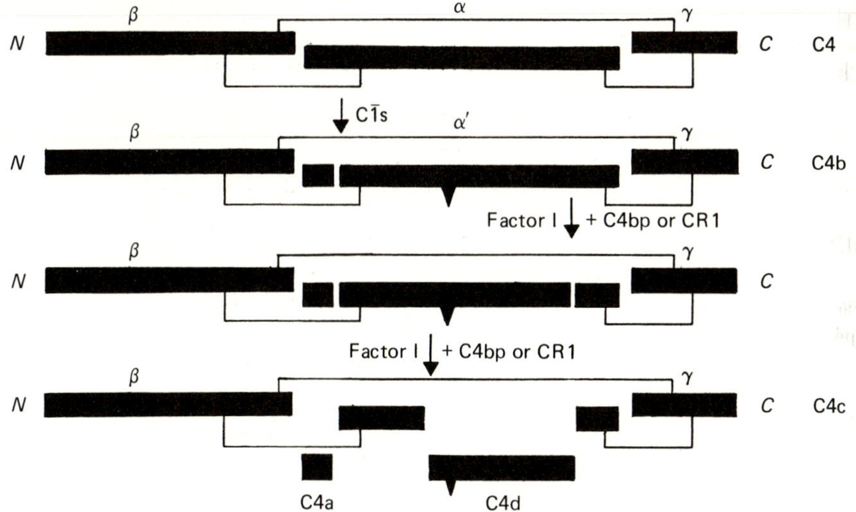



Fig. 4. *Degradation of C4*

Cleavage of the $\alpha$ chain of C4 by C$\overline{1}$s splits C4 into C4a (9000 $M_r$) and C4b (185000 $M_r$). The thiolester site of C4b, through which C4b may bind covalently to surfaces is indicated by a spike. Once C4b is formed it is rapidly cleaved in two places by factor I, to form C4c (140000 $M_r$) and C4d (45000 $M_r$) [39]. As for C3b breakdown (Fig. 3) this reaction also requires a cofactor. C4b-binding protein (C4bp) and CR1 are the physiological cofactors for this reaction [40,41]. Degradation of surface-bound C4b leaves C4d bound to the surface, while C4c dissociates.

Table 1. *Cell-surface proteins which interact with C3 or C4 fragments*

| Fragment | Cell-surface protein |
| --- | --- |
| C3b or C4b | CR1 [29,41]; factor H [43,44]; gp 45-70 (human) [45] or p65 (mouse) [46]; Decay Accelerating Factor (DAF) [47] |
| C3b | HSV-1 C glycoprotein [48]; CR2 [49]. |
| iC3b | CR3 [50]; p150,95 [51,52]; CR2 [49,53,54]; 90000 $M_r$ protein from spleen [52]; CR1 [49,55] |
| C3dg or C3d | CR2 [49,53,54]; 'CR4' [56,57] |

CR1 binds strongly to C3b or C4b, and has also a weak interaction with iC3b [49]. CR2 binds principally to iC3b or C3d, but in artificial conditions, it interacts with C3b [49,58] and may also mediate C3b-dependent rosetting in Raji cells [58]. CR3 and its homologue p150,95 may also bind to C3dg or C3d [51,59], but this is not clearly established. 'CR4' is a term first used by Myones *et al.* [56] to describe a receptor activity on granulocytes for C3d (and possibly iC3b). This receptor has not been isolated.

C3 fragments, but soluble monomeric C3 fragments which are generated abundantly during complement activation. Recognition of such fragments by cellular receptors may be important in the inflammatory response. CR1, DAF, and probably also surface-bound factor H and gp 45-70 have, in addition, roles in regulating C3b degradation. More extensive reviews of the characteristics of complement receptors have been published recently [7,59,76].

Although the proteins listed in Tables 1–3 are a diverse group, it is already clear that many of the proteins belong to one of two families. CR3 and p150,95

Table 2. *Structural features of cell-surface proteins which interact with C3 or C4 fragments*

| Cell surface protein | Structure |
| --- | --- |
| CR1 | Single polypeptide chain glycoprotein with polymorphic size variants ranging from 160000 to 260000 $M_r$ [60–62] |
| CR2 | Single polypeptide chain glycoprotein of $M_r$ 140000–145000. [49,53,54,58] |
| CR3 | Glycoprotein consisting of two non-covalently linked chains, $\alpha$ [170000 $M_r$] and $\beta$ (95000 $M_r$) [63] |
| p150,95 | Glycoprotein consisting of two non-covalently linked chains, $\alpha$ (150000 $M_r$) and $\beta$ (95000 $M_r$) [63] |
| Factor H | Single polypeptide chain glycoprotein, $M_r$ 155000 [43,64] |
| gp45-75 | Heterogeneous glycoprotein with apparent $M_r$, as estimated by SDS/polyacrylamide-gel electrophoresis, of 45000–70000. Heterogeneity reported to be due to carbohydrate [45,65] |
| Decay Accelerating Factor (DAF) | Single polypeptide chain glycoprotein of about 70000 $M_r$ [47] |

Table 3. *Cellular distribution and principal roles of proteins which interact with C3 or C4 fragments*

| Protein | Distribution | Role |
| --- | --- | --- |
| CR1 | Erythrocytes, B lymphocytes, some T lymphocytes, monocytes, granulo-cytes, mast cells, kidney glomerular epithelial cells, and possibly also peripheral nerve tissue [66–69] | Binding of immune complexes to ery-throcytes [13]<br>Regulation of C3b breakdown [29]<br>Phagocytosis [70] |
| CR2 | B lymphocytes [49,53,54] | Regulation of B cell functions (see, e.g. [71])<br>Epstein–Barr virus receptor [72] |
| CR3 | Granulocytes, monocytes, NK cells, low level on tissue macrophage [50,63,73] | Phagocytosis |
| p150,95 | Monocytes, granulocytes, NK cells; high level on tissue macrophage [50,63,73] | Unknown: probably does not mediate particle adherence [51] |
| gp 45-70 | Most leukocytes, platelets, erythrocyte precursors [45,62] | Does not mediate particle adherence<br>Possible regulation of C3b breakdown [45,62] |
| Factor H | U937 cells, Raji and tonsil B cells [43,44] | Probably does not mediate particle adherence [74]<br>Probable regulation of C3b breakdown |
| HSV1-C glyco-protein | HSV1-infected endothelial cells [48] | Unknown |
| DAF | Erythrocytes, most leukocytes, platelets [47,75] | Accelerates decay of convertases<br>Does not mediate particle adherence [47,75] |
| 'CR4' | Polymorphonuclear leukocytes [56,57] | Defined as a receptor for soluble or surface-bound C3d/C3dg: role unknown [56,57] |

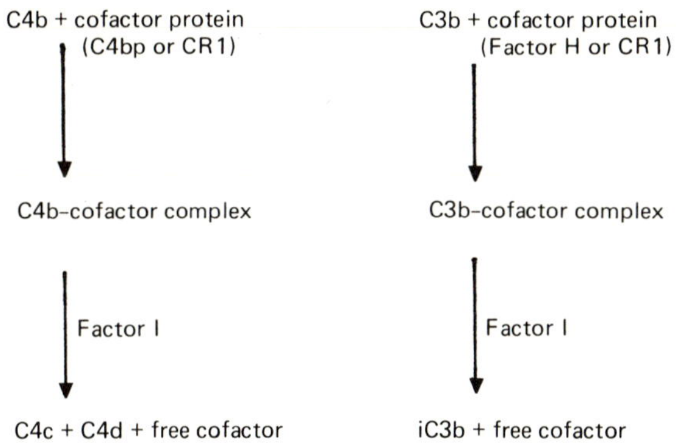



Fig. 5. *Cofactor activity*

C4bp, factor H and CR1 each serve as cofactors for the factor I-mediated degradation of C4b or C3b. Studies of distribution of cofactor activity in cells which do not express any of these proteins suggest that at least one further 'cofactor' protein must exist (R.B. Sim & E. Sim, unpublished work). This other protein may, as suggested by Holers *et al.* [62], be gp 45-70.

are closely related in structure [63,77]. Each of these glycoproteins consists of an $\alpha$ chain non-covalently linked to a $\beta$ chain. The $\beta$ chain of CR3, p150,95 and a third cell-surface protein, LFA-1, are thought to be identical. The $\alpha$ chains of CR3 and LFA-1 have been shown to have limited sequence homology [77], and so it is likely that the CR3$\alpha$, LFA-1$\alpha$ and p150,95$\alpha$ chains are closely related.

The other family consists of proteins which have striking functional similarity, but which are not clearly related in terms of their gross structure or size. These proteins are CR1, surface-bound factor H and possibly also gp 45-70 and its mouse equivalent p65. Within this group can also be considered soluble factor H, which is an abundant serum protein, and a further serum protein C4bp. DAF may also belong to this family, but there is as yet no strong evidence for its inclusion. Each of these proteins interacts with C3b or its homologue C4b, and they have particular roles associated with regulation of complement activation. The two activities which they share in particular are:

(1) the ability to act as a cofactor for factor I-mediated degradation of C3b or C4b, as summarized in Fig. 5.

(2) the capacity to destabilize or dissociate the C3 convertase enzymes of the classical or alternative pathways. Factor H, CR1 and DAF destabilize the alternative pathway C3 convertase, C3bBb, by binding to C3b and causing dissociation of Bb, while C4bp, CR1 and DAF will cause similar dissociation of the classical pathway convertase [29,40,41,47,78]

The very close similarity between CR1, factor H and C4bp in the characteristics of cofactor activity for degradation of soluble C3b or C4b [79] suggested the possibility that these proteins might share some limited structural similarities associated with substrate (C3b/C4b) binding. The gross structure of these proteins is however quite different (Table 4), so the idea of a homologous substrate-binding domain of limited size might be all that could be expected in terms of similarity. The idea of a common substrate-binding domain was strengthened by studies from several laboratories on limited proteolysis of CR1, factor H and C4bp, where in initial stages of breakdown there is similarity in

Table 4. *General properties of cofactor proteins*

The $M_r$ given for CR1 represents the commonest polymorphic variant [62]. Solubilized CR1 and factor H appear to be monomeric, while C4b is an oligomer consisting of seven disulphide-linked subunits each of 70000 $M_r$

| Property | CR1 | Factor H | C4bp |
|---|---|---|---|
| $M_r$ | 220000 [80] | 155000 [64] | 70000 ($\times$ 7) [81] |
| No. of amino acids | 2030 [80] | 1275 [64] | 549 ($\times$ 7) [82] |
| Carbohydate | 6–9% [69,80,83] | 9% [64] | 12% [82] |
| Sequence data | None | Partial [84,85] | Complete [82] |

Fig. 6. *Limited proteolysis of CR1, factor H and C4bp by trypsin.*

Initial cleavage destroys interaction with surface bound ligand, while cofactor activity for degradation of soluble ligand is retained, as discussed in the text. Disulphide bridges are represented in arbitrary positions. For C4bp and factor H, orientation is *N*- to *C*-terminal from left to right. For CR1 the orientation of fragments is unknown.

the behaviour of these proteins. As shown in Fig. 6, trypsin cleavage of factor H and C4bp results in formation of an *N*-terminal fragment of about 35000 $M_r$, which remains disulphide-linked to the rest of the molecule [64,86]. For factor H, it is known that the C3b binding site is in the 35000 $M_r$ *N*-terminal fragment [87]. The C4b-binding site of C4bp also resides in the *N*-terminal region [88–90]. Cleavage of this site in either protein occurs with retention of cofactor activity for factor I-mediated degradation of soluble C3b or C4b, but the ability of H or C4bp to bind to surface-attached C3b or C4b is lost [64,86–90]. CR1 also forms a fragment of about 35000 $M_r$ in the course of trypsin degradation (Fig. 6) [80] and retains cofactor activity for cleavage of soluble C3b, while the ability to bind to solid-phase substrate is lost [80]. However, the orientation of the fragments formed from CR1, and the location of the C3b binding site in CR1, are still unknown. The differential loss of interaction with solid-phase or soluble substrate seen on cleavage of these proteins may reflect merely a large decrease of binding affinity after proteolysis, or may indicate that recognition of solid-phase and soluble substrate is qualitatively different.

Two recent important aspects of research on these proteins have greatly strengthened this idea of a close relationship. These are gene linkage studies, and amino acid or cDNA sequencing.

Studies by Rodriguez de Cordoba, Rubinstein, Atkinson, Nussenzweig and others [91–94] on the inheritance of known polymorphic variants of CR1, factor

Table 5. *Polymorphic variants of CR1, C4bp and factor H*

Inheritance patterns are all autosomal co-dominant. Information on CR1 is from [60–62], for factor H [91], and for C4bp [92].

| Protein | Polymorphism | Variants | Gene frequency |
|---|---|---|---|
| CR1 | Size | 160 kDa | Very rare |
| | | 220–240 kDa | 0.83 |
| | | 250–260 kDa | 0.16 |
| | | 280 kDa | Very rare |
| C4bp | Charge | C4bp 1 | 0.98 |
| | | C4bp 2 | 0.018 |
| Factor H | Charge | FH 1 | 0.691 |
| | | FH 2 | 0.302 |
| | | FH 3 | 0.006 |

H and C4bp have shown that the structural genes for these proteins are closely linked. The polymorphic variation of these proteins is summarized in Table 5.

Sequencing studies on these proteins have shown up further similarities. C4bp, which has the shortest polypeptide chain length of these proteins, has been completely sequenced at the cDNA and protein level by Chung, Bentley & Reid [82]. C4bp was found to contain eight internally homologous regions, each of about 60 amino acids long, with 25–35% homology between them, based on identity and conservative replacement. Comparison of this repetitive sequence with other known sequences showed that the Ba region of complement factor B also contained three of these internal homology units [85], and it is now known that the *N*-terminal region of C2, a homologue of factor B, also contains three such units [5]. Factor B and C2 also interact with C3b and C4b, and so some functional homology between B, C2 and C4bp does exist. However, it has been observed, as discussed by Chung *et al.* [82] that another, non-complement protein, $\beta_2$I glycoprotein, also contains six internally homologous units [96], which are closely related to those found in C4bp, C2 and factor B. This protein, so far as is known, does not interact with C4b or C3b.

Incomplete sequence data at the amino acid and cDNA level is also available for human factor H [84,85]. More extensive data are available for mouse factor H. (T. Kristensen & B. F. Tack, unpublished work). It is evident that factor H also contains internal homology units which are similar to those in C4bp, Ba and $\beta_2$I. Fig. 7 shows the homology between a region of factor H [85] and regions of C4bp, Ba and $\beta_2$I. Fig. 8 shows internal homology in various peptide and cDNA-derived sequences in factor H [85]. From the internal homologies, it is clear that factor H contains several internal repeat units of the type found in C4bp. Information on mouse factor H, noted above, suggests that factor H contains at least 18 internal units of this type. From the information in Fig. 8, a 'consensus' sequence for the internal repeat of factor H can be constructed, and this is shown in Fig. 9, and compared with consensus sequences for Ba, C4bp and $\beta_2$I. The consensus sequence represents only residues which are completely, or very highly, conserved between repeat units, and is based principally on the spacing of Cys and Trp residues.

(a)

```
         1                    10                   20                   30
C4BP  C  E  -  P  -  N  S  C  I  N  L  P  D  I  -  P  H  A  S  W  E  T  Y  P  R  P  T  K  E  D
H     M  E  T  G  R  N  H  -  L  N  -  A  K  I  N  P  P  -  I  V  Q  N  A  T  I  V  S  R  Q  -

         31                   40                   50                   60
C4BP  V  -  -  Y  V  V  G  T  V  L  R  Y  R  C  H  -  P  G  Y  K  P  T  T  D  E  P  T  T  V  I
H     M  S  K  Y  P  S  G  E  R  V  R  Y  Q  C  R  S  P  -  Y  E  M  F  G  D  E  -  -  V  M

         61                   70
C4BP  C  -  Q  K  N  L  R  W  T  P  Y  Q  G  C      42% HOMOLOGY (CONSERVATIVE REPLACEMENTS)
H     C  L  N  G  N  W  T  E  P  P  -  Q  -  C      27% HOMOLOGY (IDENTITIES)
```

(b)

```
         1                    10                   20                   30
β₂I   L  P  E  C  -  R  E  V  K  C  P  F  P  S  R  -  P  D  N  G  F  V  -  N  -  Y  P  A  -  K
H     M  -  E  T  G  R  -  N  H  -  L  N  A  K  I  N  P  -  -  P  T  V  Q  N  A  T  I  V  S  R

         31                   40                   50                   60
β₂I   P  T  L  -  Y  Y  -  -  K  D  K  A  T  F  G  C  H  D  G  Y  S  L  D  G  P  E  E  I  E  C
H     -  Q  M  S  K  Y  P  S  G  E  R  V  R  Y  Q  C  R  S  P  Y  E  M  F  G  D  E  E  V  M  C

         61                   70
β₂I   T  K  L  G  N  W  S  A  M  P  S  C      42% HOMOLOGY (CONSERVATIVE REPLACEMENTS)
H     -  L  N  G  N  W  T  E  P  P  Q  C      24% HOMOLOGY (IDENTITIES)
```

(c)

```
         1                    10                   20                   30
Ba    R  W  S  G  -  Q  T  A  I  C  D  N  G  A  G  Y  C  S  N  P  G  -  I  P  I  G  T  -  -  -
H     M  E  I  G  R  N  H  -  -  -  L  N  A  K  I  -  -  -  N  P  P  T  V  Q  N  A  T  I  V  S

         31                   40                   50                   60
Ba    R  K  V  G  S  Q  Y  R  L  E  D  S  V  T  Y  H  C  -  S  R  G  L  T  L  R  G  S  Q  R  R
H     R  Q  M  -  S  K  Y  P  S  G  E  R  V  R  Y  Q  C  R  S  -  P  Y  E  M  F  G  D  E  E  V

         61                   70
Ba    T  C  Q  E  G  G  S  W  S  G  T  E  -  P  S  C      39% HOMOLOGY (CONSERVATIVE REPLACEMENTS)
H     M  C  L  N  -  G  N  W  -  -  T  E  P  P  Q  C      26% HOMOLOGY (IDENTITIES)
```

Fig. 7. *Homology between a segment of factor H sequence and regions of C4bp, Ba and $\beta_2 I$*

Factor H sequence is from [85]. The C4bp sequence shown (*a*) corresponds to the sequence beginning at residue 246 [82]. The $\beta_2 I$ sequence shown (*b*) begins at residue 179 [96], and the Ba sequence begins at residue 125 in the numbering of Morley & Campbell [95]. Dashes indicate gaps introduced to maximize homology. Identical residues are boxed. Conservative replacements (Dayhoff groupings) are underlined.

A summary of the similarities between factor H, CR1 and C4bp is shown in Table 6. As discussed above, all of these proteins possess factor I cofactor and convertase decay acceleration activity. Limited proteolytic cleavage of these proteins appears to destroy their interaction with surface-bound ligands, while

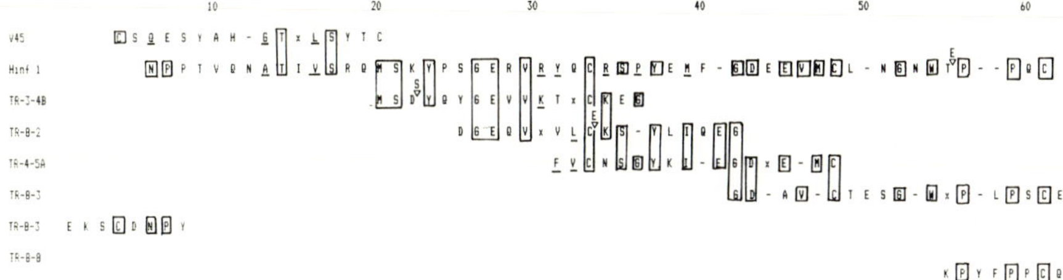

Fig. 8. *Internal homology between segments of factor H sequence*

Dashes indicate gaps inserted to maximize homology. X represents unidentified residues. Other details are as in Fig. 7.

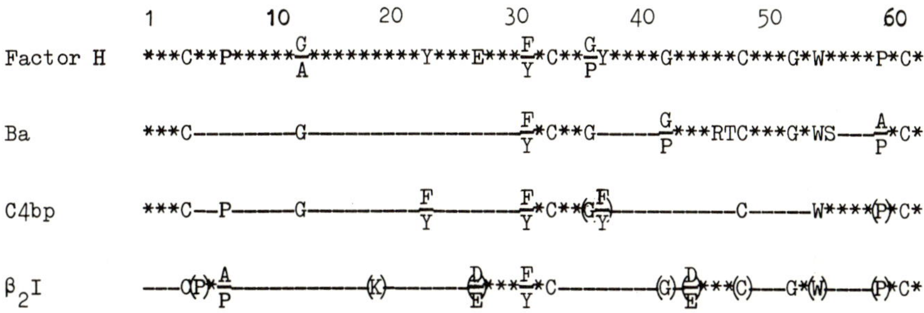

Fig. 9. *Consensus sequences for the internal repeat units of C4bp, factor H, Ba and $\beta_2I$*

Residues shown in parentheses are present in most, but not all, repeat units. Other residues shown are completely conserved. Asterisks (*) indicated residues which are not highly conserved. No account is taken here of conservative replacement. Dashes indicate regions of variable length.

Table 6. *Comparison of properties of CR1, Factor H and C4bp*

| Property | CR1 | Factor H | C4bp |
|---|---|---|---|
| Cofactor activity | C4b<br>C3b | C3b | C4b<br>(C3b) |
| Decay acceleration | C4b<br>C3b | C3b | C4b |
| Sequence homology<br>(also with Ba, $\beta_2I$ and C2) | ? | + | + |
| Membrane-bound form | + | + | ? |
| Soluble form | + | + | + |
| Elongated structure | ? | + | + |
| Oligomer formation | ? | + | + |
| Different recognition of<br>soluble and bound ligand | + | + | + |
| Zn²⁺ binding | + | + | + |

preserving their capacity to interact with soluble ligands. It is likely therefore, particularly for factor H [97], that they recognize these two forms of ligands differently. CR1 and factor H both exist in soluble and membrane-bound forms [29,36,43,78], but it is not known if this is the case for C4bp. Factor H and C4bp have very elongated structures [64,81; S. J. Perkins & R. B. Sim, unpublished work]. C4bp exists as an oligomeric species, as discussed above, and factor H can also form non-covalent oligomers [98, S. J. Perkins & R. B. Sim, unpublished work]. These features have not been investigated for CR1. All three proteins also interact strongly with $Zn^{2+}$ ions [99], a feature which is likely to reflect further structural similarity.

The very close functional and structural similarities between CR1, C4bp and factor H, and the linkage of their structural genes, indicate that these proteins may represent a family evolved from a single precursor. As discussed above, gp 45-70, and possibly also DAF, may belong to this family. C2 and factor B, as well as the non-complement protein $\beta_2 I$, also possess structural feature in common with this group.

Structural similarities between CR1, factor H and C4bp were predictable on the basis of function, but the extensively repetitive structure of C4bp and factor H is surprising. Since the C3b or C4b binding site of factor H or C4bp is likely to involve, at most, three or four of the internal repeat units, it is clear that the extensive homology between the proteins and the repeat structure cannot all be related to substrate binding. The repeat units, which are rich in cysteine residues, may however be associated with the curious elongated structure of the proteins. Factor H and $\beta_2 I$ (also an elongated protein, with estimated $f/f_0 = 1.6$) have unusual and similar c.d. spectra [100,101]. The spectrum of factor H was interpreted by DiScipio and colleagues [100] as having little or no $\alpha$ or $\beta$ structure. Structure predictions, based on the amino acid sequence of C4bp, [102] suggest that this protein, unlike H and $\beta_2 I$, contains mainly $\beta$, and some $\alpha$ structure. DiScipio and colleagues did however observe that partial reduction of factor H altered the c.d. spectrum to a form similar to that which would be expected for the $\alpha/\beta$ ratios predicted for C4bp. It seems likely that the conserved cysteines in the factor H and C4bp repeat units are maintaining an unusual secondary and tertiary structure associated with the elongated shape of the molecules.

We are very grateful to Drs. R. D. Campbell, D. R. Bentley, K. B. M. Reid, S. J. Perkins, B. F. Tack and L. Tack for helpful discussion and access to unpublished information. We thank Professor R. R. Porter for his advice and encouragement.

## References


1. Reid, K. B. M. (1983) *Biochem. Soc. Trans.* **11**, 1–12
2. Whaley, K. & Ferguson, A. (1981) *Mol. Aspects Med.* **4**, 209–273
3. Fothergill, J. E. & Anderson, W. H. K. (1978) *Curr. Top. Cell. Regul.* **13**, 259–311
4. Porter, R. R. (1980) *Proc. R. Soc. London Ser. B* **210**, 477–498
5. Bentley, D. R. & Campbell, R. D. (1985) *Biochem. Soc. Symp.* **51**, 7–18
6. Mayer, M. M. (1981) *Johns Hopkins Med. J.* **148**, 243–258
7. Fearon, D. T. & Wong, W. W. (1983) *Annu. Rev. Immunol.* **1**, 243–271
8. Takahashi, M. & Takahashi, S. (1981) *Clinics Immunol. Allergy* **1**, 261–279
9. Schifferli, J. A., Woo, P. & Peters, D. K. (1982) *Clin. Exp. Immunol.* **47**, 555–562
10. Sim, E., Gill, E. W. & Sim, R. B. (1984) *Lancet* **ii**, 422–424
11. Medof, M. E., Prince, G. M. & Oger, J. J.-F. (1982) *Clin. Exp. Immunol.* **48**, 715–725
12. Horgan, C. & Taylor, R. P. (1984) *Arthritis Rheum.* **27**, 320–329
13. Cornacoff, J. B., Hebert, L. A., Smead, W. L., Vanaman, M. E., Birmingham, D. J. & Waxman, F. J. (1983) *J. Clin. Invest.* **71**, 236–247
14. Law, S.-K. A., (1983) *Ann. N.Y. Acad. Sci.* **421**, 246–258

15. Law, S. K., Lichtenberg, N. A. & Levine, R. P. (1980) *Proc. Natl. Acad. Sci. U.S.A.* **77**, 7194–7198
16. Sim, R. B., Twose, T. M., Paterson, D. S. & Sim, E. (1981) *Biochem. J.* **193**, 115–127
17. Davies, S. G. & Sim, R. B. (1981) *Biosci. Rep.* **1**, 461–468
18. Janatova, J. (1983) *Ann. N.Y. Acad. Sci.* **421**, 218–234
19. Tack, B. F. (1983) *Springer Semin. Immunopathol.* **6**, 259–282
20. Sottrup-Jensen, L., Stepanik, T. M., Kristensen, T., Lønblad, P. B. Jones, C. M., Wierzbicki, D. M., Magnusson, S., Domdey, H., Wetsel, R. A., Lundwall, A., Tack, B. F. & Fey, G. H. (1985) *Proc. Natl. Acad. Sci. U.S.A.* **82**, 9–13
21. von Zabern, I., Nolte, R. & Vogt, W. (1981) *Scand. J. Immunol.* **13**, 413–431
22. Jacobs, R. J. & Reichlin, M. (1983) *J. Immunol.* **130**, 2775–2781
23. Arnaout, M. A., Melamed, J., Tack, B. F. & Colten, H. R. (1981) *J. Immunol.* **127**, 1348–1354
24. Johnson, U. & Holmström, E. (1982) *Acta Pathol. Microbiol. Immunol. Scand. Sect. C* **90**, 321–326
25. Siersted, H. C., Jensensius, J. C., Svehag, S.-E. & Brandslund, I. (1982) *J. Clin. Lab. Immunol.* **12**, 201–208
26. Harrison, R. A. & Lachmann, P. J. (1980) *Mol. Immunol.* **17**, 9–20
27. Sim, E., Wood, A. B., Hsiung, L.-M. & Sim, R. B. (1981) *FEBS Lett.* **132**, 55–60
28. Pangburn, M. K., Schreiber, R. D. & Muller-Eberhard, H. J. (1977) *J. Exp. Med.* **146**, 257–270
29. Fearon, D. T. (1979) *Proc. Natl. Acad. Sci. U.S.A.* **76**, 5867–5871
30. Gaither, T. A., Hammer, C. H. & Frank, M. M. (1978) *J. Immunol.* **123**, 1195–1204
31. Malhotra, V. & Sim, R. B. (1984) *Biochem. Soc. Trans.* **12**, 781–782
32. Medof, M. E., Iida, K., Mold, C. & Nussenzweig, V. (1982) *J. Exp. Med.* **156**, 1739–1754
33. Medicus R. G. & Arnaout, M. A. (1983) *Eur. J. Immunol.* **13**, 465–470
34. Ross, G. D., Lambris, J. D., Cain, J. A. & Newman, S. L. (1982) *J. Immunol.* **129**, 2051–2060
35. Davis, A. E., Harrison, R. A. & Lachmann, P. J. (1984) *J. Immunol.* **132**, 1960–1965
36. Yoon, S. H. & Fearon, D. T. (1985) *J. Immunol.* **134**, 3332–3338
37. De Bruijn, M. H. L. & Fey, G. (1985) *Proc. Natl. Acad. Sci. U.S.A.* **82**, 708–712
38. Kinoshita, T., Lavoie, S. & Nussenzweig, V. (1984) *J. Immunol.* **134**, 2564–2570
39. Press, E. M. & Gagnon, J. (1981) *Biochem. J.* **199**, 351–357
40. Fujita, T., Gigli, I. & Nussenzweig, V. (1978) *J. Exp. Med.* **148**, 1044–1051
41. Iida, K. & Nussenzweig, V. (1981) *J. Exp. Med.* **153**, 1138–1150
42. Huey, R., Fukuoka, Y., Hoeprich, P. D., Jr. & Hugli, T. E. (1985) *Biochem. Soc. Symp.* **51**, 69–81
43. Malhotra, V. & Sim, R. B. (1985) *Eur. J. Immunol.* **15**, 935–941
44. Schulz, T. F., Schreiner, D., Alsenz, J., Lambris, J. D. & Dierich, M. P. (1984) *J. Immunol.* **132**, 392–398
45. Cole, J. E., Housley, G. A., Dykman, T. R., MacDermott, R. P. & Atkinson, J. P. (1985) *Proc. Natl. Acad. Sci. U.S.A.* **82**, 859–863
46. Wong, W. W. & Fearon, D. T. (1985) *J. Immunol.* **134**, 4048–4056
47. Nicholson-Weller, A., Burge, J., Fearon, D. T., Weller, P. F. & Austen, K. F. (1982) *J. Immunol.* **129**, 184–189
48. Friedman, H. M., Cohen, G. H., Eisenberg, R. J., Seidel, C. A. & Cines, D. B. (1984) *Nature (London)* **309**, 633–635
49. Micklem, K. J., Sim, R. B. & Sim, E. (1984) *Biochem. J.* **224**, 75–86
50. Beller, D. I., Springer, T. A. & Schreiber, R. D. (1982) *J. Exp. Med.* **156**, 1000–1009
51. Malhotra, V., Hogg, N. & Sim, R. B. (1985) *Eur. J. Immunol.*, in the press
52. Micklem, K. J. & Sim, R. B. (1985) *Biochem. J.* **231**, 233–236
53. Iida, K., Nadler, L. & Nussenzweig, V. (1983) *J. Exp. Med.* **158**, 1021–1033
54. Weis, J. J., Tedder, T. F. & Fearon, D. T. (1984) *Proc. Natl. Acad. Sci. U.S.A.* **81**, 881–885
55. Medof, M. E., Iida, K., Mold, C. & Nussenzweig, V. (1982) *J. Exp. Med.* **156**, 1739–1754
56. Myones, B. L., Frade, R. & Ross, G. D. (1984) *Complement* **1**, 165
57. Vik, D. P. & Fearon, D. T. (1985) *J. Immunol.* **134**, 2571–2579
58. Barel, M., Charriaut, C. & Frade, R. (1981) *FEBS Lett.* **136**, 111–114
59. Ross, G. D. (1982) *Fed. Proc. Fed. Am. Soc. Exp. Biol.* **41**, 3089–3093
60. Wong, W. W., Wilson, J. G. & Fearon, D. T. (1983) *J. Clin. Invest.* **72**, 685–693
61. Dykman, T. R., Hatch, J. A.., Aqua, M. S. & Atkinson, J. P. (1984) *J. Immunol.* **134**, 1787–1789
62. Holers, V. M., Cole, J. L., Lublin, D. M., Seya, T. & Atkinson, J. P. (1985) *Immunol. Today* **6**, 188–192
63. Sanchez-Madrid, F., Nagy, J. A., Robbins, E., Simon, P. & Springer, T. A. (1983) *J. Exp. Med.* **158**, 1785–1803
64. Sim, R. B. & DiScipio, R. G. (1982) *Biochem. J.* **205**, 285–293

65. Cole, J. L., Ballard, L. L., Jones, E. A., Yu, G. & Atkinson, J. P. (1985) *Fed. Proc. Fed. Am. Soc. Exp. Biol.* **44**, 987

66. Fearon, D. T. (1980) *J. Exp. Med.* **152**, 20–30

67. Kazatchkine, M. D., Fearon, D. T., Appay, M. D., Mandet, C. & Bariety, J. (1982) *J. Clin. Invest.* **69**, 900–912

68. Nyland, H., Matre, R. & Tonder, O. (1979) *Acta Pathol. Microbiol. Scand. Sect. C* **87**, 7–10

69. Vranian, G., Conrad, D. H. & Ruddy, S. (1981) *J. Immunol.* **126**, 2302–2311

70. Ehlenberger, A. G. & Nussenzweig, V. (1977) *J. Exp. Med.* **145**, 357–371

71. Frade, R., Crevon, M. C., Barel, M., Vasquez, A., Krikorian, L., Charriaut, C. & Galanaud, P. (1985) *Eur. J. Immunol.* **15**, 73–76

72. Fingeroth, J. D., Weis, J. J., Tedder, T. F., Strominger, J. L., Biro, A. P. & Fearon, D. T. (1984) *Proc. Natl. Acad. Sci. U.S.A.* **81**, 4510–4514

73. Hogg, N., Takacs, L., Palmer, D. G., Salvendran, Y. & Allen, C. (1986) *Eur. J. Immunol.*, in the press

74. Malhotra, V. (1985) D. Phil Thesis, Oxford University

75. Kinoshita, T., Medof, M. E., Silber, R. & Nussenzweig, V. (1985) *J. Exp. Med.* **162**, 75–92

76. Fearon, D. T. (1984) *Immunol. Today* **5**, 105–110

77. Springer, T. A. & Anderson, D. C. (1986) *Biochem. Soc. Symp.* **51**, 47–57

78. Whaley, K. & Ruddy, S. (1976) *J. Exp. Med.* **144**, 1147–1163

79. Sim, E. & Sim, R. B. (1983) *Biochem. J.* **210**, 567–576

80. Sim, R. B. (1985) *Biochem. J.* **232**, 883–889

81. Dahlbäck, B., Smith, C. A. & Muller-Eberhard, H. J. (1983) *Proc. Natl. Acad. Sci. U.S.A.* **80**, 3461–3465

82. Chung, L. P., Bentley, D. R. & Reid, K. B. M. (1985) *Biochem. J.* **230**, 133–141

83. Atkinson, J. P. & Jones, E. A. (1984) *J. Clin. Invest.* **74**, 1649–1657

84. Kristensen, T., Wetsel, R. A. & Tack, B. F. (1985) *Fed. Proc. Fed. Am. Soc. Exp. Biol.* **44**, 1531

85. Ripoche, J., Day, A. J., Willis, A. C., Belt, K. T., Campbell, R. D. & Sim, R. B. (1986) *Biosci. Rep.*, in the press

86. Hong, K., Kinoshita, T., Dohi, Y. & Inoue, K. (1982) *J. Immunol.* **129**, 647–652

87. Alsenz, J., Lambris, J. D., Schultz, T. F. & Dierich, M. P. (1984) *Biochem. J.* **224**, 389–398

88. Chung, L. P. & Reid, K. B. M. (1985) *Biosci. Rep.* **5**, 855–865

89. Reid, K. B. M. & Gagnon, J. (1982) *FEBS Lett.* **137**, 75–79

90. Fujita, T., Kamato, T. & Tamura, N. (1985) *J. Immunol.* **134**, 3320–3324

91. Rodriguez de Cordoba, S. & Rubinstein, P. (1984) *J. Immunol.* **132**, 1906–1908

92. Rodriguez de Cordoba, S., Ferreira, A., Nussenzweig, V. & Rubinstein, P. (1983) *J. Immunol.* **131**, 1565–1569

93. Rodriguez de Cordoba, S., Dykman, T. R., Ginsberg-Fellner, F., Ercilla, G., Aqua, M., Atkinson, J. P. & Rubinstein, P. (1984) *Proc. Natl. Acad. Sci. U.S.A.* **81**, 7890–7892

94. Rodriguez de Cordoba, S., Lublin, D. M., Rubinstein, P. & Atkinson, J. P. (1985) *J Exp. Med.* **161**, 1189–1195

95. Morley, B. J. & Campbell, R. D. (1984) *EMBO J.* **3**, 153–158

96. Lozier, J., Takahashi, N. & Putnam, F. W. (1984) *Proc. Natl. Acad. Sci. U.S.A.* **81**, 3640–3644

97. Sim, E., Palmer, M. S., Puklavec, M. & Sim, R. B. (1983) *Biosci Rep.* **3**, 1119–1131

98. Ripoche, J., Al-Salihi, A., Rousseaux, J. & Fontaine, M. (1984) *Biochem. J.* **221**, 89–96

99. Day, A. J. & Sim, R. B. (1986) *Biochem. Soc. Trans.* **14**, 73–74

100. DiScipio, R. G. & Hugli, T. E. (1982) *Biochim. Biophys. Acta* **709**, 58–64

101. Osborne, J. C., Lee, N. S. & Brewer, B. B. (1982) *Fed. Proc. Fed. Am. Soc. Exp. Biol.* **41**, 1021

102. Chung, L. P. (1985) D. Phil. Thesis, Oxford University

103. World Health Organisation (1968) *Bull. W.H.O.* **39**, 935–936

104. World Health Organisation (1981) *Bull. W.H.O.* **59**, 489–490

*Biochem. Soc. Symp.* **51**, 97–112
*Printed in Great Britain*

# Lysosomal Membrane Glycoproteins: Properties of LAMP-1 and LAMP-2

JEFF W. CHEN, GREGORY L. CHEN,* M. PATRICIA D'SOUZA,
THERESA L. MURPHY and J. THOMAS AUGUST

*Department of Pharmacology and Experimental Therapeutics, The Johns Hopkins University
School of Medicine, Baltimore, MD 21205, U.S.A.*

## Synopsis

Several properties of the lysosomal membrane glycoproteins LAMP-1 and LAMP-2 have been analysed. Each molecule was strongly associated with lysosome membranes and was extracted only in the presence of detergent. Studies of the biosynthesis and processing of the glycoproteins showed that each contained a polypeptide core of approx. 43000 Da as identified by use of tunicamycin and endoglycosidase H. Nascent glycoproteins pulse-labelled for 5 min with [$^{35}$S]methionine were approx. 92000 Da. These precursor molecules were processed in 30 min to highly heterogenous mature glycoproteins of approx. 110000 Da (LAMP-1) and 105000 Da (LAMP-2). Concomitant with the increase in apparent $M_r$ the molecules became endoglycosidase H resistant and acquired sialic acid residues, indicating that they were converted to complex-type oligosaccharides. The final maturation of the glycoproteins was blocked by monensin. Immunohistochemical analysis of tissues from Balb/c and Beige/J mice showed that the molecules were present on many types of cells, consistent with their presence in lysosomes. The patterns of tissue expression of LAMP-1 and LAMP-2 in the two mouse strains were the same except that the intensity of staining of LAMP-2 was less than that of LAMP-1. LAMP-2, but not LAMP-1, gave a decreased immunofluorescent staining intensity in transformed HaNIH as compared with NIH/3T3 cells. The marked similarities between the LAMP proteins raise the consideration of common functions, possibly associated with the high oligosaccharide content of the molecules.

## Introduction

Lysosomes have a central role in cellular homeostasis as sites for digestion of foreign materials and for degradation of intracellular components undergoing autolytic processing (de Duve, 1983). Recently, several reports have described glycoproteins associated with lysosomal membranes. These molecules are of interest for their possible role in the biogenesis of lysosomes or in certain specialized functions, such as fusion with other vesicles, selective recognition and

* Present address: University of California, Berkeley, CA, U.S.A.

transport of molecules, vesicle acidification, and resistance to lysosomal hydrolytic enzymes.

We previously identified two glycoproteins of mouse cells specifically localized in the lysosomal membrane, LAMP-1 of 105000–115000 Da and LAMP-2 of 100000–110000 Da (Chen *et al.*, 1985*a,b*). Electron microscopy with ferritin bridge labelling showed that both were localized just beneath the limiting lysosomal membrane of large 'dense body' lysosomes and smaller multivesicular lysosomes. LAMP-1 and LAMP-2 appeared to be different polypeptides as indicated by tryptic peptide mapping, sequential immunoprecipitation (Chen *et al.*, 1985*a*), and *N*-terminal amino acid sequence analysis (J. W. Chen, unpublished work). Additional studies of LAMP-1 suggested that the molecule contained a large number of *N*-linked oligosaccharides, as the glycoprotein pulse-labelled with [$^{35}$S]methionine and treated with endoglycosidase H yielded a core peptide of 45000 Da (Chen *et al.*, 1985*b*).

LAMP-1 and LAMP-2 were compared with other recently described lysosomal membrane glycoproteins. Reggio *et al.* (1984) and Tougard *et al.* (1985) identified a protein of about 100000 Da present both in lysosomes and prelysosomal acidic vesicles. Polyclonal antibodies against this protein reacted with a gastric mucosal H$^+$/K$^+$ATPase and it was suggested that the protein may be a component of the proton pump involved in vesicle acidification. Lewis *et al.* (1985) described glycoproteins of 120000, 100000, and 80000 Da located in lysosomal membranes. The 120000-Da component was found to contain a 42000-Da polypeptide core and at least 18 *N*-linked oligosaccharides rich in sialic acid. In addition, Lippincott-Schwartz & Fambrough (1986) have identified a glycoprotein of approx. 100000 Da in chick embryo fibroblasts that yielded a core polypeptide of 48000 Da, this molecule was located predominantly in lysosomal membranes with smaller fractions present in endosomes, the Golgi apparatus and the plasma membrane.

In this report we describe the further characterization of LAMP-1 and LAMP-2. The molecules were very similar in their biosynthesis and expression *in vivo* and both contained a large number of *N*-linked, highly sialylated oligosaccharide chains. The similarity in oligosaccharide composition of these glycoproteins and those described by Lewis *et al.* (1985) and Lipincott-Schwartz & Fambrough (1985) raise the possibility of common functions that could be related to the biogenesis or function of lysosomes.


**Materials and Methods**

*Antibodies*

Monoclonal antibody anti-LAMP-1 IgG$_{2a}$, was derived from spleen cells of a rat immunized with a membrane fraction of mouse embryo 3T3 cells (Hughes & August, 1981). Monoclonal antibody anti-LAMP-2, IgG$_{2a}$, was derived from spleen cells of a rat immunized with a lentil lectin membrane fraction of Balb/c 3T3 cells (Chen *et al.*, 1985*a*).

## Cells

NIH 3T3 cells (Jainchill *et al.*, 1969) were obtained from Dr. Don Blair of the Frederick Cancer Institute, MD, U.S.A., and were used at passage 4. HaNIH cells (Scolnick & Parks, 1974) and P388D1 (Shevach *et al.*, 1972) were obtained as previously described (Hughes & August, 1981) and maintained as described (Chen *et al.*, 1985*b*).

## Biosynthetic labelling of cells

HaNIH cells grown to 80–90% confluence in 75 cm² or 150 cm² flasks were washed with warm Hank's buffered salt solution (HBSS) (Gibco, Chicago Falls, OH, U.S.A.), and cultured with 6 ml of methionine-free medium (Gibco) for 2 h at 37 °C. The cells were then pulse-labelled with 6 ml of labelling medium {methionine-free medium supplemented with 125 $\mu$Ci of [$^{35}$S]methionine/ml (Amersham, Arlington Heights, IL, U.S.A.)} for 5 min or as indicated in the text. Cells were then washed twice with warm HBSS and collected immediately or incubated with Dulbecco's minimal essential medium containing 10% fetal bovine serum for the times indicated in the text. The inhibitors tunicamycin or monesin, when present as indicated in the text, were added with each of the media. Cells to be harvested were washed three times with phosphate-buffered saline at 4 °C, removed from the plate by scraping, and collected by centrifugation.

## Cell extraction and protein immunoprecipitation

Metabolically labelled cells were extracted with a lysis buffer [10 mM-Tris/HCl (pH 7.5)/0.5% Nonidet-P40 (NP-40; Particle Data Inc., Elmhurst, IL, U.S.A.)/5 mM-EDTA/1 mM phenylmethanesulphonyl fluoride/0.15 M-NaCl (Hughes & August, 1982)]. After 30 min on ice, the lysate was freeze-thawed three times, and the detergent-insoluble material was removed by centrifugation at 100000 $g$ for 60 min at 4 °C. The soluble extract, containing (1–5 × 10⁶ acid-precipitable d.p.m., was incubated with 200 $\mu$l of monoclonal antibody from tissue culture supernatants for 1 h at 4 °C followed by addition of a titred amount of goat anti-rat second antibody and incubation on ice for 6 h. The antibody–antigen complexes were washed twice with 20 mM-Tris/HCl, pH 7.6, containing 2.5 M-KCl, 100 mM-NaCl, 1 mM-EDTA and 0.5% NP-40, and once with 20 mM-Tris/HCl, pH 7.6. The precipitates were analysed by SDS/polyacrylamide-gel electrophoresis under reducing conditions (Laemmli, 1970) followed by fluorography (Bonner & Laskey, 1974), unless otherwise indicated. $M_r$ standards were: myosin ($M_r$ 200000), $\beta$-galactosidase ($M_r$ 116000), phosphorylase $b$ ($M_r$ 97400), bovine serum albumin ($M_r$ 68000), RNA polymerase subunit ($M_r$ 43000) and chymotrypsin ($M_r$ 23000).

## Light microscopic immunohistochemistry

Light microscopic immunohistochemistry was performed as described (McMillan *et al.*, 1981). Tissue obtained from Balb/c (Charles River) or Beige-J

(Jackson Laboratory, Bar Harbor, MA, U.S.A.) mice, were frozen in liquid $N_2$ in OCT mounting compound (Lab-Tek Products Division, Miles Laboratories, Naperville, IL, U.S.A.) on brass chucks. Frozen sections (4 $\mu$m) were applied to room-temperature slides precoated with 0.5% gelatin and 0.05% chromium potassium sulphate (Fischer Scientific, Pittsburgh, PA, U.S.A.). The specimens were immediately fixed in cold acetone for 10 s, air dried and refrigerated until use. Immediately before staining, sections were fixed in cold acetone for 3 min, air dried, rinsed in phosphate-buffered saline for 10 min, and processed as follows: (1) incubated in 0.3% $H_2O_2$ in methanol for 10 min at room temperature and rinsed for 10 min in phosphate-buffered saline; (2) incubated with 100 $\mu$l of a 1:20 dilution of normal human serum (type AB+) in diluent buffer (phosphate-buffered saline) containing 3% normal human serum for 20 min at room temperature in a humid chamber and then rinsed in phosphate-buffered saline for 5 min; (3) incubated overnight at 4 °C with 100 $\mu$l of a 1:10 dilution of either $\alpha$-LAMP-1, $\alpha$-LAMP-2, 5D227 or P3 $\times$ 63Ag8 hybridoma supernatant and washed twice in phosphate-buffered saline for 10 min; (4) treated with 100 $\mu$l of horseradish peroxidase-conjugated goat anti-(rat IgG) IgG (Kirkegaard and Perry Laboratories, Gaithersburg, MD, U.S.A.), 10 $\mu$l/ml in diluent buffer, for 30 min at room temperature and washed twice for 10 min with phosphate-buffered saline; (5) covered with phosphate-buffered saline containing 0.8 $\mu$g of 3,3'-diaminobenzidine tetrahydrochloride/ml and 0.03% $H_2O_2$ for 4–5 min, washed with water, counterstained with Mayer's hematoxylin for 30 s, dehydrated, cleared in xylene, and mounted in Permount (Fischer Scientific).

## Results

### Amphiphilic properties of the glycoproteins

The association of LAMP-1 and LAMP-2 with lysosomal membranes was compared by use of a variety of detergents and chaotropic agents to extract the glycoproteins into a soluble fraction. As described in Fig. 1, aliquots of cells were incubated with the various extraction reagents and centrifuged at 100000 $g$. The supernatants were retained and the pellets extracted by a lysis buffer containing 0.5% Nonidet P-40. Glycoproteins in the supernatant ('supernatant') and extracted pellet ('pellet') fractions were then immunoprecipitated and analysed by SDS/polyacrylamide-gel electrophoresis (Fig. 1).

LAMP-1 and LAMP-2 were each extracted only in the presence of detergent. Triton X-100 or Nonidet P-40. The glycoproteins both remained in the cell pellet when cells were extracted with either 1.0 M-KI, 0.5 M-guanidinium chloride, 1.0 M-urea, or 5 mM-EDTA. These results indicated that both LAMP-1 and LAMP-2 had properties of amphiphilic molecules with strong hydrophobic associations with lysosomal membranes.

### Biosynthesis of the glycoproteins

Precursor forms of LAMP-1 and LAMP-2 and processing of the molecules were examined by pulse–chase labelling and immunoprecipitation. HaNIH cells were pulse-labelled for 5 min with [$^{35}$S]methionine and then incubated with

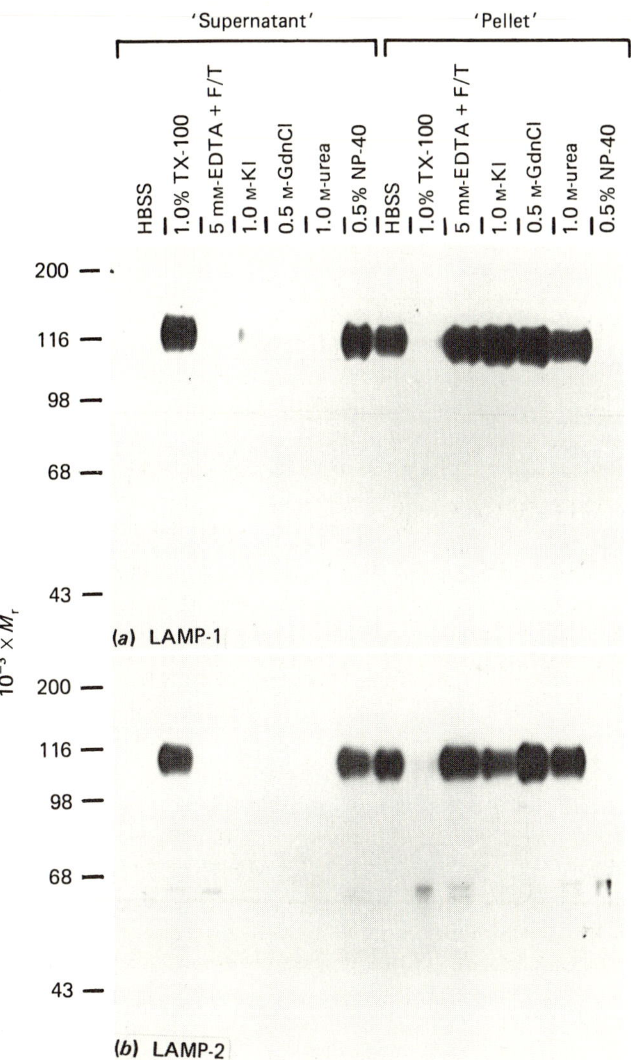


Fig. 1. *LAMP-1 and LAMP-2 are integral membrane proteins*

HaNIH cells were metabolically labelled with [$^{35}$S]methionine as described under 'Materials and Methods'. The washed cells were dispersed into seven pre-weighed conical tubes and collected by centrifugation. Various extraction agents were added to the pellets at a ratio of extraction agent: cell pellet of 10:1 (w/w), as follows: (a), Hank's buffered saline (HBSS); (b), 1.0% Triton X-100 (TX-100); (c), 5 mM-EDTA with three cycles of freeze-thaw (F/T); (d), 1.0 M-KI; (e), 50 mM guanidinium chloride (GdnCl); (f), 1.0 mM-urea; (g) 0.5% Nonidet P-40 (NP-40). To each of these, phenylmethanesulphonyl fluoride was added to a final concentration of 1 mM. After incubation at room temperature for 30 min the cells were centrifuged at 100000 g for 1 h at 4 °C. The supernatants were dialysed exhaustively against lysis buffer (0.5% Nonidet P-40, 5 mM-EDTA, 150 mM-NaCl, 10 mM-Tris/HCl, pH 7.5) and centrifuged at 100000 g for 1 h at 4 °C ('Supernatant'). The pellets from the first 100000 g centrifugation were extracted a second time with lysis buffer and the suspension was centrifuged at 100000 g. The supernatant was retained ('Pellet'). Samples (1 × 10$^6$ acid-precipitable d.p.m.) of the supernatant and the pellet extracts were incubated with the α-LAMP-1 or α-LAMP-2 monoclonal antibodies and the immune complexes were processed and analysed by SDS/polyacrylamide-gel electrophoresis as described under 'Materials and Methods'.

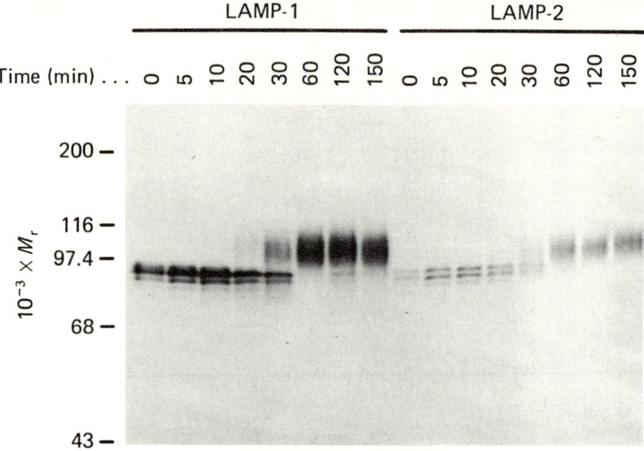

Fig. 2. *Biosynthetic processing of LAMP-1 and LAMP-2 in cells pulse-labelled with*
[$^{35}$S]*methionine*

HaNIH cells were pulse-labelled for 5 min, chased for the indicated time, and extracted and analysed
by immunoprecipitation with anti-LAMP-1 and anti-LAMP-2 monoclonal antibodies followed by
SDS/polyacrylamide-gel electrophoresis, as described under 'Materials and Methods'.

unlabelled methionine for different times. Immunoprecipitates obtained by use
of the anti-LAMP-1 or anti-LAMP-2 antibodies were analysed by SDS/poly-
acrylamide-gel electrophoresis.

LAMP-1 and LAMP-2 present in the 5 min pulse-labelled fraction (0 chase)
and the early chase fractions appeared as multiple forms ranging in apparent
$M_r$ from about 88000 to 92000 (Fig. 2). These precursor forms were processed
to mature molecules of $M_r$ approx. 110000 (LAMP-1) or 105000 (LAMP-2),
beginning at about 30 min after pulse-labelling. There consistently was reduced
labelling of LAMP-2 by [$^{35}$S]methionine.

## Effect of tunicamycin

Asparagine-linked oligosaccharides are formed via a lipid-linked high-manose
precursor (Glc$_3$Man$_9$GlcNAc$_2$) which is transferred co-translationally to nascent
peptide chains during their transport across membranes of the rough endoplasmic
reticulum (reviewed by Kornfeld & Kornfeld, 1985). Tunicamycin inhibits the
formation of this lipid-linked precursor oligosaccharide, thereby preventing
$N$-glycosylation of proteins (Tkacz & Lampen, 1975). Analysis of the effect of
tunicamycin may therefore indicate the presence of $N$-linked oligosaccharides
on the mature molecule and reveal the nature of the core polypeptide.

HaNIH cells incubated in the presence of tunicamycin were pulse-labelled with
[$^{35}$S]methionine and LAMP-1 and LAMP-2 were immunoprecipitated from
detergent extracts of cells. Molecules were present as low-$M_r$ core polypeptides,
$M_r$ 42000 for LAMP-1 (Fig. 3) and $M_r$ 44000 for LAMP-2 (results not shown
due to the difficulty in photographically reproducing the lightly labelled
LAMP-2).

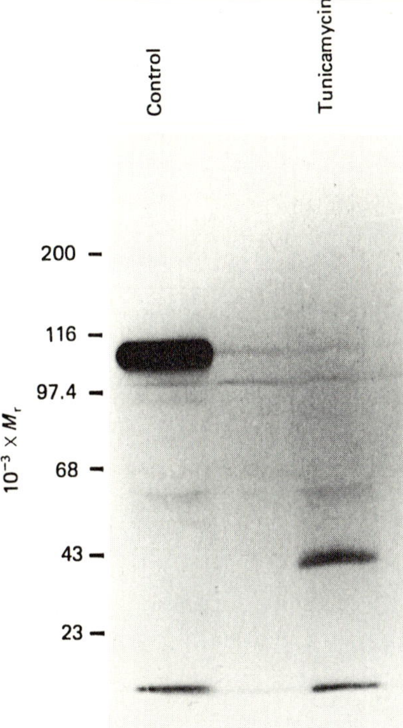


Fig. 3. *Identification of a core precursor polypeptide in cells treated with tunicamycin*

HaNIH cells were preincubated and pulse-labelled for 15 min with [35S]methionine in the presence or absence (control) of 2 $\mu$g of tunicamycin/ml (Calbiochem, La Jolla, CA, U.S.A.). The cells were extracted and $10^6$ d.p.m. of acid-precipitable radioactivity was incubated with anti-LAMP-1 and the immunoprecipitates were analysed by SDS/polyacrylamide-gel electrophoresis as described under 'Materials and Methods'.

## Effect of endoglycosidase H

The $Glc_3Man_9GlcNAc_2$ oligosaccharides transferred to polypeptides are quickly processed by a series of enzymes present in the rough endoplasmic reticulum and Golgi apparatus, resulting in the removal of the glucose and several mannose residues. For those glycoproteins that traverse the Golgi, this is followed by the addition of other sugars characteristic of complex oligosaccharides. Prior to removal of the mannose residues, the oligosaccharide is sensitive to endo-$\beta$-$N$-acetylglucosaminidase H, which acts specifically on oligosaccharides that contain four or more mannose residues, cleaving the carbohydrate chain between the two proximal $N$-acetylglucosamine residues (Tarentino & Maley, 1979). Treatment with endoglycosidase H should therefore confirm the presence of high-mannose oligosaccharides and yield a product that closely resembles the core polypeptide synthesized in the presence of tunicamycin.

HaNIH cells were pulse-labelled for 5 min with [35S]methionine, LAMP-1 and LAMP-2 were immediately immunoprecipitated from cell extracts, and equal aliquots of the immunoprecipitates were treated with endoglycosidase H or

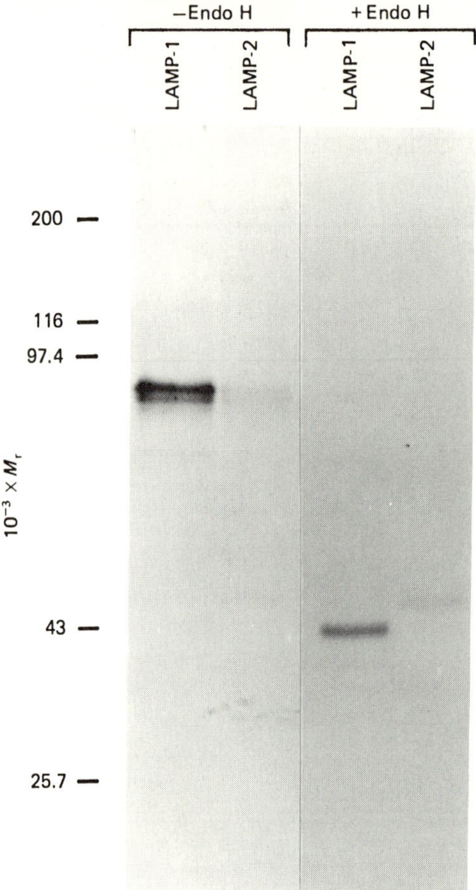



Fig. 4. *Identification of the products of pulse-labelled LAMP-1 and LAMP-2 treated with endoglycosidase H*

HaNIH cells labelled with [$^{35}$S]methionine for 5 min were harvested immediately. The immune complexes obtained with anti-LAMP-1 and anti-LAMP-2 as described under 'Materials and Methods' were solubilized in 40 $\mu$l of Endo H buffer [0.1 M-sodium citrate (pH 5.6)/0.1% SDS/0.2% 2-mercaptoethanol/1 mM-phenylmethanesulphonyl fluoride] by boiling for 2 min. Immunoprecipitates were divided into two equal portions. Endoglycosidase H (2 m-unit; Boeringher Mannheim, Indianapolis, IN, U.S.A.) was added to one sample and both were incubated at 37 °C for 12 h. The proteins were analysed by SDS/polyacrylamide-gel electrophoresis followed by fluorography.

control buffer. The enzyme had a marked effect on both LAMP-1 and LAMP-2. The pulse-labelled, unprocessed glycoproteins appeared as single polypeptide bands of approx. 43000 and 45000 Da for LAMP-1 and LAMP-2, respectively (Fig. 4). The untreated, control aliquots contained the expected precursor forms of about 90000 Da, as seen in Fig. 2.

The effect of endoglycosidase H was also studied with cells pulse-labelled and chased for varying times. Post-translational processing of LAMP-1 and LAMP-2 was accompanied by an increase in the apparent $M_r$ of the glycoproteins at about 30 min after pulse labelling of the nascent polypeptides, as seen in Fig. 2. These changes can be attributed to the removal of mannose residues from the high-mannose core and the addition of $N$-acetylglucosamine, galactose, fucose and sialic acid residues characteristic of complex oligosaccharides, and possibly the addition of $O$-linked oligosaccharides. Support for this model was provided by the effect of endoglycosidase H on samples immunoprecipitated at different times after pulse–chase labelling. The precursor glycopeptides remained sensitive to endoglycosidase H after 5, 10, 15, and 20 min of chase incubation following the 5 min pulse labelling. Both molecules became resistant to the enzyme between 30 and 40 min after labelling, corresponding to the time when the proteins showed the apparent increase in $M_r$ from the 90000-Da precursor glycoproteins to the 105000–110000-Da forms (M. P. D'Souza, unpublished work).

*Effect of monensin*

Further evidence of the role of the Golgi apparatus in the terminal processing of LAMP-1 and LAMP-2 was obtained by use of monensin. Monensin, a monovalent ionophore, markedly effects a number of eukaryotic cell functions, including the post-translational modification of glycoproteins in the Golgi complex (Tartakoff & Vassalli, 1978; Strous & Lodish, 1980; Johnson & Spear, 1983).

In the absence of monensin, LAMP-1 and LAMP-2 demonstrated the expected 90000-$M_r$ form at $t = 0$, and were procesed at $t = 120$ min to the mature forms of 140000 and 125000 Da in P388 and HaNIH cells, respectively (Fig. 5). In the presence of monensin, at concentrations as low as 0.2 $\mu$M, both LAMP-1 and LAMP-2 at $t = 120$ min were both present as diffuse bands of about 85000 Da. The same number of acid-precipitable counts were included in each immunoprecipitation reaction, and there was no apparent decrease in the incorporation of [${}^{35}$S]methionine even at higher doses of monensin, suggesting that there was no significant effect of the drug on the rate of synthesis or metabolic degradation of these antigens. The experiment repeated with NIH cells gave similar results.

*Expression* in vivo *of LAMP-1 and LAMP-2*

The expression *in vivo* of LAMP-1 and LAMP-2 was examined by light microscopic immunohistochemistry of tissues from Balb/c and Beige-J mice (Fig. 6). Beige-J mice have an inherited defect on lysosomal function leading to a Chediak–Higashi-like syndrome (Chi *et al.*, 1978; Frankel *et al.*, 1978). These mice were examined to ascertain whether or not there were any gross alterations in tissue expression of LAMP-1 or LAMP-2 accompanying the Beige phenotype which is characterized by granulation anomalies of leukocytes and gigantism of cytoplasmic organelles (Witkop *et al.*, 1983). LAMP-1 and LAMP-2 displayed essentially identical tissue staining patterns in the two mouse strains. Increased staining of LAMP-1 as compared with LAMP-2 was consistently observed. This was in accord with evidence that LAMP-1 is more abundant than LAMP-2. Both antigens were notably found in macrophages and epithelial cells. Staining was predominantly intracellular. At the light microscopic level we were not able to

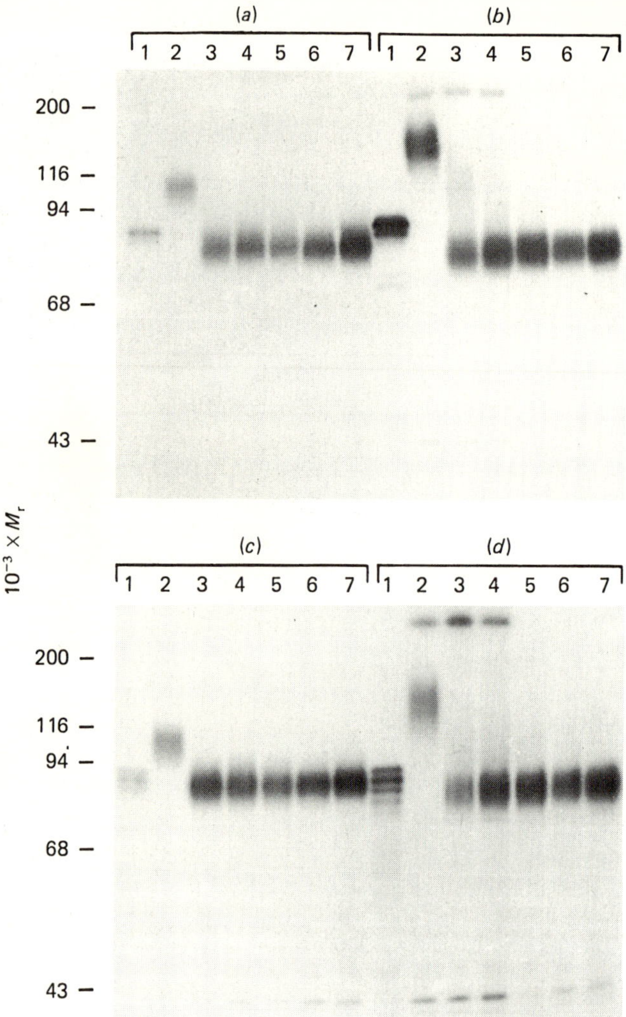

Fig. 5. *Monesin affects the post-translational processing of LAMP-1 and LAMP-2*

HaNIH and P388 cells were pulse-labelled with [$^{35}$S]methionine for 5 min and incubated in the presence of unlabelled methionine as described in the text with the exception that 0.2–10.0 $\mu$M-monensin (Calbiochem) was included in the preincubation, labelling, and chase media. (*a,b*), LAMP-1; (*c,d,*), LAMP-2; (*a,c*), HaNIH cells; (*b,d*), P388 cells. Control cells without monensin were pulse-labelled for 5 min and harvested immediately (lanes 1) or chased for 120 min (lanes 2). Other cells were all preincubated, labelled for 5 min and chased for 120 min in the presence of different concentrations of monensin: 0.2 $\mu$M (lanes 3); 0.5 $\mu$M (lanes 4); 1.0 $\mu$M (lanes 5); 2.0 $\mu$M (lanes 6); 5.0 $\mu$M (lanes 7). The cells were extracted, immunoprecipitated and examined by SDS/polyacrylamide-gel electrophoresis.

resolve any differences in cytoplasmic distribution of LAMP-1 or LAMP-2 in Balb/c as compared with Beige-J cells. The control immunoglobulins in these experiments were the 5D227 monoclonal antibody, IgG2a, which reacts with polymorphic determinant not expressed in Balb/c mice (Hughes & August, 1981) and P3 × 63Ag8 (Kohler *et al.*, 1976). Both gave minimal nonspecific antibody binding.

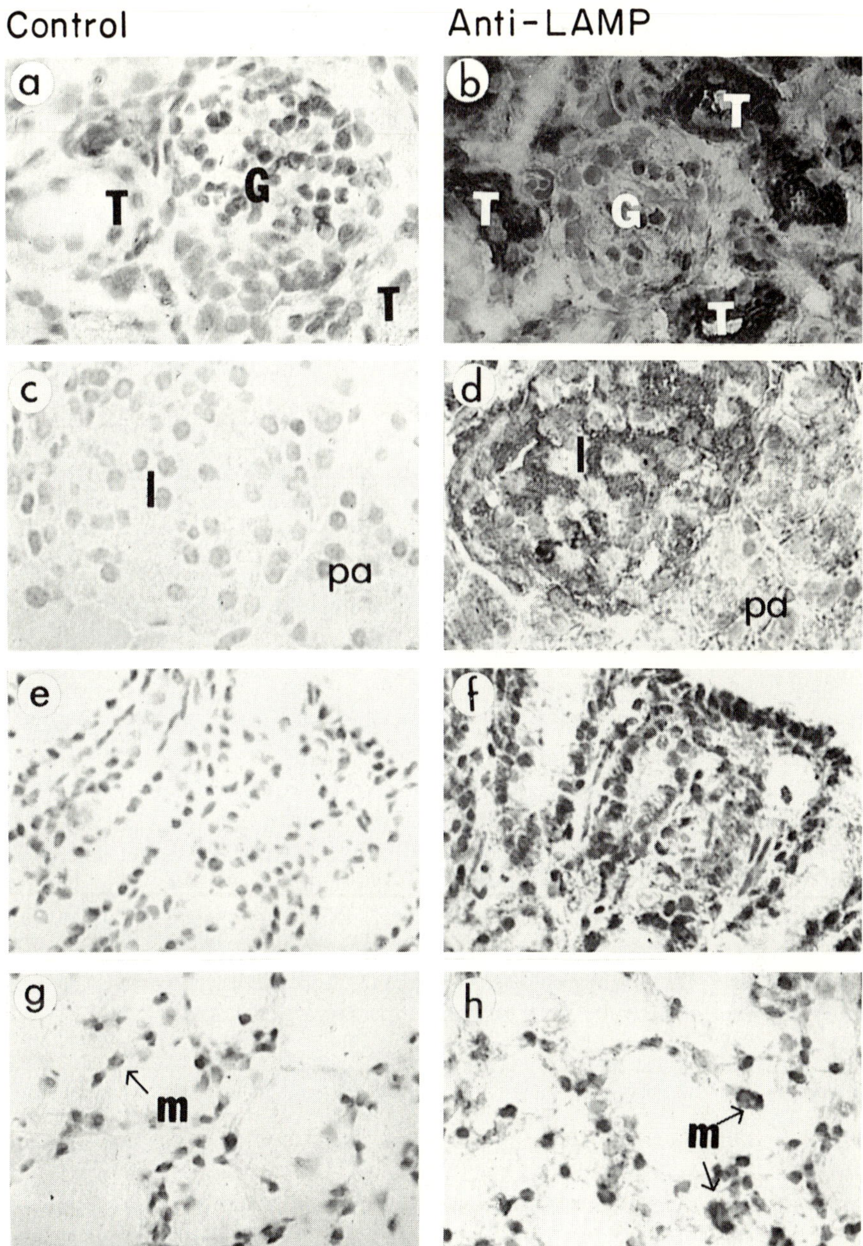


Fig. 6. *Expression in vivo of LAMP-1 and LAMP-2*

Frozen sections of mouse tissue were incubated with anti-LAMP-1 (f,h), anti-LAMP-2 (b,d), 5D227 (e,g) or p3 × 63A8 (a,c), and then processed as described under 'Materials and Methods'. For each tissue examined, the sections treated with control antibodies were devoid of reaction product (a,c,e,g). Kidney of Beige-J mouse (a,b): tubules (T) stained intensely while glomeruli (G) show negligible staining. Pancreas from Beige-J mouse (c,d): the pancreatic acinar (pa) cells show moderate granular cytoplasmic staining, while the islets of Langerhans (I) stained intensely. Large intestine from Balb/c mouse (e,f): staining of epithelial cells of intestinal villi and crypts appeared to be concentrated in the apices of the cells. Lung from Balb/c mouse (g,h): intense intracellular and surface staining of alveolar macrophages (M). Magnification × 1000.

A principal site of expression of LAMP-1 and LAMP-2 was in epithelial cells. Staining was granular and cytoplasmic. In most simple epithelia, staining was polarized with respect to the cell nucleus. Intense intracellular staining was observed in the viable epithelium of the tongue surface, apical cytoplasm of epithelial cells of intestinal villi and crypts (Fig. 6*f*), basal cytoplasm of cells lining bronchi and bronchioles, and the efferent ductuli of the epididymis. Kidney tubules contained a high concentration of these antigens, while glomeruli were negative (Fig. 6*b*). Pancreatic acinar cells were positive in a granular cytoplasmic pattern, while the islets of Langerhans were intensely stained (Fig. 6*d*). Staining in hepatocytes was concentrated in a juxtanuclear pattern. Macrophages in most organs, including the lung (Fig. 6*h*), connective tissue, spleen and lymph nodes, stained intensely. Lymphocytes were not stained by either antibody. The antigens were also diffusely located throughout most of the grey and white matter of the brain, with a slight increase of staining in cerebellar Purkinje cells as compared with the granule cell and molecular layers. Other regions of the nervous system were not stained. Staining was absent in stromal tissues, including smooth, cardiac and skeletal muscle, and collagen although there was minor staining in the blood vessel walls.

*Expression of LAMP-1 and LAMP-2 in transformed cells*

There have been reports that lysosomes and lysosomal integrity may be involved in cell transformation (Zajac-Kaye & Ts'o, 1984). For this reason we have compared the expression of LAMP-1 and LAMP-2 in NIH 3T3 mouse embryo cells and Harvey sarcoma virus transformed NIH 3T3 cells.

The immunofluorescence localization of the antigen was the same in both types of cell, with a vesicular perinuclear staining pattern consistent with that of the lysosomal localization of these antigens (Chen *et al.*, 1985*a,b*) (Fig. 7). However, while the pattern of distribution of LAMP-1 and LAMP-2 was indistinguishable between the two cell types, the intensity of staining of LAMP-2 was greater in NIH as compared with HaNIH. This difference between the two cell types was not detected in the intensity of staining of LAMP-1.

**Discussion**

In this study we have compared some of the properties of LAMP-1 and LAMP-2, two glycoproteins specifically localized in lysosomal membranes and absent from the plasma membrane or from endocytotic vacuoles. The molecules were markedly similar in many of their properties, particularly in oligosaccharide structure, and in cell and tissue expression. As previously described, the membrane distributions of the molecules analysed by electron microscopy were identical (Chen *et al.*, 1985*a*). LAMP-1 was a major component of the membrane; it constituted 0.1% of total detergent-extracted cell protein and, when labelled with [$^3$H]glucosamine, the immunoprecipitated glycoprotein accounted for about 15% of the total acid-insoluble radioactivity fractionated by two-dimensional gel electrophoresis (Hughes & August, 1982). The concentration of LAMP-2 has not been determined, but the antigenic reactivity of the

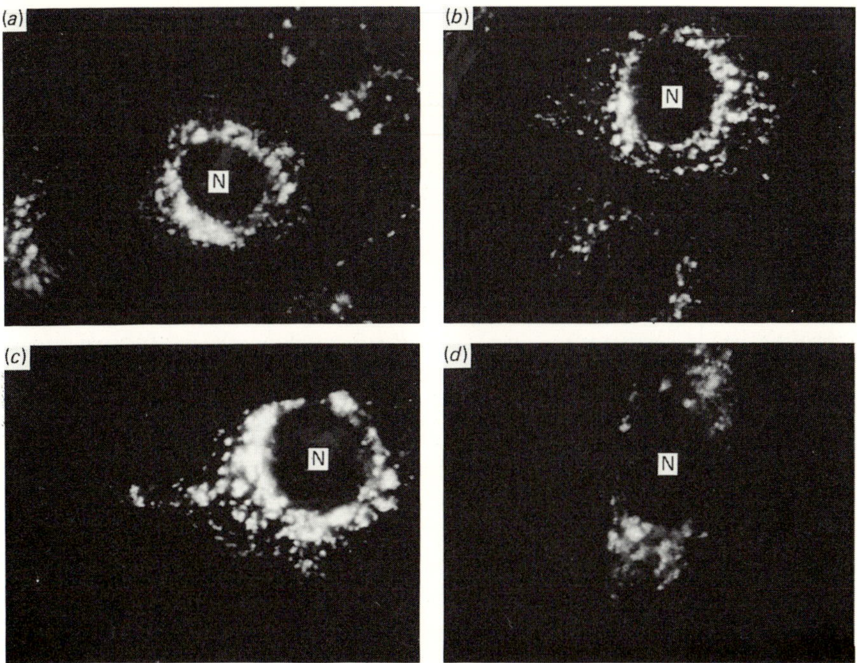

Fig. 7. *Immunocytochemical localization of LAMP-1 and LAMP-2 in NIH and HaNIH cells*

NIH (*a,b*) and HaNIH (*c,d*) cells were grown to 50% confluency in 35-mm plastic dishes and fixed *in situ* with 3.0% paraformaldehyde in phosphate-buffered saline for 10 min. The cells were washed several times with phosphate-buffered saline followed by Dulbecco's modified Eagle's medium. They were then incubated for 30 min at room temperature with anti-LAMP-1 (*a,c*) or anti-LAMP-2 (*b,d*) culture supernatant diluted 1:10 with phosphate-buffered saline containing 0.1% bovine serum albumin and 0.1% saponin (Sigma, St. Louis, MO, U.S.A.). The cells were then washed with phosphate-buffered saline and incubated for an additional 30 min in phosphate-buffered saline containing 0.1% bovine serum albumin and with affinity-purified goat anti-(rat IgG) conjugated to rhodamine (Kirkegard & Perry Laboratories) in the same solution containing 0.1% saponin. Cells were photographed using a Zeiss microscope equipped with epi-illumination and a 100×, N.A. 1.4 oil objective. Photographs were made with Kodak Ektachrome P800 film and processed by the E-6P procedure. All photographs were made at the same exposure. N, nucleus. Magnification ×1200.

molecule and the incorporation of sugar substrates were comparable with that found with LAMP-1. Biosynthetic labelling of LAMP-2 with [$^{35}$S]methionine was markedly less than that of LAMP-1, but this could be attributed either to the concentration of the protein, methionine content, or rate of synthesis. Glyco-proteins with similar properties were described by Burnside & Schneider (1982) as major components of lysosomal membranes and by Kato *et al.* (1984) as constituting 5% of the total protein and 10% of the membrane protein of tritosomes. The tissue expression of the molecules in the mouse, including the Beige-J mouse genetically deficient in lysosomal function, also appeared to be identical. Both proteins were present in cells known to contain high concentrations of lysosomes. In addition, intense staining of pancreatic islet cells and a granular pattern of staining in acinar cells suggested that both glycoproteins were also components of secretory and storage granules. Possible evidence for independent expression of LAMP-1 and LAMP-2 was obtained by comparing the immunohistochemical staining of the protein in different cell lines. In repeated experiments the intensity of immunofluorescent labelling of LAMP-2 in Harvey sarcoma virus transformed NIH 3T3 cells was reduced as compared with the untransformed cells, whereas there was no difference in LAMP-1 expression between the two cells.

Distinctive features of LAMP-1 and LAMP-2 were the high concentrations of oligosaccharides. A model suggested by the data is that the glycoproteins contain a large number of asparagine-linked high-mannose oligosaccharides added cotranslationally and then processed to chain sequences found in complex-type oligosaccharides. Studies with tunicamycin, which blocks assembly of the lipid-linked oligosaccharide precursor, indicated that the core polypeptides of LAMP-1 and LAMP-2 were synthesized as molecules of 42000 and 44000 Da, respectively. These results were substantiated by the effect of endoglycosidase H, which acts specifically on high-mannose oligosaccharides. The approx. 92000 Da precursor glycoproteins pulse-labelled with [$^{35}$S]methionine for 5 min and isolated by immunoprecipitation were converted by the enzyme to poly-peptide bands of 43000 and 45000 Da for LAMP-1 and LAMP-2, respectively. The results also indicate that the multiple bands of the precursor glycoproteins found by SDS/polyacrylamide-gel electrophoretic analysis of the unprocessed, pulse-labelled samples can be attributed to heterogeneity in the asparagine-linked high-mannose core oligosaccharides of these glycoproteins. The core polypeptides obtained after treatment with tunicamycin or endoglycosidase H were present as single bands. The difference in apparent $M_r$ between the approx. 43000-Da core and 92000-Da precursor molecules would be sufficient for as many as 20–25 high-mannose chains/polypeptide. Lewis *et al.* (1985) directly demonstrated at least 18 asparagine-linked oligosaccharides on the lpg120. Subsequent processing of the mannose-rich precursor to mature molecules of 110000 Da and 105000 Da for LAMP-1 and LAMP-2, respectively, can be attributed to the formation of the complex-type, highly sialylated oligosaccharides. The 5 min [$^{35}$S]methionine pulse-labelled molecules become resistant to endoglycosidase H after about 30 min of chase incubation, corresponding to the appearance of the mature glycoproteins. Simultaneously, the molecules became highly sialylated, with a large number of acidic isoelectric variants revealed by two-dimensional gel electrophoresis, particularly with LAMP-1 which contained more than 16 distinct fractions between pH 4.1 and 7.0 (Chen *et al.*, 1985*a*). Neuraminidase eliminated these acidic forms, yielding molecules with an alkaline pH (Chen *et al.*, 1985*b*). The basis for the apparent increase in $M_r$ during terminal process-ing is unknown. The molecules were markedly heterogeneous in that the apparent $M_r$ of [$^3$H]glucosamine-labelled glycoprotein appeared greater than that of [$^{35}$S]methionine-labelled molecules or protein stained with silver or Coomassie Blue (Chen *et al.*, 1985*a*). It thus appeared that a small fraction of the protein contained a majority of the sugar residues and that this heavily glycosylated fraction correspondingly migrated at higher apparent $M_r$ value. The effect was not attributed to sialic acid, as there was no comparable effect of neuroaminidase on $M_r$. Treatment with monensin blocked the increase in apparent $M_r$ and resulted in molecules of about 85000 Da, slightly smaller than the normal pulse-labelled precursor of about 92000 Da. Ongoing studies (S.

Mane, unpublished work) show that monensin markedly reduces the rate of processing of the high-mannose oligosaccharides and it is speculated that some process occurring during terminal processing in the Golgi and related to the apparent change in $M_r$ was blocked by the drug. For example, the reduced rate of migration in SDS/polyacrylamide-gel electrophresis of the low-density lipoprotein receptor has been related to $O$-linked residues (Cummings *et al.*, 1983) and it is possible that there is $O$-linked glycosylation of the LAMP proteins in the Golgi.

These results provide evidence that the biosynthesis of both LAMP-1 and LAMP-2 involves passage through the Golgi apparatus and processing of the $N$-linked high-mannose chains to complex-type oligosaccharide. This likely differs from the processing of lysosomal enzymes which do not appear to traverse the Golgi system and are modified by a highly specific mannose 6-phosphate recognition system catalysed by enzymes believed to reside in the *cis*-Golgi cisternae (reviewed by Kornfeld & Kornfeld, 1985). In repeated attempts, we have not detected phosphorylation of the LAMP glycoproteins. Moreover, we recently have identified a molecule in human cells that is homologous to LAMP-2 and find this glycoprotein markedly enriched in cells derived from patients with I-cell disease (MLII) or pseudo-Hurler polydystrophy (MLIII) (J. W. Chen, unpublished work). These diseases are characterized by the absence of the mannose phosphate recognition marker on lysosomal enzymes and diminished cellular concentrations of the enzymes.

The recently described lysosomal membrane glycoproteins are similar in several properties. One remarkable correlation is the high content of $N$-linked oligosaccharide. The rat protein lpg120 of Lewis *et al.* (1985), the chicken CV24 antigen of Lippencott-Schwartz & Fambrough, and LAMP-1 and LAMP-2, all contain polypeptide cores of 40000–50000 Da that are modified by high-mannose oligosaccharides to precursor glycoproteins of about 90000 Da and processed to mature glycoproteins of about 110000–120000 Da. It can be speculated that this high carbohydrate composition is important to the function of the glycoproteins. Another relationship is the presence of CV24 protein and 100000 Da glycoprotein identified by Reggio *et al.* (1984) on endosome and plasma membranes in addition to the lysosomal localization, whereas the lpg120 (Lewis *et al.*, 1985) and LAMP-1 and LAMP-2 are restricted to lysosomes. Moreover, the Reggio antigen and CV24 were found on the ruffled border plasmalemma of osteoclasts, whereas the lpg120 was absent (Baron *et al.*, 1985). It is likely that some of these glycoproteins are homologous. Unfortunately, direct immunological comparison is difficult because of the different species of origin of the antigens. Definitive comparison of the proteins awaits peptide sequencing.

This research was supported in part by the National Institutes of Health (RO1 GM31168) and the Office of Naval Research (N00014-82-K-0221). J. W. C. is the recipient of Medical Scientist Training Program grant 5T 32-TM-GM07309-11 and T. L. M. of a postdoctoral fellowship award by the National Cancer Institute, grant CA-09243. The authors thank Dr. Mette Strand for helpful comments and critical reading of this manuscript. We gratefully acknowledge Joni Cambias for technical assistance and Linda Poole and Dana Lawrence for assistance in preparing the manuscript.

## References

Baron, R., Neff, J., Lippincott-Schwartz, J., Louvard, D., Mellman, I., Helenius, A. & Marsh, M. (1985) *J. Cell Biol.* **101**, 53a

Bonner, W. M. & Laskey, R. A. (1974) *Eur. J. Biochem.* **46**, 83–88

Burnside, J. & Schneider, D. L. (1982) *Biochem. J.* **204**, 525–534

Chen, J. W., Murphy, T. L., Willingham, M. C., Pastan, I. & August, J. T. (1985a) *J. Cell Biol.* **101**, 85–95

Chen, J. W., Pan, W., D'Souza, M. P. & August, J. T. (1985b) *Arch. Biochem. Biophys.* **239**, 574–586

Chi, E. Y., Ignacio, E. & Lagunoff, D. (1978) *J. Histochem. Cytochem.* **26**, 131–137

Cummings, R. D., Kornfeld, S., Schneider, W. J., Hobgood, K. K., Tolleshaug, H., Brown, M. S. & Goldstein, J. L. (1983) *J. Biol. Chem.* **258**, 15261–15273

de Duve, C. (1983) *Eur. J. Biochem.* **137**, 391–397

Frankel, F. R., Tucker, R. W., Bruce, J. & Stenberg, R. (1978) *J. Cell Biol.* **79**, 401–408

Hughes, E. N. & August, J. T. (1981) *J. Biol. Chem.* **256**, 664–671

Hughes, E. N. & August, J. T. (1982) *J. Biol. Chem.* **257**, 3970–3977

Jainchill, J. L., Aaronson, S. A. & Todaro, G. J. (1969) *J. Virol.* **4**, 549–553

Johnson, D. C. & Spear, P. G. (1983) *Cell* **32**, 987–997

Kato, K., Osumi, Y., Akazaki, K. & Okada, K. (1984) *Abstr. 3rd Int. Congr. Cell Biol.* 100

Köhler, G., Howe, S. C. & Milstein, C. (1976) *Eur. J. Immunol.* **6**, 292–295

Kornfeld, R. & Kornfeld, S. (1985) *Annu. Rev. Biochem.* **54**, 631–664

Laemmli, U. K. (1970) *Nature (London)* **227**, 680–685

Lewis, V., Green, S. A., Marsh, M., Vihko, P., Helenius, A. & Mellman, I. (1985) *J. Cell Biol.* **100**, 1839–1847

Lippincott-Schwartz, J. & Fambrough, D. M. (1986) *J. Cell Biol.*, in the press

McMillan, E. M., Wasik, R. & Everett, M. A. (1981) *Am. J. Clin. Pathol.* **76**, 737–744

Reggio, H., Bainton, D., Harms, E., Coudrier, E. & Louvard, D. (1984) *J. Cell Biol.* **99**, 1511–1526

Scolnick, E. M. & Parks, W. P. (1974) *J. Virol.* **13**, 1211–1219

Shevach, E. M., Stobo, J. D. & Green, I. (1972) *J. Immunol.* **108**, 1146–1151

Strous, G. J. A. & Lodish, H. F. (1980) *Cell* **22**, 709–717

Tarentino, A. L. & Maley, F. (1979) *J. Biol. Chem.* **249**, 811–817

Tartakoff, A. & Vassalli, P. (1978) *J. Cell Biol.* **79**, 694–707

Tkacz, J. S. & Lampen, J. O. (1975) *Biochem. Biophys. Res. Commun.* **65**, 248–257

Tougard, C., Louvard, D., Picart, R. & Tixier-Vidal, A. (1985) *J. Cell Biol.* **100**, 786–793

Witkop, C. J., Quevedo, W. C. & Fitzpatrick, T. B. (1983) In *The Metabolic Basis of Inherited Disease* (Stanbury, J. B., Wyngaarden, J. B., Fredrickson, D. S., Goldstein, J. L. & Brown, M. S., eds.), pp. 301–346, McGraw-Hill Book Co., New York

Zajac-Kaye, M. & Ts'o, P. O. P. (1984) *Cell* **39**, 427–437

*Biochem. Soc. Symp.* **51**, 113–115
*Printed in Great Britain*

# Structure and Function of the Receptor for Polymeric Immunoglobulins

KEITH E. MOSTOV,* MARTIN FRIEDLANDER and GÜNTER BLOBEL

*Rockefeller University, 1230 York Avenue, New York, NY 10021, U.S.A.*

Polymeric IgA and IgM (polymeric immunoglobulins, pIg) are specifically taken up by a variety of glandular epithelial cells and transported across the cell into external secretions, such as milk, bile, and intestinal fluid, where they form the first immunological defense against infection. During this transport, the epithelial cell adds an extra polypeptide called secretory component (SC) to the pIg. SC has been proposed to be the receptor that mediates the transcellular transport of pIg (Brandtzaeg, 1981). On the basis of cytochemical, cell fraction-ation, and pIg uptake studies, the pathways of SC and pIg in the cell have been shown to be unusually complex. SC is synthesized in the rough endoplasmic reticulum (RER), and transported through the Golgi apparatus, to the baso-lateral cell surface. Here it binds pIg, and SC–pIg complex is endocytosed. However, unlike many instances of receptor-mediated endocytosis, the ligand is not degraded in lysosomes. Rather the SC–pIg complex is rapidly carried in vesicles across the cell, exocytosed at the apical surface, and discharged into the secretion.

A second remarkable aspect of SC is its apparent dual character. SC functions first as a plasma membrane receptor for pIg. Normally such receptors are integral membrane proteins which are anchored in the lipid bilayer by a hydrophobic portion(s). Yet SC isolated from secretions is hydrophilic and water soluble.

To explain these observations, we have proposed (Mostov *et al.*, 1980) that SC is made as a transmembrane precursor. This precursor is the pIg receptor, and a portion of it is cleaved off and becomes SC which is released into secretions bound to pIg. In the accompanying model of an epithelial cell (Fig. 1), the apical surface is at the top and the basolateral surface at the bottom. In step 1, the pIg receptor is synthesized at the RER with a portion in the RER lumen (open circle), a membrane spanning segment (bar) and a cytoplasmic tail (filled circle). The pIg receptor is transported through the Golgi (step 2) to the basolateral surface (step 3) where its extracellular protein binds pIg (step 4). The receptor–ligand complex is endocytosed via coated pits and transported across the cell in uncoated vesicles (step 5). These vesicles fuse with the apical surface (step 6) and the complex is released into the secretion. The site of cleavage of SC from the pIg receptor is not known, but may be either at the apical surface, as depicted in step 7 of Fig. 1, or during transcytosis.

* Present address: Whitehead Institute, Nine Cambridge Center, Cambridge, MA 02142, U.S.A.

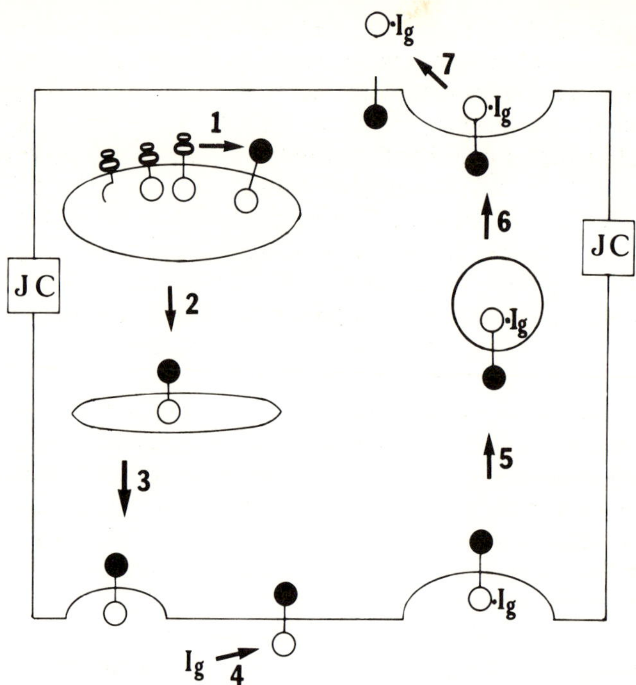

Fig. 1. *Model of transport of pIg by its receptor*

The apical surface is at the top and the basolateral surface is at the bottom. The open circle is the ligand-binding portion of the pIg receptor and the filled circle is the cytoplasmic domain. [Reprinted, with permission, from Mostov *et al.* (1980).]

When we began our investigations only the mature form of SC found in secretions had been identified. The precursor of SC, i.e. the pIg receptor, had not been detected. We began by investigating the earliest steps in the biosynthesis of rabbit pIg receptor (Mostov *et al.*, 1980). Mature rabbit SC had been well characterized and was known to be heterogenous. Four immunologically related glycoproteins were known: two of 50–55 kDa and two of 70–75 kDa. We first translated mRNA from the rabbit liver and lactating mammary gland in a cell-free protein synthesizing system and immunoprecipitated the products with antibody to rabbit SC. Four primary translation products were found: two of 70–75 kDa and two of 90–95 kDa. All four of these proteins are rabbit pIg receptors and are the precursors of the 50–55 kDa and 70–75 kDa forms of rabbit SC, respectively. We showed by affinity chromatography that all four forms bind to pIg. Next we used a cell-free protein synthesis and membrane integration system to demonstrate directly that all four of these pIg receptor forms are transmembrane proteins and have large (10–15 kDa) cytoplasmic tails.

We then studied the complete pathway of pIg receptor, including its cleavage to SC, in a human colon adenocarcinoma cell line, HT29 (Mostov & Blobel, 1982). When the cells are pulse-labelled with [$^{35}$S]cysteine, a single polypeptide of 95 kDa can be immunoprecipitated with antibody to human SC. This 95 kDa protein, the human pIg receptor, is a transmembrane protein and has a cytoplasmic tail of about 10–15 kDa. This form is then converted in the Golgi to a 100 kDa form which has complex oligosaccharides. This 100 kDa form is then cleaved to the 80 kDa mature human SC, which is then released from the cells. By protein sequencing, we found that the SC is cleaved from the *N*-terminal portion of pIg receptor and that the cytoplasmic tail is at its *C*-terminus.

The next step was to determine the complete structure of the pIg receptor (Mostov *et al.*, 1984). We cloned cDNA from rabbit liver, selected clones coding for pIg receptor, and sequenced a clone which contained the complete coding sequence. The pIg receptor consists of 773 amino acids, including an 18-residue-long signal sequence at the *N*-terminus. Residues 630–652 comprise the predominantly hydrophobic membrane-spanning segment and are followed by the cytoplasmic tail, consisting of 103 amino acids. The extracellular pIg-binding portion of the molecule has five homologous repeating units of 100–115 amino acids. Each unit contains a conserved pair of cysteines separated by 60–70 amino acids. These five repeating units have striking homology to the variable region of immunoglobulin *κ* chains. A computer-assisted alignment also suggested that the pIg receptor contained a sixth region of homology, immediately following the first five homology units. Considering these sequence homologies and the conserved pairs of cysteines, it is very likely that the pIg receptor domains are probably folded in a similar conformation to immunoglobulin domains.

The similarity of the pIg receptor domains to immunoglobulin variable regions suggests that the pIg receptor may have evolved from an immunoglobulin which had a variable region with specificity for a site on pIg. Alternatively, the interaction of the pIg receptor domains with the pIg domains may resemble the interaction of the pIg domains with each other.


# References

Brandtzaeg, P. (1981) *Clin. Exp. Immunol.* **44**, 221–232
Mostov, K. E., Kraehenbuhl, J.-P. & Blobel, G. (1980) *Proc. Natl. Acad. Sci. U.S.A.* **77**, 7257–7261
Mostov, K. E. & Blobel, G. (1982) *J. Biol. Chem.* **257**, 11816–11821
Mostov, K. E., Friedlander, M. & Blobel, G. (1984) *Nature (London)* **308**, 37–43

*Biochem. Soc. Symp.* **51**, 117–129
*Printed in Great Britain*

# Structure and Function of Transferrin Receptors and their Relationship to Cell Growth

IAN S. TROWBRIDGE and DEBORAH A. SHACKELFORD

*The Salk Institute for Biological Studies, Post Office Box 85800, San Diego, CA 92138, U.S.A.*

## Synopsis

The transferrin receptor binds the major serum iron-transport protein, transferrin, and mediates cellular iron uptake. The receptor is a major immuno-dominant cell surface glycoprotein of cultured cells and its expression on the cell surface is co-ordinately regulated with cell growth. Recent structural and functional studies of the transferrin receptor are reviewed. The properties of monoclonal antibodies against the transferrin receptor that inhibit transferrin-mediated iron uptake are described. Studies with these antibodies establish that the transferrin receptor plays an important role in cell growth and suggest that monoclonal antibodies that interfere with the function of growth-related receptors may be useful in regulating tumour cell growth.

## Introduction

Transferrin receptors bind the major iron transport protein, transferrin, and mediate cellular iron uptake. A large fraction of serum iron is utilized in the bone marrow for haem synthesis and specific receptors for transferrin were first identified on immature erythroid cells by Jandl & Katz in 1963 [1]. Subsequently, as the growth requirements of mammalian cells in tissue culture were defined, it was found that transferrin was an essential growth factor for virtually all cells [2]. This observation stimulated interest in the role of transferrin in cell growth, and transferrin-binding studies carried out in the late 1970s showed that transferrin receptors could be detected on a wide variety of cultured human tumour cell lines but were only found on normal lymphocytes after they had been stimulated to proliferate by a mitogenic plant lectin [3–6].

At about this time, in studies aimed at defining changes in the cell surface of human haematopoietic cells during differentiation and malignant transformation, an immunodominant cell surface antigen was identified in our laboratory that was found on many cultured haematopoietic cell lines, on a variable fraction of leukaemic cells, but not on normal peripheral blood lymphocytes [7]. The expression of this cell surface glycoprotein, defined by the monoclonal antibody B3/25, was strongly correlated with cell growth. Immunoprecipitation studies showed that the glycoprotein was induced on peripheral blood lymphocytes after mitogenic stimulation. Further, the antigen was lost from the surface of the promyelocytic cell line, HL-60, during terminal differentiation along the myeloid pathway induced by chemical inducers such as dimethyl sulphoxide. This occurred prior to morphological differentiation of the cells and their arrest in G1 of the cell cycle and was the result of a large decrease in the rate of biosynthesis of the glycoprotein that occurred within 24–48 h of exposure of the cells to inducer.

We were initially led to the idea that the cell surface glycoprotein may be the transferrin receptor because of evidence that it was associated with a serum-derived protein that was tightly bound to the surface of cultured cells. This was confirmed by showing that monoclonal antibodies against the glycoprotein could indirectly coprecipitate radiolabelled transferrin as a specific receptor–ligand complex [8]. Independently, two other groups showed that another monoclonal antibody, OKT9, originally thought to define an early T cell differentiation antigen, was also directed against the human transferrin receptor [9,10]. More recently, monoclonal antibodies against the murine and rat transferrin receptors have also been identified on the basis of their capacity to coprecipitate specifically transferrin–transferrin-receptor complexes [11–14].

The isolation of the human transferrin receptor as a result of the immunological studies described provided striking confirmation that the expression of the receptor is co-ordinately regulated with cell growth. Further, monoclonal antibodies against the transferrin receptor have proved to be useful not only for structural studies of the receptor, but also to investigate its role in cell growth.


## Structural Characterization of the Human Transferrin Receptor

The general structural features of the human transferrin receptor inferred from biochemical studies of the receptor of cultured human leukaemic cell lines are shown in Fig. 1. The receptor is a transmembrane glycoprotein consisting of two disulphide-bonded subunits of 95000 $M_r$ [15–18]. The bulk of the receptor is exposed on the cell surface, as treatment of human cells with trypsin quantitatively releases into the medium a large soluble fragment ($M_r$ 70000) lacking inter-molecular disulphide bonds [16,17]. On the basis of proteinase K treatment of newly synthesized receptor in microsomes, it has been estimated that the cyto-plasmic domain of the receptor is of $M_r$ 5000 [17]. Chemical crosslinking experiments are consistent with each subunit of the receptor having a single transferrin-binding site [17].

The transferrin receptor undergoes three kinds of post-translational modifi-cation. Biosynthetic labelling studies and treatment with endoglycosidase H suggest that the receptor has three asparagine-linked carbohydrate chains including at least one high-mannose type and one complex oligosaccharide. The receptor is also modified by the covalent attachment of fatty acid, as shown by metabolic labelling experiments with [³H]palmitate [16,18]. Tryptic cleavage of the palmitate-labelled receptor localizes the acylation site(s) to the region of the receptor proximal to the cell membrane [18]. Only the mature form of the receptor is acylated and there is evidence that the fatty acid moiety of the receptor may turn over more rapidly than the receptor itself [18]. It is not known which amino acid residue is modified by the addition of fatty acid, but by

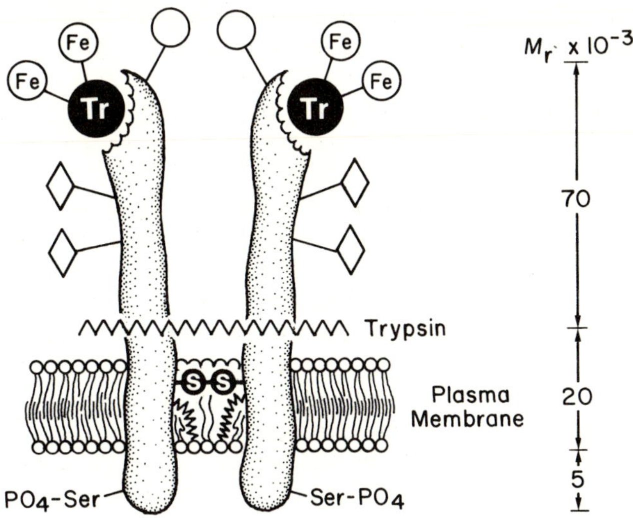



Fig. 1. *Schematic representation of the human transferrin receptor*

The model is based on the biochemical studies in [16–18]. ◇, high-mannose oligosaccharide; ○, complex-type oligosaccharide; ∿∿, covalently bound fatty acid; Ⓣ, transferrin.

analogy with the vesicular stomatitis virus G protein [19] and an HLA-B heavy chain [20], the most likely candidate is a cysteine residue located either in, or proximal to, the lipid bilayer. Finally, the transferrin receptor of human cells is also phosphorylated, principally upon serine residues [17].

## Mechanism of Iron Uptake

Recent morphological and biochemical studies have established that the transferrin receptor is a member of the class of receptors that carry macro-molecular ligands into the cell by the general mechanism of receptor-mediated endocytosis [21–28]. After transferrin binds to its receptor, the receptor–ligand complex is rapidly taken up into the cell via clathrin-coated pits [21–23]. The receptor–ligand complex becomes localized predominantly in endocytic vesicles, which become acidified as a result of the action of a proton pump [26,27]. Under these acidic conditions, iron dissociates from transferrin, but apotransferrin remains bound to the receptor and is recycled back to the cell surface. At pH 7, apotransferrin rapidly dissociates from the receptor [24,25] leaving the receptor to undergo another round of endocytosis. From studies of several different cell types, it has been estimated that 50% of the transferrin receptors on the cell surface are internalized within 2–5 min and that the transit time of receptors inside the cell is 5–15 min [21,23,29]. As the half-life of transferrin receptors has been shown to be 2–3 days in a human T leukaemic cell line [16] and 14 h in HeLa cells [30], it is clear that each receptor molecule can undergo many rounds of recycling in its lifetime.

Several important problems remain to be resolved, including the nature of the molecular signals that direct the internalization and intracellular routing of

Table 1. *Monoclonal antibodies against the transferrin receptor*

| Monoclonal antibody | Species specificity | Class | Growth inhibition | Reference |
|---|---|---|---|---|
| 42/6 | Human | IgA | + | [33] |
| RI7 208 | Mouse | IgM | + | [11,12] |
| REM 17 | Mouse | IgM | + | [12] |
| B3/25 | Human | $IgG_1$ | − | [7,8] |
| T56/14 | Human | $IgG_1$ | − | [8] |
| OKT9 | Human | $IgG_1$ | − | [9] |
| 5E9 | Human | $IgG_1$ | − | [64] |
| RI7 217 | Mouse | $IgG_{2a}$ | − | [12,40] |
| RR24 | Mouse | $IgG_{2b}$ | − | [12] |
| R234-14 | Mouse | $IgG_{2a}$ | − | [12] |

receptors during recycling. Recently, convincing evidence has been reported supporting the widely-held view that transferrin receptors recycle spontaneously in the absence of bound ligand [31]. However there is also conflicting data supporting the notion that transferrin binding is required to trigger the internalization of receptors [32]. Both sets of studies were carried out using the same human cell line, K562, and the reason for the discrepancy is unclear. Finally, after its release from transferrin, iron has still to be translocated across an intracellular membrane in order to reach the cytoplasm, and nothing is known about how this is achieved.

## Monoclonal Antibodies that Block Transferrin Receptor Function

The co-ordinate expression of transferrin receptors with cell growth and the fact that transferrin is an essential growth factor for most cultured cells implies that transferrin receptors play an important role in cell proliferation. Direct evidence for this idea has been obtained from the use of monoclonal antibodies against the transferrin receptor that interfere with its function. As shown in Table 1, antibodies against both the human and murine transferrin receptors have been derived that inhibit transferrin-mediated iron uptake and block the growth of cells in tissue culture.

The first such antibody to be obtained was a murine monoclonal antibody against the transferrin receptor of human cells, designated 42/6 [33]. This antibody was identified on the basis of its capacity to inhibit the binding of [125]I-labelled human transferrin to the human leukaemic cell line CCRF-CEM. It was shown that monoclonal antibody 42/6 inhibited the growth of CCRF-CEM cells in a dose-dependent manner, a concentration of 5 $\mu$g of purified antibody/ml being sufficient to inhibit growth completely. Subsequently, two rat antibodies, RI7 208 and REM 17, against the murine transferrin receptor were derived that similarly blocked the growth of cultured murine cell lines [11,12]. These latter antibodies were selected by directly testing their effects on growth of a panel of mouse lymphoma cell lines.

Studies with these three antibodies have established several points. First, the antibodies do not discriminate between normal and malignant cells. Monoclonal antibody 42/6 inhibits the proliferation of normal human lymphocytes stimulated with mitogens [34] and bone marrow myeloid progenitor cells [35] as effectively as it inhibits the growth of CCRF-CEM cells. However, there is a wide variation in the sensitivity of different cell types to growth inhibition by the antibodies. For example, even after exposure to saturating amounts of monoclonal antibody RI7 208 for several weeks, mouse L cells continue to grow at the same rate as in untreated control cultures [11]. Similarly, monoclonal antibody 42/6 has little or no effect on the growth of several human carcinoma, melanoma and fibroblast cell lines [36]. The explanation for these differences in sensitivity to the inhibitory effects of the antibodies remains obscure, as there are no clear differences in the number of receptors that resistant and sensitive cells display nor in the affinity of antibody binding. However, it is known from the characterization of resistant mutant cells derived from murine lymphoma cell lines sensitive to growth inhibition by anti-(transferrin receptor) antibodies [12] that a 50% reduction in antibody binding to receptors is sufficient to confer resistance. Thus, subtle differences in the density of transferrin receptors displayed on the cell surface or their arrangement in the membrane may account, in part, for the differences in the sensitivity of various cell types to growth inhibition by anti-(transferrin receptor) antibodies. It is also possible that cells differ in their ability to adapt to iron deprivation.

Second, it was found that the ability of ferric complexes to overcome the growth inhibitory effects of anti-(transferrin receptor) antibodies varies widely depending upon the cells studied. Earlier studies showed that Chinese hamster V79 lung fibroblasts [37] and human embryonic fibroblasts [38] could be grown in iron-supplemented media in the absence of transferrin. Consequently, if the effects of the anti-(transferrin receptor) antibodies on cell growth were a direct consequence of iron starvation, then their effects might be reversed by soluble iron complexes such as ferric-fructose or ferric nitrilotriacetic acid. Inhibition of the growth of KG-1 cells, a human acute leukaemia cell line, by monoclonal antibody 42/6 was completely reversed by ferric nitrilotriacetic acid [35]. However, antibody-induced inhibition of the growth of normal peripheral blood lymphocytes was only partially overcome by the addition of soluble iron [34], and in the case of other cell lines, including CCRF-CEM cells [33], the growth inhibitory effects of anti-transferrin antibodies could not be reversed by the addition of soluble iron. The results with KG-1 cells show that antibody-induced iron deprivation is sufficient to inhibit cell growth. Consistent with this conclusion is the observation that cell lines adapted for growth in transferrin-free medium containing ferric nitrilotriacetic acid become resistant to growth inhibition by anti-(transferrin receptor) antibodies [39]. It is likely that the failure of soluble ferric complexes to overcome the effects of anti-(transferrin receptor) antibodies on the growth of other cells reflects the fact that some cells cannot efficiently utilize iron provided in this form. It cannot be excluded, however, that in the case of some cells that are sensitive to the anti-receptor antibodies even in the presence of ferric chelates, the antibodies may interfere with additional processes required for growth, such as receptor-mediated uptake of other nutrients or growth factors.

Finally, as shown in Table 1, there is a striking correlation between the immunoglobulin class of anti-receptor antibodies and their biological effects. The three antibodies that inhibit iron uptake and block cell growth are polymeric, monoclonal antibody 42/6 is an IgA, whereas the two antibodies that block the growth of murine cells are IgM antibodies. Antibodies of the IgG class against either the murine or human transferrin receptor have little or no effect on cell growth. This suggests that extensive crosslinking of receptors by the polymeric antibodies is an important factor in the inhibition of transferrin receptor function. Further support for this idea comes from the observation that cell growth is inhibited by IgG antibodies if they are crosslinked on the cell surface by the addition of anti-immunoglobulin antibodies [12]. Although monoclonal antibody 42/6 was selected on the basis of its ability to inhibit transferrin binding to human cells [33], the two IgM antibodies against the murine transferrin receptor do not inhibit transferrin binding. These antibodies appear to interfere primarily with the internalization of transferrin receptors [12]. Receptors on cells treated with IgG antibodies are rapidly internalized, whereas in the presence of IgM anti-(transferrin receptor) antibodies, transferrin receptors and bound antibody remain localized on the cell surface. Interestingly, treatment of cells with IgG, but not IgM, anti-(transferrin receptor) antibodies leads to a 5-fold increase in the rate of degradation of receptors and a concomitant down-regulation of receptors on the cell surface.

## Expression of Transferrin Receptors on Normal and Malignant Tissues

Most of the initial evidence that the expression of transferrin receptors is co-ordinately regulated with cell growth is based on studies of cultured cells. In contrast, the early work on transferrin receptors in freshly isolated tissues focused upon their abundance on tissues with high iron requirements unrelated to growth, such as placental trophoblast which provides iron to the fetal circulation, and maturing erythroid cells. Nevertheless, there are now clear indications that the expression of transferrin receptors is also influenced by the proliferative status of cells *in vivo*.

The distribution of transferrin receptors on haematopoietic stem cells in man [40] and in the mouse [41] has been analysed by sorting bone marrow cells into receptor-positive and receptor-negative populations and then enumerating the different kinds of restricted stem cells in each cell fraction by the appropriate colony assay *in vitro*. The results in both species are generally concordant and show that transferrin receptors are found on most mature erythroid precursors (CFU-E), but only on a minority of myeloid precursors (CFU-C) and early erythroid stem cells (BFU-E). In the mouse, treatment of bone marrow with a cytotoxic ricin A–anti-(transferrin receptor) conjugate killed CFU-E but not the pluripotential stem cell (CFU-S), suggesting that the latter express few transferrin receptors [42]. This is consistent with results of cell sorting experiments in the rat, which show that both CFU-S and lymphopoietic stem cells are in the transferrin receptor-negative population of bone marrow and fetal liver [14]. In the mouse, the proportion of stem cells expressing transferrin receptors correlates well with estimates of the fraction of actively cycling cells within each haematopoietic stem cell compartment [43,44].

Studies of non-Hodgkin's lymphomas have provided further evidence that the expression of transferrin receptors is indicative of the proliferative status of tumours [45,46]. In one case, a direct correlation was found between DNA synthesis by B cell lymphoma populations and the proportion of cells expressing transferrin receptors [46]. Interestingly, it appears from these studies that the high expression of transferrin receptors on such lymphomas may be a negative prognostic indicator. A survey of normal human tissues and solid tumour using immunohistological staining of frozen tissue sections to detect transferrin receptors showed that receptors were only detected at a limited number of sites in normal tissues, but were more widely distributed on a random sampling of human tumour biopsies [47]. Generally, similar results were obtained from immunohistological studies on normal rat tissues [14].

In summary, the pattern of transferrin receptor expression *in vivo* is complex, reflecting both the increased expression of transferrin receptors on proliferating cells and cellular needs for iron unrelated to growth. The expression of transferrin receptors on tumours is presumably a consequence of their active proliferation.

## Immunotherapy with Monoclonal Antibodies that Block Receptor Function

Monoclonal antibodies that block the function of the surface receptors required for tumour cell growth have several potential advantages for immuno-therapy [48]. The availability of monoclonal antibodies against the murine transferrin receptor that block function has provided the opportunity to investigate their potential as anti-tumour agents *in vivo* in a syngeneic mouse model system [49,50]. Initial experiments have been carried out using the transplantable AKR mouse T cell leukaemia SL-2 which was previously used by Bernstein and his colleagues to investigate the requirements for effective serotherapy with anti-Thy-1 monoclonal antibodies [49,50]. Preliminary studies showed that the growth *in vivo* of SL-2 cells was inhibited by the rat monoclonal antibody, RI7 208, at a concentration of 5–10 $\mu$g/ml.

As described elsewhere [48,51], intravenous or peritoneal injection of the monoclonal antibody RI7 208 had a significant effect on the growth *in vivo* of SL-2 leukaemic cells. This was manifested not only by the prolonged survival of the tumour-bearing mice given antibody compared with untreated controls, but also in a marked inhibition of the growth of the primary tumour at the subcutaneous site of inoculation. The anti-(transferrin receptor) antibody was at least as effective in prolonging survival of tumour-bearing mice as the Thy-1 antibody, 19E12, given on the same schedule (3 mg of antibody on days 0, 4 and 7). Long-term survival of the majority of treated mice could be achieved by either more prolonged immunotherapy with the anti-(transferrin receptor) antibody or by administration of anti-Thy-1 and anti-(transferrin receptor) antibodies together. It is likely that the anti-tumour activity of antibody RI7 208 is mediated by its direct effects on transferrin receptor function and not by activation of host immunological effector mechanisms. Mutant SL-2 cells have been derived that still bind the anti-(transferrin receptor) antibody but are resistant to its inhibitory effects on growth in tissue culture. These cells give rise to tumours in AKR/J mice that are also resistant to the effects of the anti-(transferrin receptor) antibody.

Despite the fact that transferrin receptors are expressed at various sites in normal tissues, there has been little evidence of toxicity associated with the administration of the RI7 208 monoclonal antibody during immunotherapy trials. However, in other studies, it has been found that daily injection of 1 mg of antibody for a week depresses erythropoiesis in the bone marrow of treated mice as measured by a 2-fold decrease in the number of erythroid progenitor cells, CFU-E (J. Lesley, R. Schulte & I. S. Trowbridge, unpublished work). Concomitantly, the number of CFU-E in the spleens of treated mice was increased more than 10-fold on a per spleen basis. Similar but smaller changes in the number of myeloid progenitor cells (CFU-C) in bone marrow and spleen were also observed. The fact that erythroid progenitor cells appear to be the primary target for anti-(transferrin receptor) antibodies *in vivo* is consistent with their high expression of receptors.

## Molecular Cloning of the Human Transferrin Receptor

Two approaches have been used to isolate the human transferrin receptor gene. Genomic sequences that encode the human transferrin receptor were isolated by Kuhn and his colleagues using the method of DNA-mediated gene transfer [52]. Mouse L(tk$^-$) cells were cotransformed with the Herpes simplex thymidine kinase gene and total human DNA. Transformants expressing the human transferrin receptor gene were isolated by selection in hypoxanthine/aminopterin/thymidine (HAT) medium followed by fluorescence-activated cell sorting of HAT-resistant cells for expression of the human transferrin receptor. From a genomic library prepared from secondary transformants of mouse L cells expressing the human transferrin receptor, recombinant phage containing inserts of the human transferrin receptor gene were obtained. A probe from the 5′ end of the gene was then used to isolate a cDNA clone containing the entire coding region of the transferrin receptor gene. Independently, Schneider and his colleagues [53] were able to isolate cDNA clones from libraries constructed from mRNA isolated from human placenta and enriched for transferrin receptor mRNA by immunoselection of polysomes.

The primary sequence of the human transferrin receptor deduced from the cDNA sequence data [54,55] confirms the general structural features of the receptor inferred from previous biochemical studies. The receptor, which lacks an *N*-terminal signal sequence, is a polypeptide of 760 amino acid residues with an $M_r$ of nearly 85000. There is a single hydrophobic transmembrane region that is located proximal to the *N*-terminus of the receptor (residues 62–89) and is preceded by a cluster of basic residues (Lys-Pro-Lys-Arg). It is likely, therefore, that the human transferrin receptor is oriented with its *N*-terminus on the cytoplasmic side of the plasma membrane and has a cytoplasmic domain of 61 amino acids and an extracellular domain of 671 amino acids with three potential sites for the attachment of asparagine-linked oligosaccharides. Of a total of eight cysteines, four (residues 62, 67, 89 and 98) are clustered within, or close to, the membrane-spanning region. One or more of these cysteine

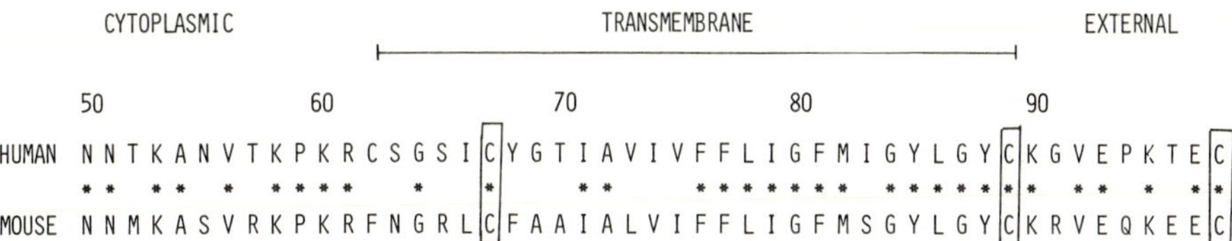

Fig. 2. *Comparison of the amino acid sequences of the human and mouse transferrin receptors flanking the membrane-spanning region*

The numbering refers to the human sequence. Amino acid residues conserved between species are indicated by asterisks. The data are from [54–56].

residues must be involved in disulphide bonding between the transferrin receptor subunits, in order to account for the fact that the external tryptic fragment of the transferrin receptor is not dimeric. These cysteines are also candidates for the site of attachment of covalently-bound lipid. The primary structure of the human transferrin receptor gives little insight into its function. No clear homology between the human transferrin receptor and other proteins, including transferrin or other cell surface receptors, is discernible.

Recently, cDNA clones of the mouse transferrin receptor have been isolated and one of these encoding the transmembrane and adjacent flanking regions of the receptor has been partially sequenced [56]. There is a high degree of homology between the mouse and human receptors in this region and three of the four cysteines are conserved (Fig. 2).

## Transferrin Receptor is a Substrate *in vivo* and *in vitro* for Protein Kinase C

The transferrin receptor has been previously shown to be phosphorylated [17]. Recently, it was demonstrated that incubation of HL-60 cells with a tumour-promoting phorbol diester causes a rapid increase in the level of receptor phosphorylation and a concomitant decrease in the number of receptors expressed on the cell surface [57]. The increased phosphorylation of the transferrin receptor in response to phorbol esters appears to be a general phenomenon. As shown in Fig. 3(*a*), 12-*O*-tetradecanoyl phorbol 13-acetate (TPA) induces the phosphorylation of the transferrin receptor of two human T leukaemic cell lines, CCRF-CEM and HUT 102B2. It is known that a major effect of TPA is the activation of the $Ca^{2+}$- and phospholipid-dependent protein kinase C which probably serves as a receptor for tumour-promoting phorbol esters (reviewed in [58]). Protein kinase C specifically catalyses the phosphorylation of proteins on serine and threonine residues and several growth-related cell membrane receptors, including the receptors for epidermal growth factor [59,60], insulin [61], somatomedin C [61] and IL-2 [62] have been directly or indirectly shown to be substrates for the enzyme.

As shown in Fig. 3(*b*), the transferrin receptor isolated from both human and mouse cells can be phosphorylated *in vitro* in the presence of protein kinase C. Phosphoamino acid analysis of the transferrin receptor from two human cell lines phosphorylated *in vitro* in the presence of protein kinase C revealed predominantly phosphoserine and trace amounts of phosphothreonine (Fig. 4).

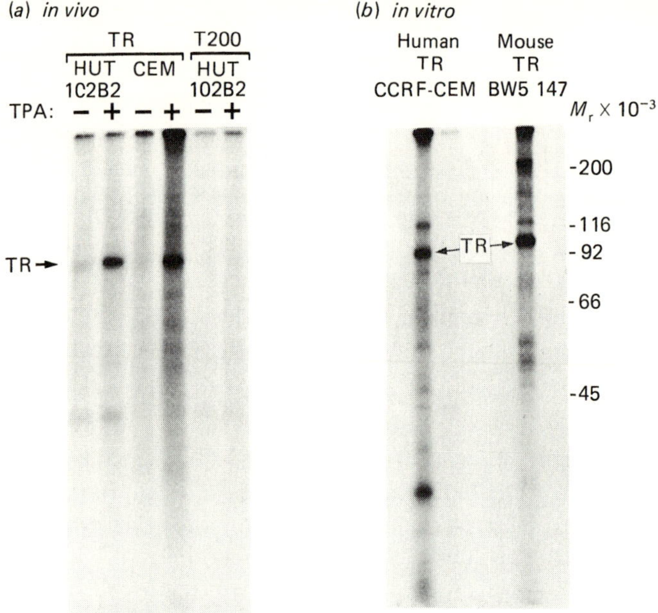


Fig. 3. *Phosphorylation in vivo and in vitro of the transferrin receptor*

(*a*) HUT102B2 ($2.4 \times 10^6$/ml) and CCRF-CEM ($4 \times 10^6$ ml) cells were labelled with 0.5 mCi of $H_3{}^{32}PO_4$ (ICN)/ml for 4 h at 37 °C as described previously [62]. After the 4 h incubation, TPA (50 ng/ml) was added to half of each sample (+) and the incubation continued for 15 min. The cells were washed twice with phosphate-buffered saline and lysed in 1 % NP40 in phosphate-buffered saline containing 0.1 mM-$Na_3VO_4$ and 5 mM-sodium pyrophosphate. The transferrin receptor (TR) and T200 glycoprotein were immunoprecipitated with the monoclonal antibodies B3/25 [8] and T29/33 [65], respectively, which had been covalently coupled to Sepharose beads. The immuno-precipitates were analysed by SDS/polyacrylamide-gel electrophoresis on 10 % acrylamide gels. (*b*) The human transferrin receptor was isolated from $1.4 \times 10^8$ unlabelled CCRF-CEM cells by using monoclonal antibody B3/25–Sepharose beads. The mouse transferrin receptor was precipitated from $0.7 \times 10^8$ unlabelled BW5147 cells by using monoclonal antibody RI7 217–Sepharose beads [12]. The immunoprecipitates were washed as described [62] with the final wash in 5 mM-$MgCl_2$/10 mM-Tris/HCl (pH 7.4). Phosphorylation *in vitro* of the transferrin receptor by kinase C was performed as described [59]. Protein kinase C was generously provided by Dr. Gordon Gill and Gary Heiserman (University of California, San Diego). The enzyme was purified from rat brain as described [59] including the final purification on a phenyl-Sepharose column. After phosphorylation, the transferrin receptor immunoprecipitates were washed again and analysed by SDS/polyacrylamide-gel electrophoresis.

The significance of the phosphorylation of the transferrin receptor by protein kinase C is unclear. It has been suggested that phosphorylation may act as a signal for the internalization of transferrin receptors [32]. However, as described earlier, whether transferrin receptors recycle spontaneously or are only endo-cytosed after ligand binding is unresolved. Further, it has been reported that transferrin does not stimulate phosphorylation of its receptor [57]. Although the transferrin receptors of both mouse and human cells are phosphorylated by protein kinase C, a comparison of their reported sequences shows no likely sites of phosphorylation that are conserved between species. By analogy with the EGF receptor [63], serine-63 may be a site of phosphorylation in the human sequence, but is unlikely to be accessible to protein kinase C *in vivo*, whereas

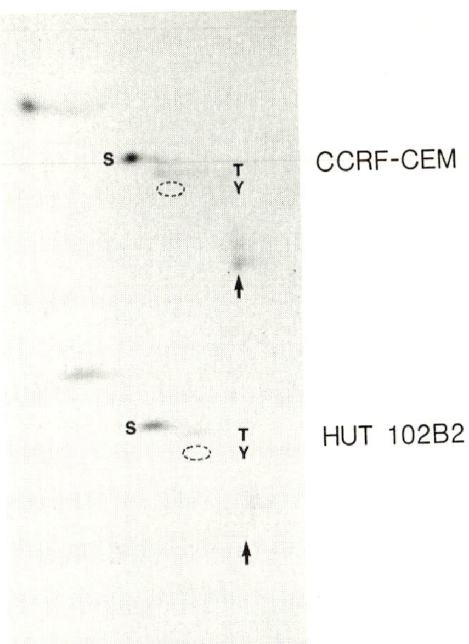


Fig. 4. *Phosphoamino acid analysis of human transferrin receptor*

The human transferrin receptor from CCRF-CEM and HUT102B2 cells was purified by immuno-precipitation and phosphorylated by protein kinase C *in vitro* as described in the legend to Fig. 3(*b*). The transferrin receptor was purified from SDS/polyacrylamide gels as described [62] and hydrolysed in 5.7 M-HCl for 2 h at 110 °C. The acid hydrolysate was resolved in two dimensions by electrophoresis toward the anode (left side) at pH 1.9 followed by electrophoresis toward the anode (top) at pH 3.5 [66]. The origin is designated by the arrow. The positions of phosphoserine (S), phosphothreonine (T), and phosphotyrosine (Y) standards mixed with the labelled sample are indicated.

serine-55 of the murine sequence as aligned in Fig. 2 is a probable site. The phosphorylation of residues that are proximal to the transmembrane region could affect the interaction of the receptor with components of the cell membrane.

We would like to acknowledge the contributions made by our colleagues in the Department of Cancer Biology to the work described. We thank Ami Koide for her help in preparation of this manuscript. Our work was supported by Grants CA 34787 and CA 17733 from the National Cancer Institute. D. A. S. is a recipient of Junior Fellowship J-52-83 from the California Division–American Cancer Society.

# References

1. Jandl, J. H. & Katz, J. H. (1963) *J. Clin. Invest.* **43**, 314–326
2. Barnes, D. & Sato, G. (1980) *Cell* **22**, 649–655
3. Larrick, J. W. & Cresswell, P. (1979) *J. Supramol. Struct.* **11**, 421–428
4. Galbraith, G. M. P., Galbraith, R. M. & Faulk, W. P. (1980) *Cell Immunol.* **49**, 215–222
5. Hamilton, T. A., Wada, H. G. & Sussman, H. H. (1979) *Proc. Natl. Acad. Sci. U.S.A.* **75**, 6406–6410
6. Galbraith, R. M. & Galbraith, G. M. P. (1981) *Immunology* **44**, 703–710
7. Omary, M. B., Trowbridge, I. S. & Minowada, J. (1980) *Nature (London)* **286**, 888–891
8. Trowbridge, I. S., & Omary, M. B. (1981) *Proc. Natl. Acad. Sci. U.S.A.* **78**, 3039–3043
9. Sutherland, R., Delia, D., Schneider, C., Newman, R., Kemshead, J. & Greaves, M. (1981) *Proc. Natl. Acad. Sci. U.S.A.* **78**, 4515–4519

10. Goding, J. W. & Burns, G. F. (1981) *J. Immunol.* **127**, 1256–1258
11. Trowbridge, I. S., Lesley, J. & Schulte, R. (1982) *J. Cell Physiol.* **112**, 403–410
12. Lesley, J. F. & Schulte, R. J. (1985) *Mol. Cell. Biol.* **5**, 1814–1821
13. Van Agthoven, A., Goridis, C., Naquet, P., Pierres, A. & Pierres, M. (1984) *Eur. J. Biochem.* **140**, 433–440
14. Jefferies, W. A., Brandon, M. R., Williams, A. F. & Hunt, S. V. (1985) *Immunology* **54**, 333–341
15. Seligmann, P. A., Schleicher, R. B. & Allen, R. H. (1979) *J. Biol. Chem.* **254**, 9943–9946
16. Omary, M. B. & Trowbridge, I. S. (1981) *J. Biol. Chem.* **256**, 12888–12892
17. Schneider, C., Sutherland, R., Newman, R. A. & Greaves, M. F. (1982) *J. Biol. Chem.* **257**, 8516–8522
18. Omary, M. B. & Trowbridge, I. S. (1981) *J. Biol. Chem.* **256**, 4715–4718
19. Rose, J. K., Adams, G. A. & Gallione, C. J. (1984) *Proc. Natl. Acad. Sci. U.S.A.* **81**, 2050–2054
20. Kaufman, J. F., Krangel, M. S. & Strominger, J. L. (1984) *J. Biol. Chem.* **259**, 7230–7328
21. Bleil, J. D. & Bretscher, M. S. (1982) *EMBO J.* **1**, 351–355
22. Harding, C., Heuser, J. & Stahl, P. (1983) *J. Cell Biol.* **97**, 329–339
23. Hopkins, C. R. & Trowbridge, I. S. (1983) *J. Cell Biol.* **97**, 508–521
24. Dautry-Varsat, A., Ciechanover, A. & Lodish, H. F. (1983) *Proc. Natl. Acad. Sci. U.S.A.* **80**, 2258–2262
25. Klausner, R. D., Ashwell, G., van Renswoude, J., Harford, J. B. & Bridges, K. R. (1983) *Proc. Natl. Acad. Sci. U.S.A.* **80**, 2263–2266
26. van Renswoude, J., Bridges, K. R., Hartford, J. B. & Klausner, R. D. (1982) *Proc. Natl. Acad. Sci. U.S.A.* **79**, 6186–6190
27. Yamashiro, D. J., Tycko, B., Fluss, S. R. & Maxfield, F. R. (1984) *Cell* **37**, 789–800
28. Hopkins, C. R. (1983) *Cell* **35**, 321–330
29. Klausner, R. D., van Renswoude, J., Ashwell, G., Kempf, C., Schechter, A. N., Dean, A. & Bridges, K. R. (1983) *J. Biol. Chem.* **258**, 4715–4724
30. Ward, J. H., Kushner, J. P. & Kaplan, J. (1982) *J. Biol. Chem.* **257**, 10317–10323
31. Watts, C. (1985) *J. Cell. Biol.* **100**, 633–637
32. Klausner, R. D., Harford, J. & van Renswoude, J. (1984) *Proc. Natl. Acad. Sci. U.S.A.* **81**, 3005–3009
33. Trowbridge, I. S. & Lopez, F. (1982) *Proc. Natl. Acad. Sci. U.S.A.* **79**, 1175–1179
34. Mendelsohn, J., Trowbridge, I. & Castagnola, J. (1983) *Blood* **62**, 821–826
35. Taetle, R., Honeysett, J. M. & Trowbridge, I. (1983) *Int. J. Cancer* **32**, 343–349
36. Trowbridge, I. S. & Newman, R. A. (1984) in *Antibodies to Receptors: Probes for Receptor Structure and Function* (Greaves, M. F., ed.), pp. 235–261, Academic Press/Chapman and Hall, London
37. Messmer, T. O. (1973) *Exp. Cell. Res.* **77**, 404–408
38. Walthall, B. J. & Ham, R. G. (1981) *Exp. Cell. Res.* **134**, 303–311
39. Taetle, R., Rhyner, K., Castagnola, J., To, D. & Mendelsohn, J. (1985) *J. Clin. Invest.* **75**, 1061–1067
40. Sieff, C., Bicknell, D., Caine, G., Robinson, J., Lam, G. & Greaves, M. F. (1982) *Blood* **60**, 703–713
41. Lesley, J., Hyman, R., Schulte, R. & Trotter, J. (1984) *Cell Immunol.* **83**, 14–25
42. Lesley, J., Domingo, D. L., Schulte, R. & Trowbridge, I. S. (1984) *Exp. Cell. Res.* **150**, 400–407
43. Becker, A. J., McCulloch, E. A., Siminovitch, L. & Till, J. E. (1965) *Blood* **26**, 296–308
44. Gregory, C. J. & Eaves, A. C. (1978) *Blood* **51**, 527–537
45. Habeshaw, J. A., Lister, T. A., Stansfeld, A. G. & Greaves, M. F. (1983) *Lancet* **i**, 498–500
46. Kvaloy, S., Langholm, R., Kaalhus, O., Michaelsen, T., Funderud, S., Foss Abrahamsen, A. & Godal, T. (1984) *Int. J. Cancer* **33**, 173–177
47. Gatter, K. C., Brown, G., Trowbridge, I. S., Woolston, R.-E. & Mason, D. Y. (1983) *J. Clin. Pathol.* **36**, 539–545
48. Trowbridge, I. S. (1983) in *Monoclonal Antibodies and Cancer* (Boss, B. D., Langman, R., Trowbridge, I. & Dulbecco, R., eds.), pp. 53–61, Academic Press
49. Bernstein, I. D. & Nowinski, R. C. (1982) in *Hybridomas in Cancer Diagnosis and Treatment* (Mitchell, M. S. & Oettgen, H. F., eds.), pp. 97–112, Raven Press, New York
50. Badger, C. C. & Bernstein, I. D. (1983) *J. Exp. Med.* **157**, 828–842
51. Trowbridge, I. S., Newman, R. A., Domingo, D. L. & Sauvage, C. (1984) *Biochem. Pharmacol.* **33**, 925–940
52. Kuhn, L. C., McClelland, A. & Ruddle, F. H. (1984) *Cell* **37**, 95–103
53. Schneider, C., Kurkinen, M. & Greaves, M. (1983) *EMBO J.* **2**, 2259–2263
54. McClelland, A., Kuhn, L. & Ruddle, F. H. (1984) *Cell* **39**, 267–274
55. Schneider, C., Owen, M. J., Banville, D. & Williams, J. G. (1984) *Nature (London)* **311**, 675–678

56. Stearne, P. A., Pietersz, G. A. & Goding, J. W. (1985) *J. Immunol.* **134**, 3474–3479
57. May, W. S., Jacobs, S. & Cuatrecasas, P. (1984) *Proc. Natl. Acad. Sci. U.S.A.* **81**, 2016–2020
58. Nishizuka, Y. (1984) *Nature (London)* **308**, 693–698
59. Cochet, C., Gill, G. N., Meisenhelder, J., Cooper, J. A. & Hunter, T. (1984) *J. Biol. Chem.* **259**, 2553–2558
60. Fox, C. F. & Iwashita, S. (1984) *J. Biol. Chem.* **259**, 2559–2567
61. Jacobs, S., Sahyoun, N. E., Saltiel, A. R. & Cuatrecasas, P. (1983) *Proc. Natl. Acad. Sci. U.S.A.* **80**, 6211–6213
62. Shackelford, D. A. & Trowbridge, I. S. (1984) *J. Biol. Chem.* **259**, 11706–11712
63. Hunter, T., Ling, N. & Cooper, J. A. (1984) *Nature (London)* **311**, 1311–1315
64. Haynes, B. F., Hemler, M., Cotner, T., Mann, D. L., Eisenbarth, G. S., Strominger, J. L. & Fauci, A. S. (1981) *J. Immunol.* **127**, 347–351
65. Omary, M. B., Trowbridge, I. S. & Battifora, H. A. (1980) *J. Exp. Med.* **152**, 842–852
66. Hunter, T. & Sefton, B. M. (1980) *Proc. Natl. Acad. Sci. U.S.A.* **77**, 1311–1315

*Biochem. Soc. Symp.* **51**, 131–148
*Printed in Great Britain*

# Immunoglobulin G as a Glycoprotein

T. W. RADEMACHER, S. W. HOMANS, R. B. PAREKH and R. A. DWEK

*Department of Biochemistry, University of Oxford, South Parks Road, Oxford OX1 3QU, U.K.*

## Introduction

An integral feature of all normal IgG class antibodies is the conserved *N*-glycosylation site in the $C_H2$ domains at Asn-297. The combinatorial association of the two heavy chains results in the two oligosaccharide units being in direct contact with each other. These are therefore a model for both specific protein–carbohydrate and carbohydrate–carbohydrate interactions [1,2]. In addition to the conserved glycosylation sites in the Fc, *N*-linked oligosaccharides may be attached to the Fab region of IgG [3,4]. However, their frequency and location are dependent upon the occurrence of Asn-Xaa-Ser(Thr) sites in the hypervariable region [5] (Fig. 1).

Given the conservation of the polypeptide structure of the immunoglobulin molecule (the immunoglobulin fold), and that Fc *N*-glycosylation has been conserved throughout evolution, it might be expected that the Fc-oligosaccharides would show limited structural diversity [6]. Further, the primary monosaccharide sequences of these should give important information about the activity of certain glycosyltransferases in plasma cells and any structural limitations imposed by the immunoglobulin molecule. In this latter respect it should be noted that the relative sizes of an immunoglobulin domain and a fully extended *N*-linked complex biantennary oligosaccharide are similar [7] (see Fig. 9).

Recent advances have opened a new era in structural and functional studies on immunoglobulin oligosaccharides. These include (i) the use of controlled hydrazinolysis for the quantitative release of intact *N*-linked oligosaccharide chains [8]; (ii) the use of rapid and sensitive techniques for the detection, fractionation and sequencing of these oligosaccharides [7]; (iii) n.m.r. methods for the analysis of oligosaccharide solution conformations [9,10]; (iv) the interpretation of the X-ray crystallographic data from the Fc-associated oligosaccharides [7,11] and (v) the ability to isolate pure and intact aglycosyl immunoglobulins [12]. We report below some of our studies on the *N*-linked oligosaccharides of IgG.

## Distribution of *N*-Linked Oligosaccharide Chains in IgG

The *N*-linked oligosaccharides associated with human IgG can be released by controlled hydrazinolysis. Subsequent reduction, with $NaB^3H_4$, of its reducing terminal residue, radioactively labels each carbohydrate chain. By determining the maximum number of moles of oligosaccharide released from a known

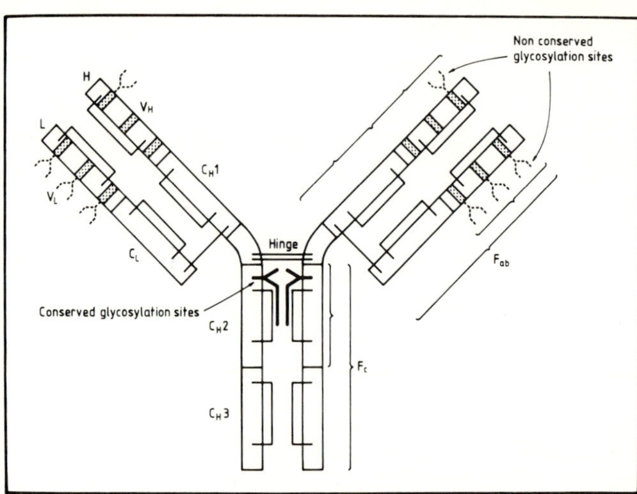


Fig. 1. *The antibody molecule consists of two heavy (H) and two light (L) chains, linked by disulphide bridges (solid lines) and is divided into homologous regions of sequence ($V_H$, $C_H1$, $C_H3$), each of which has an intra-chain disulphide bridge [19].*

The pattern of inter-chain disulphide bridging shown here is characteristic of human sub-class IgG1. In $V_H$ and $V_L$, the dotted segments represent the hypervariable regions of sequence which, in the three-dimensional structure, together form the antigen-binding site. The conserved asparagine-linked biantennary complex oligosaccharide chains are attached to Asn-297 in the $C_H2$ domains. Oligosaccharide attachment sites are found in the Fab region. Their frequency and location is dependent upon the presence of Asn-Xaa-Ser(Thr) sites in the hypervariable regions [5].

amount of human IgG the number of oligosaccharide chains on an intact IgG molecule is found to be, on average, approx. 2.8 [7]. Two of these chains are located in the Fc region and the remainder in the Fab regions [13].

## Sialylation of IgG Oligosaccharides

The radioactive oligosaccharide mixture (obtained as described in [7]) can be fractionated by subjecting it to high-voltage paper electrophoresis at pH 5.4. As shown in Fig. 2, one neutral (N) and two acidic (A1 and A2) components are obtained. Both A1 and A2 can be converted to neutral oligosaccharides by sialidase digestion, and correspond to the presence of either one (A1) or two (A2) sialic acid residues. The results in Fig. 2 show that there is a difference in the sialylation of oligosaccharides released from Fab and Fc derived from human IgG. Those from the Fab have a higher proportion of sialylated species and, of particular importance, disialylated species essentially occur only on Fab.

Since this is also true for rabbit IgG [13] and its fragments (Fig. 3), the presence of such disialylated structures may therefore be used as a marker for Fab glycosylation (see Figs. 2 and 3). It is interesting to note that the overall low level of sialic acid present on the Fc fragment [13] is inconsistent with the proposed role of desialylation in the clearance of IgG from serum [14].

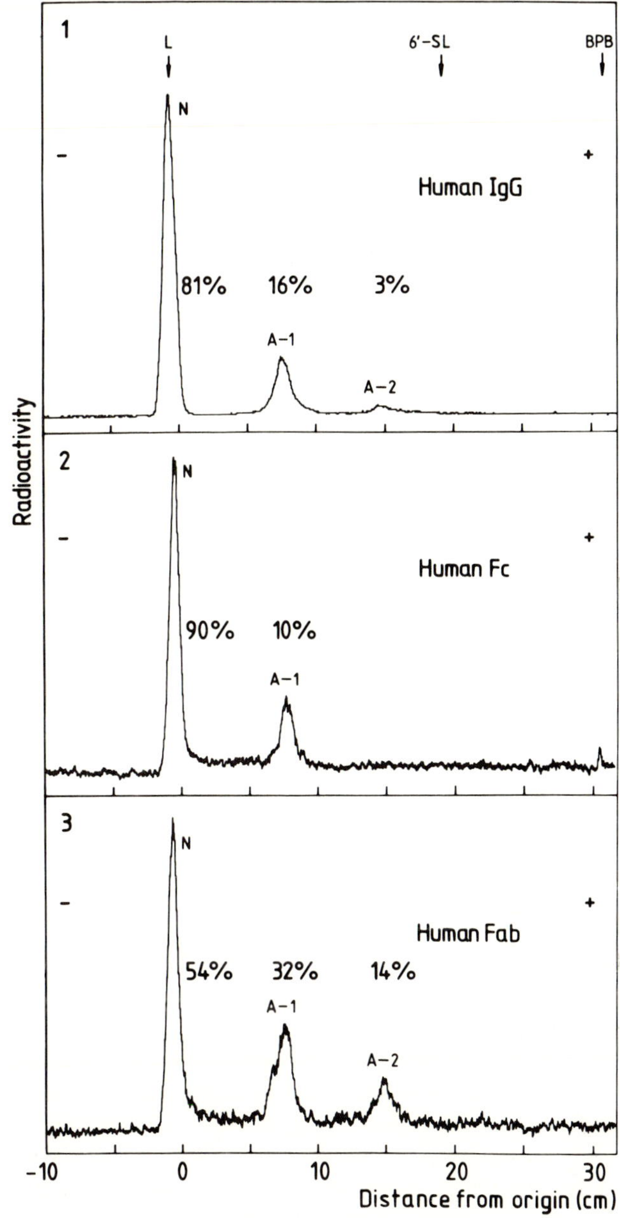


Fig. 2. *Radioelectrophoretograms of the oligosaccharides from human IgG and its fragments Fc and Fab*

Samples were subjected to high-voltage paper electrophoresis (80 V/cm) in pyridine/acetic acid/water (3:1:387, by vol.), pH 5.4. Lactose (L), 6'-sialyl-lactose (SL) and Bromophenol Blue (BPB) were used as markers. After scanning with a radiochromatogram scanner, the regions N (neutral), A-1 (monosialylated) and A-2 (disialylated) were eluted and the radioactivity in each determined, allowing the ratios in each sample to be calculated.

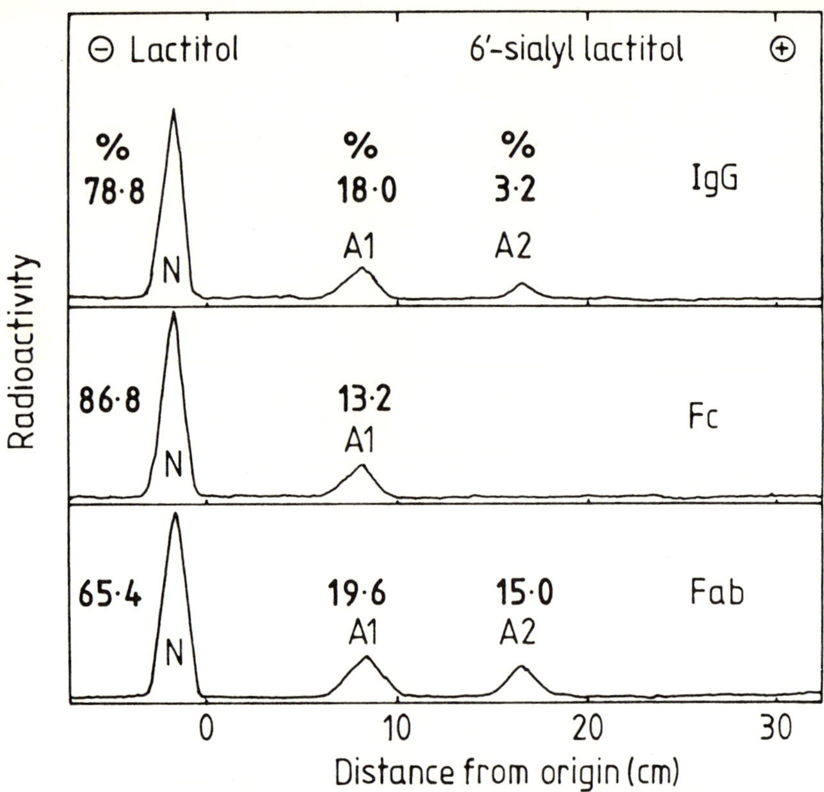


Fig. 3. *Radioelectrophoretograms of the oligosaccharides from rabbit IgG and its fragments Fc and Fab*

Analysis was performed essentially as in Fig. 2. From Mizuochi *et al.* [13].

## Fractionation and Sequencing of the 'Asialo' Oligosaccharide Fraction

The distribution of oligosaccharide structures from human IgG, and from its Fab and Fc fragments, can be analysed by subjecting the appropriate radioactive oligosaccharide mixtures to exhaustive neuraminidase digestion. The resulting 'asialo' oligosaccharide mixtures can then be fractionated on the basis of their effective hydrodynamic volumes by gel permeation chromatography using Bio-Gel P-4 (–400 mesh) as the resin (see Fig. 6, 1b–5b). The use of sequential exoglycosidase digestion in conjunction with lectin-assisted separation, paper chromatography, Bio-Gel P-4 gel permeation chromatography and g.c.–m.s. (Fig. 4) shows that at least 30 complex-type biantennary oligosaccharides are associated with human serum IgG [7]. A complete set of structures is shown in Fig. 5.

## Species-Specificity of IgG Oligosaccharides

Our studies on IgG from several species indicate that the overall sialylation pattern of the *N*-linked oligosaccharides is similar (Fig. 6, 1a–5a). Further, the

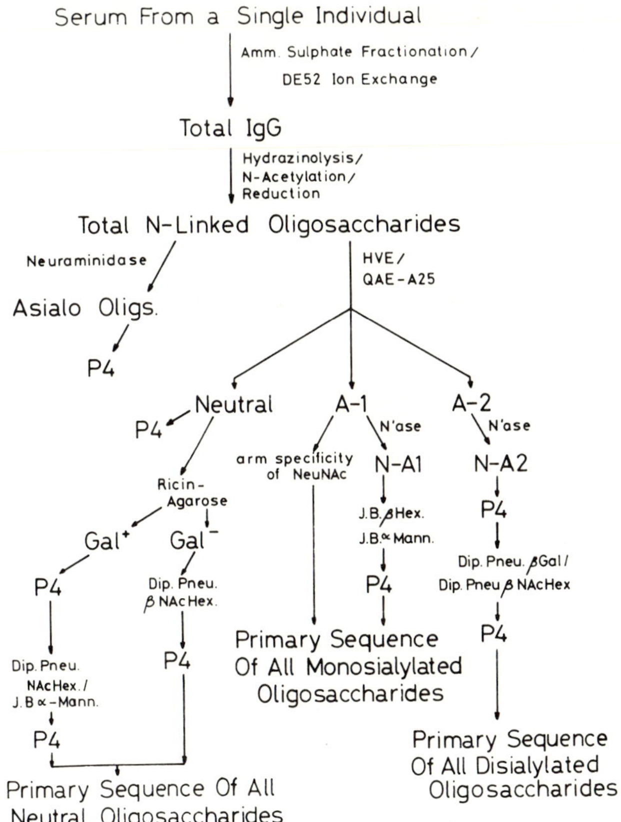


Fig. 4. *Fractionation strategy utilized to describe quantitatively the set of IgG-associated oligosaccharide structures found on a single serum sample*

Primary monosaccharide sequences were subsequently obtained by using sequential exoglycosidase digestion and methylation analysis [7].

'asialo' oligosaccharides structures can always be accommodated within essentially the same set of complex biantennary structures mentioned above (see Fig. 5). The relative incidence of each structure does however vary from species to species and a characteristic P-4 'fingerprint' is therefore observed (Fig. 6, 1b–5b).

## Oligosaccharides of Fab and Fc Moieties

Interestingly, the oligosaccharide structures associated with Fab and Fc moieties differ. Sialylation differences have already been mentioned (Figs. 2 and 3). Differences in the relative incidence of individual asialo structures also occur [13]. Since both the heavy and light chains are presumably exposed to the same biosynthetic enzymes, and since both chains are synthesized in the same cell, these differences imply that factors additional to glycosyltransferases determine the glycosylation pattern at a given site: for example, the polypeptide moiety of the nascent glycoprotein.

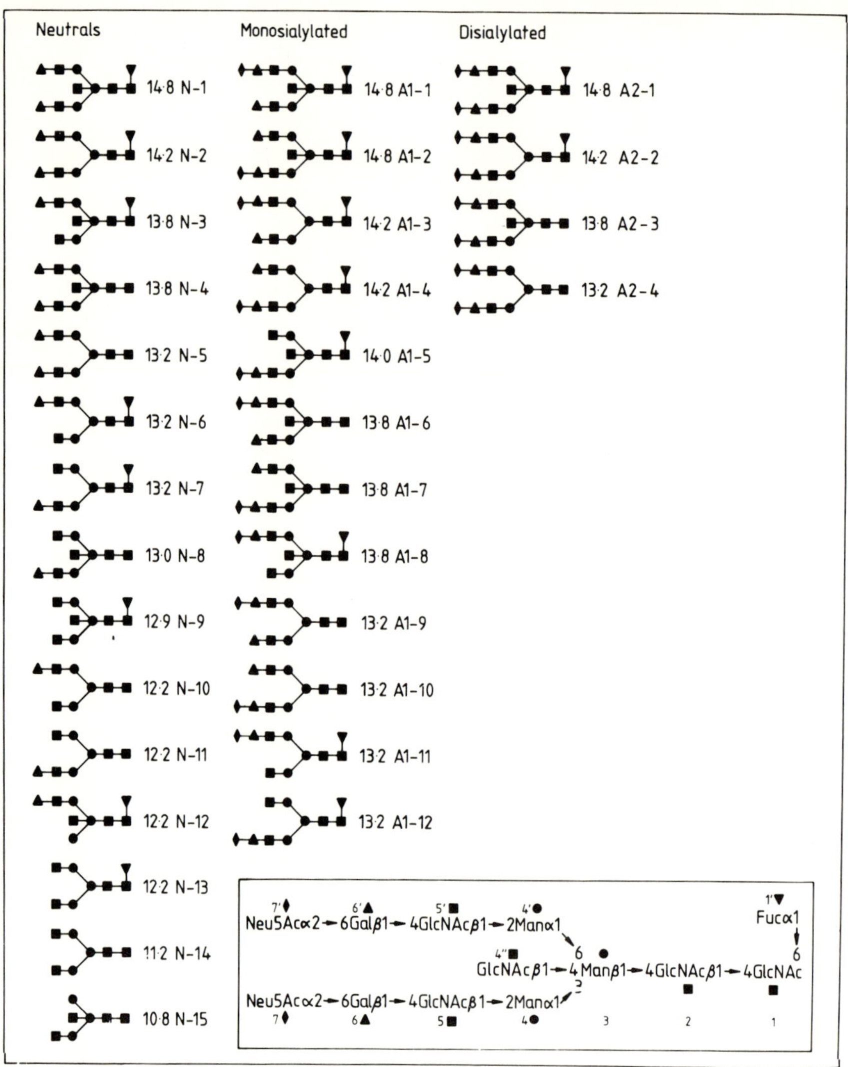



Fig. 5. *Primary sequences of the N-linked oligosaccharides associated with IgG*

The hydrodynamic volume [as measured in glucose units (g.u.)] of each structure (or of its neutral derivative in the case of those sialylated) is indicated, and was determined by comparison with $\alpha(1 \rightarrow 6)$-linked glucose oligomer standards [7,20] (see Fig. 6).

---

Fig. 6. *Analysis of oligosaccharides from mammalian IgG*

Left: representative radioelectrophoretograms of the oligosaccharides from human (1a), sheep (2a), rabbit (3a), mouse (4a) and bovine (5a) IgG. Samples were analysed as described in Figs. 2 and 3. Right: representative Bio-Gel P-4 (–400 mesh) gel-permeation chromatograms of the asialo oligosaccharides from human (1b), sheep (2b), rabbit (3b), mouse (4b) and bovine (5b) IgG are shown. Purified ³H-radiolabelled oligosaccharides were suspended in 175 μl of a 20 mg/ml partial dextran acid hydrolysate and applied to a Bio-gel P-4 (–400 mesh) gel pemeation chromatography column (1.5 cm × 200 cm). The column was maintained at 55 °C and water (200 μl/min) was used as the eluent. The eluent was monitored for radioactivity with a radioactivity monitor and for refractive index with a refractometer. Analogue signals from the monitors were digitized and subsequently analysed by using Hewlett Packard 9836 C computers. The Figure shows radioactivity (vertical axis) plotted against retention time after removal of stochastic noise by using Fourier transform techniques. The numerical superscripts refer to the elution position of glucose oligomers in glucose units (g.u.) as detected simultaneously by the refractive index monitor (results not shown). $V_0$ is the void position. Sample elution positions (in g.u.) were calculated by cubic spline interpolation between the internal standard glucose oligomer positions.

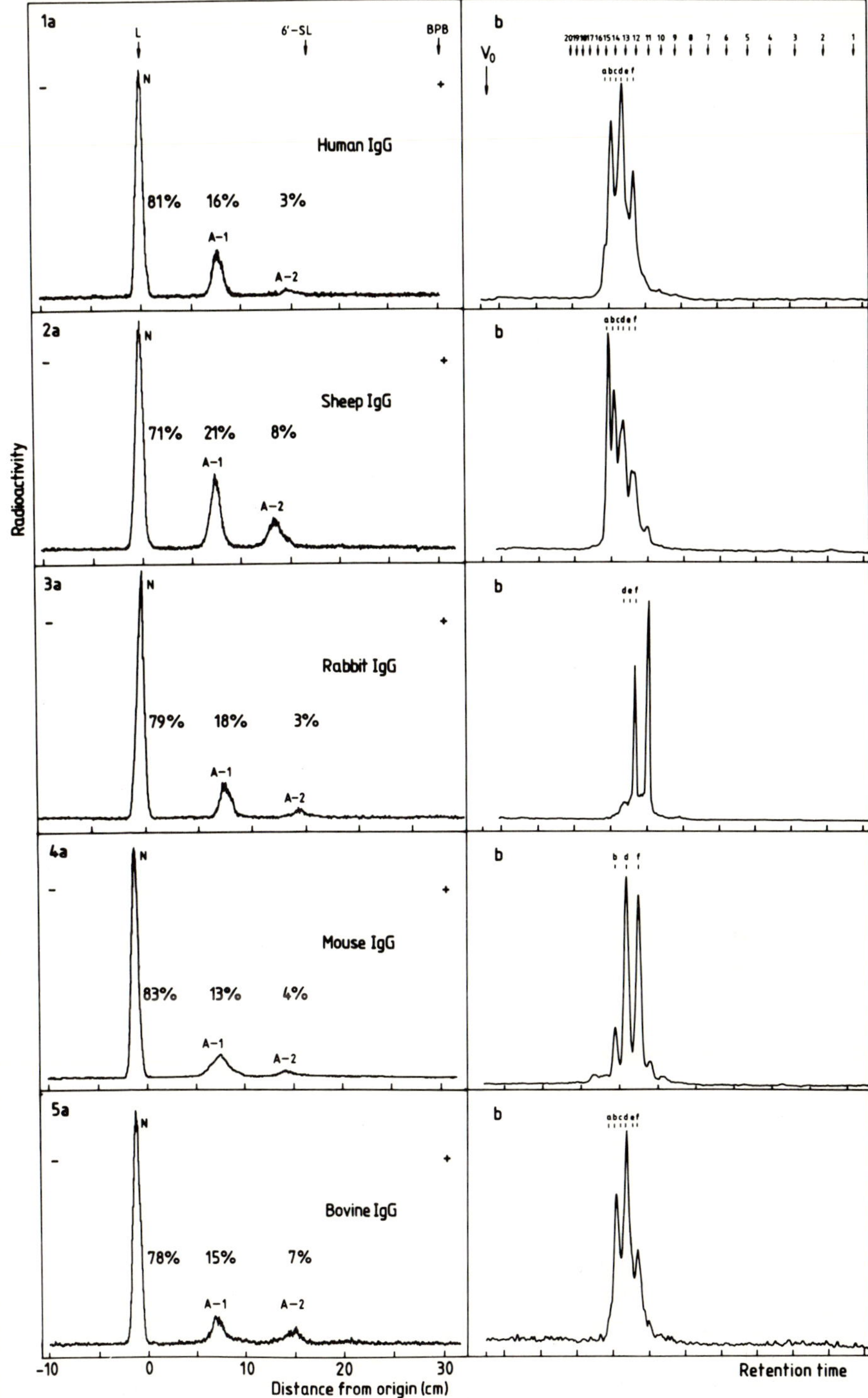

Fig. 6. *For legend see facing page.*

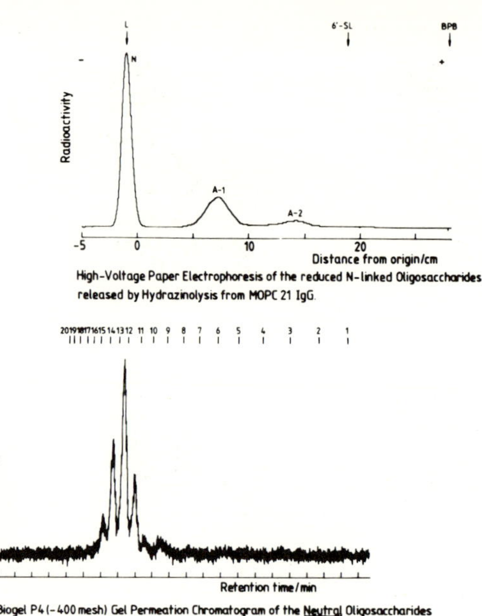



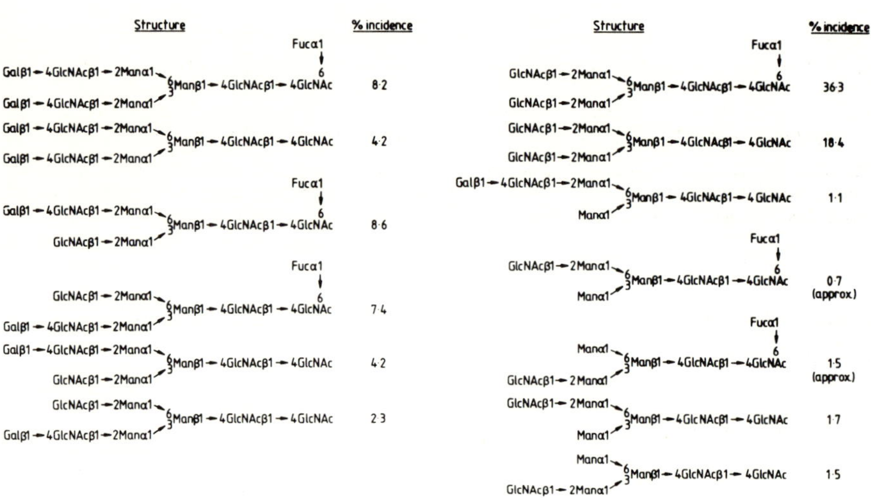



Fig. 7. *Heterogeneity of N-linked oligosaccharides found on the myeloma IgG (P3K-MOPC21)*

## IgG Oligosaccharides from Myelomas and Hybridomas

The large number of different oligosaccharides found associated with normal human IgG is not the result of using polyclonal IgG for our analysis, since this same set of structures is found on myeloma proteins [15] and mouse monoclonal antibodies (unpublished work). Some of the N-linked oligosaccharides found on a myeloma IgG (MOPC 21) are described in Fig. 7. N-Glycosylation of IgG

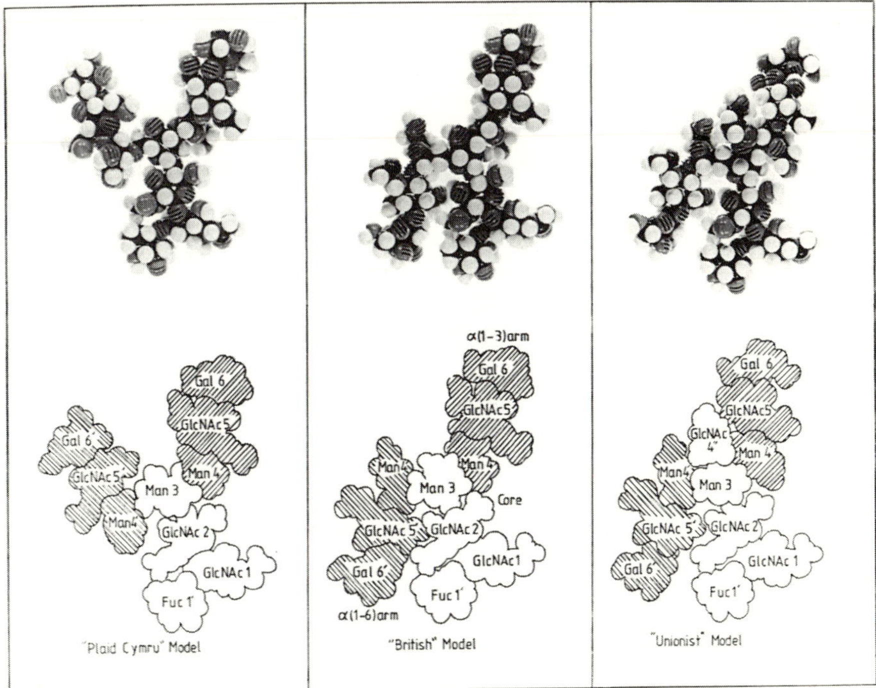



Fig. 8. *Space-filling models of the conformations of Galβ1 → 4GlcNAcβ1 → 2Manα1 → 3(Galβ1 → 4GlcNAcβ1 → 2Manα1 → 6)Manβ1 → 4GlcNAcβ1 → 4(Fucα1 → 6)GlcNAc (left and middle) and Galβ1 → 4GlcNacβ1 → 2Manα1 → 3(Galβ1 → 4GlcNAcβ1 → 2Manα1 → 6)(GlcNAcβ1→ 4)Manβ1 → 4GlcNAcβ1 → 4(Fucα1 → 6)GlcNAc (right) determined by ¹H-n.m.r. methods*

Note that the unbisected structure can adopt two orientations about the $\alpha(1 \to 6)$ linkage i.e. 'Plaid Cymru' and 'British' models, whereas the bisected analogue can only adopt a single orientation about this linkage, i.e. 'Unionist' model. Note that this substitution also uniquely changes the orientation of the $\alpha(1 \to 3)$arm.

in myelomas seems to be essentially normal, since the structures of MOPC 21 IgG are similar to those found on normal mouse IgG (results not shown).

## Conformational Analysis of IgG Oligosaccharides

The solution conformations of the monosaccharide sequences shown in Fig. 5 can be determined by using high-resolution ¹H-n.m.r. [9,10]. Work in this laboratory has shown that N-linked oligosaccharides contain regions of defined secondary structure. Despite the very large number of different primary sequences our present conclusion is that there are fewer solution structures. For example, each of the glycosidic linkages in Galβ1 → 4GlcNAcβ1 → 2Manα1 → 3 (Galβ1 → 4GlcNAcβ1 → 2Manα1 → 6) Manβ1 → 4GlcNAcβ1 → 4GlcNAc exists in solution with a single preferred conformation, with the exception of the $\alpha1 \to 6$ linkage which can be in at least two conformations. These conformations are illustrated in Fig. 8. Importantly, addition of a terminal NeuNAc residue $(2 \to 6)$ to the outer arm or removal of the Gal or GlcNAc residues does not significantly affect the orientation of the Manα1 → 3Man linkage, i.e. outer-arm heterogeneity is not associated with unique solution conformations in these cases. In contrast, alteration of the basic core structure by addition of the single 'bisecting' GlcNAc residue has a profound effect on the overall conformation. The orientation of the $\alpha(1 \to 3)$ antenna is altered with respect to the core (Fig. 8) and the orientation of the $\alpha(1 \to 6)$ antenna is restricted to the vicinity of the core and it thus appears to be 'folded back'. Once again, removal of outer-arm residues does not affect this basic structure.

For the present, we conclude that monosaccharide primary sequence heterogeneity, though extensive, can be accommodated within a *smaller* set of secondary structures. The restricted orientaton of the $\alpha 1 \to 6$ linkage in the 'bisected' structure may have an important consequence for the interaction of such oligosaccharides with the Fc (see below).

## X-Ray Crystallographic Analysis on Fc Fragments and Associated Oligosaccharides

Crystallographic studies on immunoglobulin Fc fragments have shown that, unlike other immunoglobulin domains, the two $C_H2$ domains do not form extensive lateral associations [7,11]. The resultant interstitial region is filled by the inclusion of the oligosaccharide side chains attached to Asn-297, on each heavy chain, such that the carbohydrates form a bridge across the domains (Fig. 9). The $\alpha(1 \to 6)$ antenna of each can interact along the domain surface and both hydrophobic (aromatic) and polar residues are involved. The presence of aromatic residues, often found in the combining sites of anti-saccharide antibodies [16], provides favourable interactions with the planar hydrophobic face which can be adopted by D-sugars (the $^4C_1D$ conformation). *Across* the domains there is a restriction in pairing of the two oligosaccharides which can lead to there being *different monosaccharide sequences* on the different $\alpha(1 \to 3)$ arms. One of these arms must always be devoid of galactose, thereby exposing its outer-arm $\beta(1 \to 2)$GlcNAc residue, which then interacts directly with the Man$\beta$1$(1 \to 4)$GlcNAc core segment of the opposing carbohydrate chain. The $\alpha(1 \to 3)$ arm of this latter oligosaccharide can extend outwards between the domains with no apparent steric constraints on the length of its primary sequence. Despite having identical amino acid sequences, the two heavy chains in an IgG molecule therefore often carry *N*-linked oligosaccharides of different primary sequence, rendering the molecule structurally asymmetrical.

## Does Oligosaccharide Heterogeneity lead to Selective Crystallization?

Given the oligosaccharide heterogeneity shown in Fig. 5, the description of the X-ray data could either represent a composite (or 'average') oligosaccharide structure, or a weighted 'average' structure, if selective crystallization had occurred. Sequence analysis of oligosaccharides isolated from single protein crystals from both rabbit and human Fc, which were used in the X-ray studies [11,17], are identical with those isolated from uncrystallized Fc fragments (e.g. Figs. 10, 11 and 12). While this demonstrates the absence of selective crystallization it also poses a problem. There is no apparent electron density from

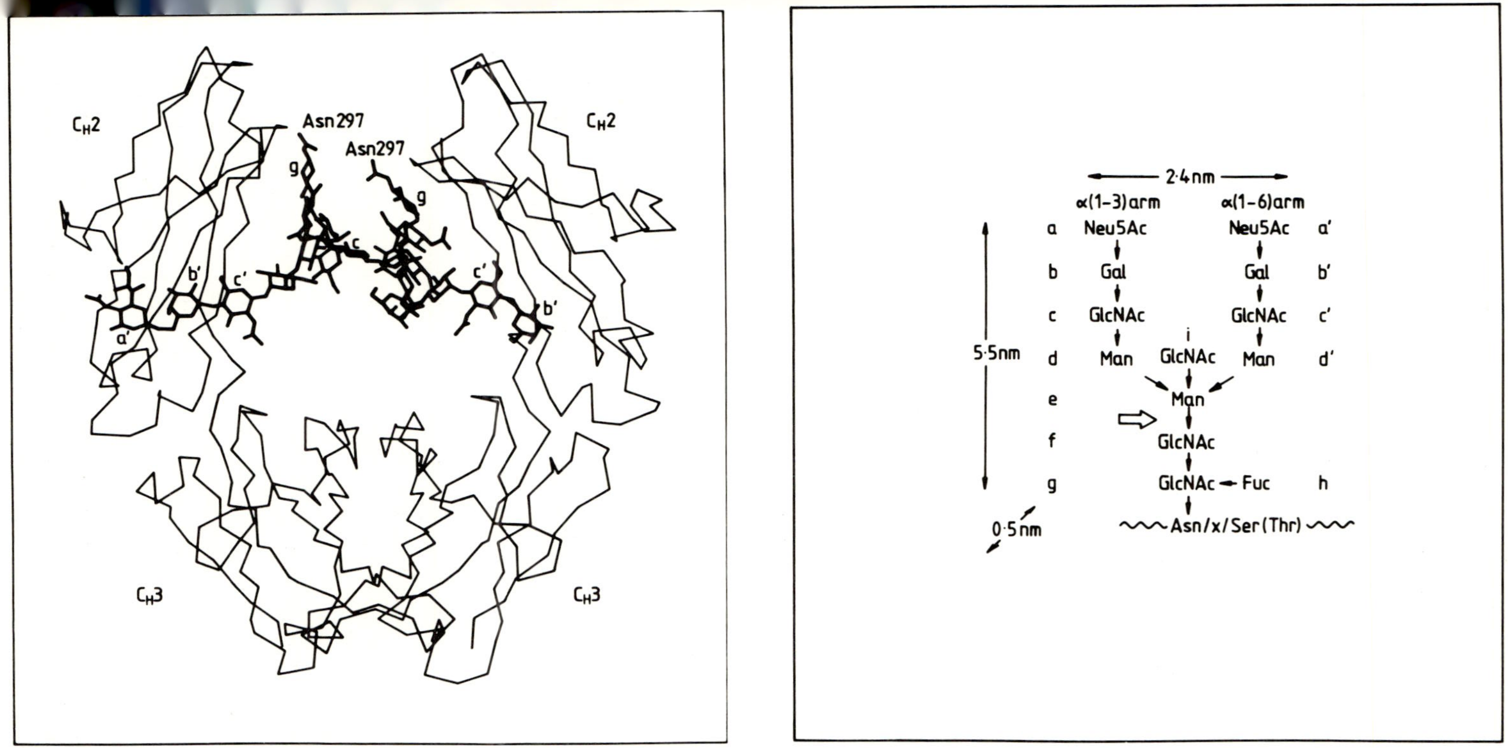

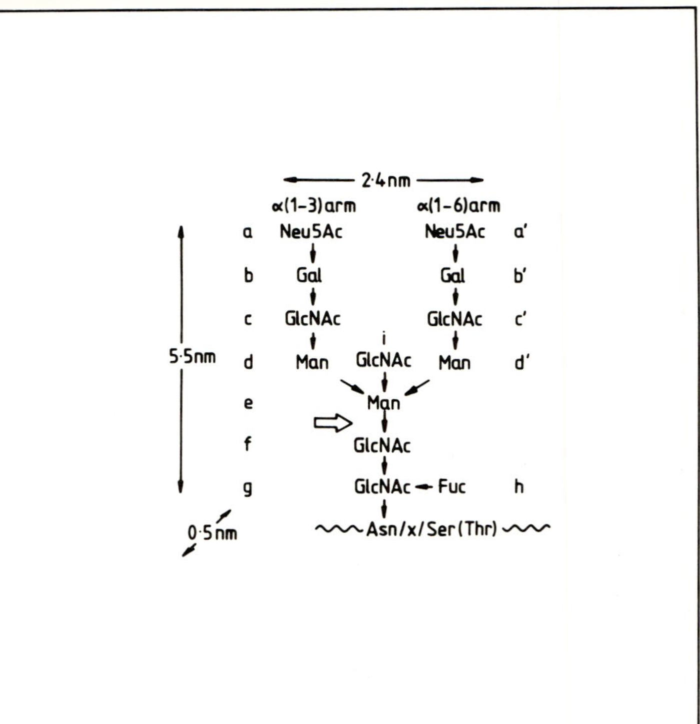


Fig. 9. *Refined structure at 2.8 Å of rabbit Fc fragment from the crystal data of Sutton & Phillips [7, 11]*

The two carbohydrate chains, each attached at Asn-297, differ in conformation and may also differ in sequence and bridge the two $C_H2$ domains. The $\alpha(1 \to 3)$ arm of the chain (left side) is always devoid of galactose and interacts through its $\beta(1 \to 2)$-linked GlcNAc residue (c) with the Man$\beta(1 \to 4)$GlcNAc segment of the opposing (right side) oligosaccharide chain (see below). The $\alpha(1 \to 3)$ arm of the right chain extends outwards between the domains with no apparent steric constraints on its length. A Neu5Ac unit (a') is shown on one $\alpha(1 \to 6)$ arm only (left). The electron density for this unit is weak and, experimentally, no disialylated oligosaccharide chains occur on the Fc (see Figs. 2 and 3). The extent of oligosaccharide heterogeneity in a single crystal is identical with that found in pooled Fc fragments (see Figs. 10, 11 and 12); consequently the X-ray data represents the composite structure. Right: the relative size of an immunoglobulin domain and a fully extended $N$-linked complex oligosaccharide [7,21] are similar. Complex-type oligosaccharides present on IgG can be subdivided into an outer-arm region (a, a', b, b', c, c'); and the core which is composed of a trimannosyl unit (d, d', e) and a $N$, $N'$-diacetylchitobiosyl unit (f, g). The 'bisect' GlcNAc (residue i) is linked $\beta(1 \to 4)$ and the fucose (residue h) is linked $\alpha(1 \to 6)$. The arrow between residues e and f indicates the site of interaction between the two oligosaccharides.

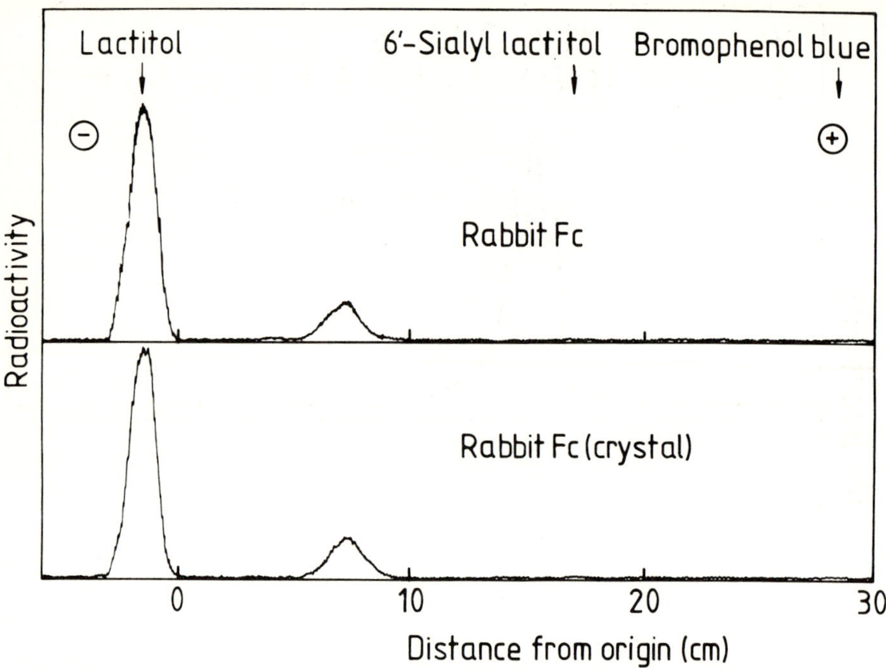


Fig. 10. *Radioelectrophoretograms of the oligosaccharides liberated by hydrazinolysis from rabbit Fc fragment and from a single crystal grown from the same Fc preparation (T. Mizuochi, A. Kobata & T. W. Rademacher, unpublished work) and used in the X-ray crystallographic study of Sutton & Phillips* [7,11]

the 'bisecting' GlcNAc residue which is present in about 30% of the Fc oligosaccharide chains. The solution conformation studies summarized above may provide some clues. Without the 'bisecting' GlcNAc the $\alpha(1\rightarrow6)$ antenna may be 'locked' in the extended conformation (Fig. 8) so as to maximize interactions with the $C_H2$ domain. In contrast the restricted (folded back) orientation of the $\alpha(1\rightarrow6)$ antenna in the 'bisected' structure would result in this arm being unable to make these protein contacts. It could therefore be 'mobile' (or disordered) with respect to the protein, and may therefore be 'absent' in the maps of the X-ray data.

## Function of Immunoglobulin Oligosaccharides

Some insight into the possible functional roles of the Fc oligosaccharides comes from studies using aglycosyl monoclonal anti-dinitrophenol mouse $IgG_{2a}$, produced in the presence of tunicamycin. The main conclusions are summarized in Fig. 13. The aglycosyl molecule retains the properties of the normal glycosylated molecule with regard to antigen binding, protein A binding, C1q binding and C1 activation [12].

There are however some important effects of removal of the oligosaccharide. Firstly, the stability of the $C_H2$ domain is reduced in the aglycosyl $IgG_{2a}$, as suggested by the 60-fold increase in its rate of cleavage by pepsin. Secondly, there

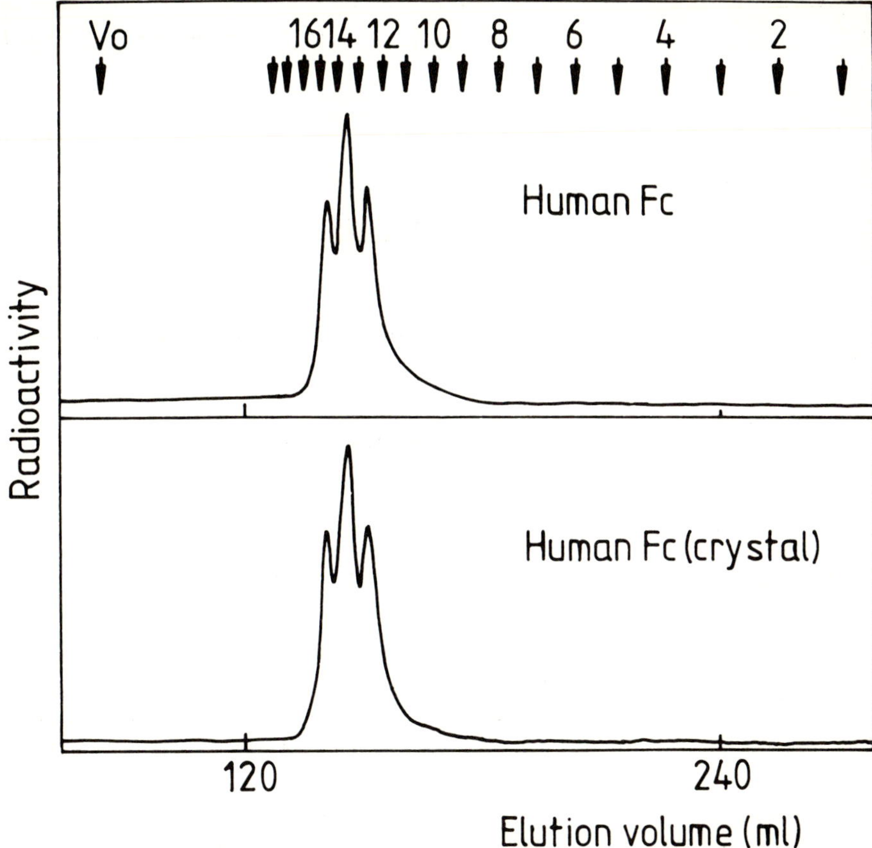



Fig. 11. *Bio-Gel P-4 (–400 mesh) gel permeation chromatograms of the hydrazine-released oligosaccharides from human Fc and from a single Fc crystal used in the X-ray analysis of Deisenhofer [17]*

The oligosaccharide mixtures were treated with neuraminidase prior to chromatography (T. Mizuochi, A. Kobata & T. W. Rademacher, unpublished work).

is a loss of binding to monocyte Fc receptors, consistent with the notion that oligosaccharide-mediated recognition is confined to events involving cell surfaces. Two additional functional features have been studied by Nose & Wigzell [18]. They found that aglycosyl IgG$_{2b}$ had lost the ability to induce antibody-dependent cellular cytotoxicity and that in its complexes with antigen it failed to be eliminated rapidly from circulation.

The inability of aglycosyl IgG to bind to Fc receptors indicates that the *N*-linked oligosaccharide must play a role in maintaining the integrity of its monocyte binding site. It has still to be established whether this role is direct or indirect. Taken with the other observations, we conclude at present that the Fc oligosaccahride does play an important role in allowing the immunoglobulin molecule to participate in cell-mediated antigen elimination reactions.

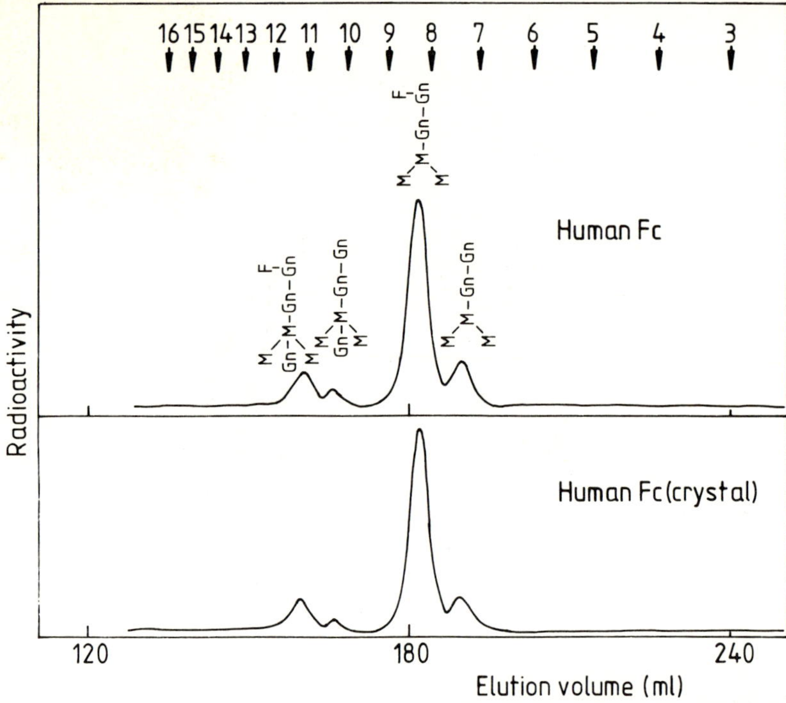



Fig. 12. *Bio-gel P-4 (–400 mesh) gel permeation chromatograms of the oligosaccharide mixtures shown in Fig. 11 after treatment with a mixture of Streptomyces pneumoniae β-galactosidase and β-N-acetylhexosaminidase*

The resulting digestion products are diagnostic for each of the four cores shown [7] (T. Mizuochi, A. Kobata & T. W. Rademacher, unpublished work).

## Changes in the IgG Oligosaccharide with Disease State

We have recently been involved in a collaborative study with the Tokyo group to compare the *N*-glycosylation pattern of serum IgG isolated from normal individuals and patients with rheumatoid arthritis [7]. The P-4 chromatograms of the asialo oligosaccharide mixtures are shown in Fig. 14 and are clearly different. These P-4 profiles for rheumatoid arthritis patients have been rationalized in terms of a 'population shift' towards structures of lower hydrodynamic volume. This shift arises solely from a reduced level of outer-arm galactosylation, so that there is an increased number of structures which have outer arms that terminate in *N*-acetylglucosamine. This decrease in galactosylation correlates highly with the presence of rheumatoid arthritis ($P < 0.001$), and much more so than the presence of rheumatoid factor or other pathological symptoms (such as subcutaneous nodules). It is not a general consequence of chronic inflammation nor of auto-immunity, since patients with the closely related disorder SLE (systemic lupus erythematosus) do not show this abnormality in their IgG (unpublished work). Nor is the decreased galactosylation in rheumatoid arthritis a generalized disorder, since the majority of the *N*-linked oligosaccharides in serum are normally galactosylated (unpublished work).

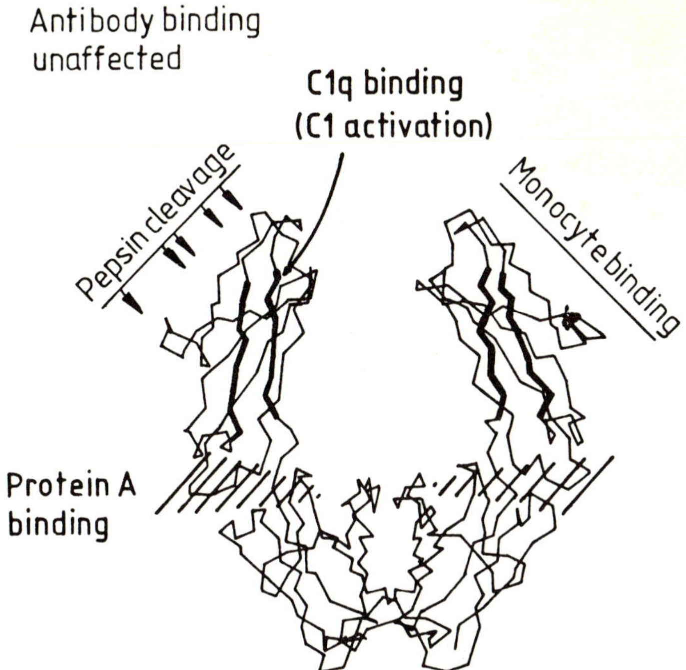



Fig. 13. *Summary of the proposed locations of a number of antibody secondary effector functions*

Aglycosylated IgG retains the properties of the normal glycosylated molecular with regard to antigen binding, protein A binding, C1q binding and C1 activation. However there is a total loss of binding to monocyte Fc receptors [12].

A consequence of this change in galactosylation may be understood by reference to the X-ray studies on Fc. A knowledge of the total population of all the oligosaccharides present, and given the steric restriction that one of the $\alpha(1 \to 3)$ arms in any individual Fc must be devoid of galactose (discussed above), a statistical calculation yields the relative incidence of all oligosaccharide pairs allowed in Fc. It turns out that, in normal serum, 10% of the Fc molecules totally lack galactose [7] whereas in the serum of patients with rheumatoid arthritis the population of the Fc molecules totally lacking galactose has risen dramatically to 40%. An immediate consequence is that the oligosaccharide binding sites on the $C_H 2$ domains, which are normally occupied by Neu5Ac$\alpha(2 \to 6)$Gal segments or galactose of the $\alpha(1 \to 6)$ arms, are now vacant (Fig. 9). This could make the IgG 'sticky' by creating a lectin-like activity, or it could result in the increased exposure of immunogenic determinants. Both these mechanism would lead to the formation of immunoglobulin complexes — although only the latter would be truly auto-immune.

## Conclusion

Several conclusions can be drawn from these studies. First, IgG irrespective of its origin always carries more or less the same limited set of *N*-linked complex

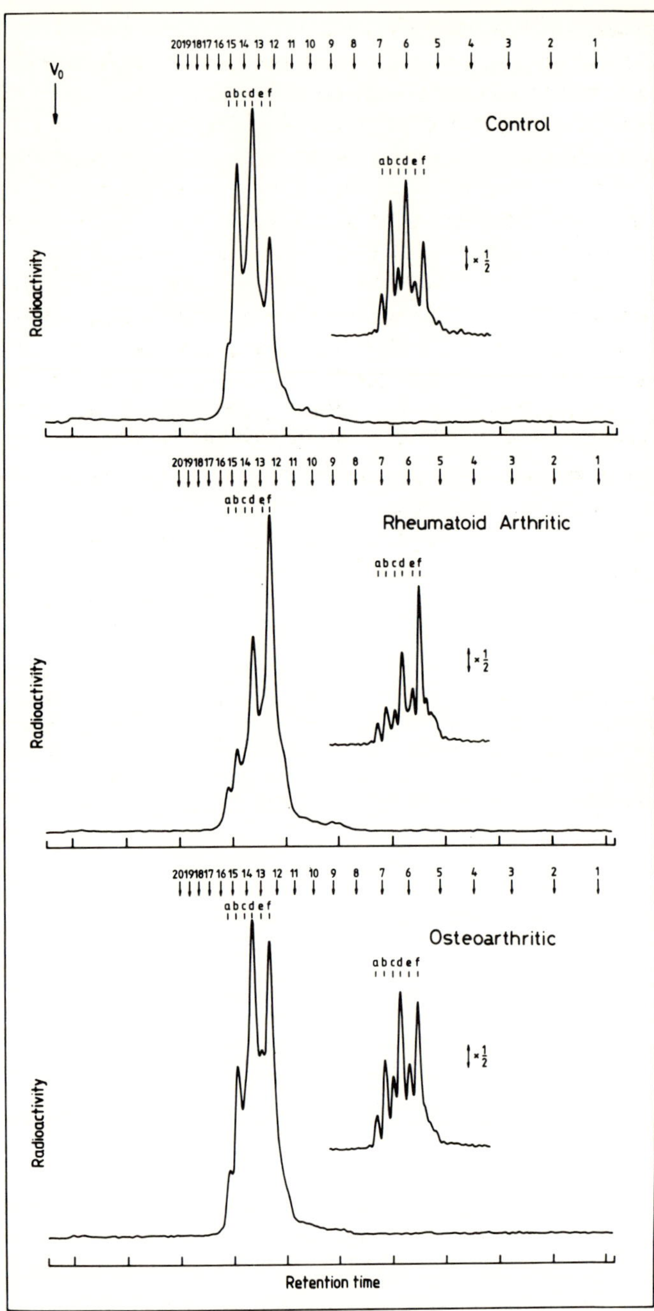


Fig. 14. *Representative Bio-Gel P-4 (–400 mesh) gel permeation chromatograms of the asialo oligosaccharides of total serum IgG from healthy individuals, patients with rheumatoid arthritis and patients with osteoarthritis*

In the disease states the observed profile differences are uniquely due to changes in the relative extent of galactosylation [7].

binatennary oligosaccharides. IgGs from different species essentially differ only with respect to the relative proportion of the individual members of this constant set of primary monosaccharide sequences. Second, while there is no obvious correlation between the occurrence of a particular monosaccharide sequence and a specific IgG function, Fc oligosaccharide *per se* is important for cell-mediated clearance of immune complexes. Third, and possibly of greater significance, the number of primary monosaccharide sequences associated with IgG is considerably greater than the number of *N*-glycosylation sites. An immunoglobulin containing molecules of identical amino acid sequences must therefore consist of many discrete sub-populations each of which will be characterized by the presence of an unique disposition of oligosaccharides. The structural consequence of what has hitherto been referred to as 'oligosaccharide heterogeneity' is therefore the diversification of a single polypeptide into a set of discrete and unique glycoproteins. Given the possibility and complexity of protein–carbohydrate and carbohydrate–carbohydrate interactions as revealed by a study of the Fc moiety, a plausible consequence of this structural diversification is a modulation of the biological function of a polypeptide. Irrespective, however, of any normal consequence of the existence of IgG glycoprotein sub-populations is the possibility that alterations in these may be important in the pathogenesis of certain diseases (e.g. rheumatoid arthritis).

The data reported here have been determined in our own laboratory and that of Professor Akira Kobata (Tokyo University). We also wish to thank our colleagues D. L. Fernandes, P. Rudd, T. Mizuochi and T. Taniguchi. The authors are members of the Oxford Oligosaccharide group. T. W. R., R. A. D. and S. W. H. are also members of the Oxford Enzyme group. This work has been supported by Monsanto.

# References


1. Rademacher, T. W. & Dwek, R. A. (1983) *Prog. Immunol.* **5**, 95–112
2. Dwek, R. A., Sutton, B. J., Perkins, S. J. & Rademacher, T. W. (1984) *Biochem. Soc. Symp.* **49**, 123–136
3. Abel, C. A., Spiegelberg, H. L. & Grey, H. M. (1968) *Biochemistry* **7**, 1271–1278
4. Spiegelberg, H. L., Abel, C. A., Fishkin, B. G. & Grey, H. M. (1970) *Biochemistry* **9**, 4217–4223
5. Sox, H. C. & Hood, B. (1970) *Proc. Natl. Acad. Sci. U.S.A.* **66**, 975–982
6. Rademacher, T. W., Homans, S. W., Fernandes, D. L., Dwek, R. A., Mizuochi, T., Taniguchi, T. & Kobata, A. (1983) *Biochem. Soc. Trans.* **11**, 132–134
7. Parekh, R. B., Dwek, R. A., Sutton, B. J., Fernandes, D. L., Leung, A., Stanworth, D., Rademacher, T. W., Mizuochi, T., Taniguchi, T., Matsuta, K., Takeuchi, F., Nagano, Y., Miyamoto, T. & Kobata, A. (1985) *Nature (London)* **316**, 452–457
8. Takasaki, S., Mizuochi, T. & Kobata, A. (1982) *Methods Enzymol.* **83**, 263–268
9. Homans, S. W., Dwek, R. A., Fernandes, D. L. & Rademacher, T. W. (1982) *FEBS Lett.* **150**, 503–506
10. Homans, S. W., Dwek, R. A., Fernandes, D. L. & Rademacher, T. W. (1983) *FEBS Lett.* **164**, 231–235
11. Sutton, B. J. & Phillips, D. C. (1982) *Biochem. Soc. Trans.* **11**, 130–132
12. Leatherbarrow, R. J., Rademacher, T. W., Dwek, R. A., Woof, J. M., Clark, A., Burton, D. R., Richardson, N. & Feinstein, A. (1985) *Mol. Immunol.* **22:4**, 407–415
13. Taniguchi, T., Mizuochi, T., Beale, M., Dwek, R. A., Rademacher, T. W. & Kobata, A. (1985) *Biochemistry* **24**, 5551–5557
14. Winkelhake, J. L. & Nicholson, G. L. (1976) *J. Biol. Chem.* **251**, 1074
15. Mizuochi, T., Taniguchi, T., Shimizu, A. & Kobata, A. (1982) *J. Immunol.* **129**, 2016–2020
16. Gettins, P., Boyd, J., Glaudemans, C. P. J., Potter, M. & Dwek, R. A. (1981) *Biochemistry* **20**, 7463–7469

17. Deisenhofer, J. (1981) *Biochemistry* **20**, 2361–2370
18. Nose, M. & Wigzell, H. (1983) *Proc. Natl. Acad. Sci. U.S.A.* **80**, 6632–6636
19. Beale, D. & Feinstein, A. (1976) *Q. Rev. Biophys.* **9**, 135–180
20. Yamashita, K., Mizuochi, T. & Kobata, A. (1982) *Methods Enzymol.* **83**, 105–126
21. Montreuil, J. (1982) *Biochem. Soc. Trans.* **11**, 134–136

*Biochem. Soc. Symp.* **51**, 149–157
*Printed in Great Britain*

# Neuronal/Lymphoid Membrane Glycoprotein MRC OX-2 is a Member of the Immunoglobulin Superfamily with a Light-Chain-Like Structure

A. NEIL BARCLAY, MELANIE J. CLARK and GEOFF W. McCAUGHAN

*MRC Cellular Immunology Unit, Sir William Dunn School of Pathology, University of Oxford, South Parks Road, Oxford OX1 3RE, U.K.*

## Synopsis

The MRC OX-2 antigen is a membrane glycoprotein of about 45000 $M_r$ present on rat neurons, thymocytes, B cells, follicular dendritic cells, endothelium and smooth muscle. Sequence of cDNA clones indicates it is a member of the Ig superfamily containing 248 amino acids organized like an Ig light chain with a V-like domain and a C-like domain followed by a transmembrane and cytoplasmic sections. There is a sequence with homologies with J-regions but analysis of the gene for human OX-2 shows that this is part of the V-domain exon and there is not a separate J-region exon as in the T cell receptor or Ig chains. The relationship of OX-2 to the other Ig-related neuronal/thymocyte antigen Thy-1 and the evolution of the Ig superfamily are discussed.

## Introduction

The MRC OX-2 monoclonal-antibody-producing hybrid cell line was produced by the fusion of spleen cells of a mouse immunized with rat thymocyte membrane glycoproteins, with the mouse myeloma NS-1 [1]. The antibody recognized thymocytes, brain, B cells [1] and on tissue sections was also shown to recognize follicular dendritic cells in lymphoid organs, endothelium and smooth muscle [2]. This unusual pattern of brain/thymocyte/blood vessels was reminiscent of the Thy-1 antigen, as summarized in Table 1, although the OX-2 antigen is about 30-fold less abundant than the Thy-1 antigen on brain and thymus. These similarities were reinforced when the OX-2 antigens were purified from brain and thymus by affinity chromatography and biochemically characterized, as summarized in Fig. 1 and Table 2 [3].

The functions of Thy-1 and OX-2 antigens are not known but it has been suggested that they are involved in recognition events at cell surfaces [3–5]. In this way the odd pattern of tissue distribution need not necessarily mean that the function is the same in the different tissues but that the molecule is a ligand of cell-surface interactions the outcome of which could vary between the tissues. The finding that Thy-1 antigen had homology with a single immunoglobulin V-domain reinforced this idea, as immunoglobulin domains are known to be involved in many recognition events [5]. It seemed possible that OX-2, like Thy-1,

Table 1. *Distribution of rat OX2 and Thy-1 antigens*

++indicates strongly positive, +positive, −negative. Data from [2,5,6,21,22,23].

| Tissue or cell type | OX-2 | Thy-1 |
| --- | --- | --- |
| Neurons | ++ | ++ |
| Thymocytes | ++ | ++ |
| Peripheral nerves | + | + |
| Neonatal brain | + | − |
| T cells | − | − |
| B cells | + | − |
| Fibroblasts and pericytes around blood vessels | − | + |
| Endothelium | + | − |
| Muscle | + | + |
| Mast cells, haematopoeitic stem cells | − | + |

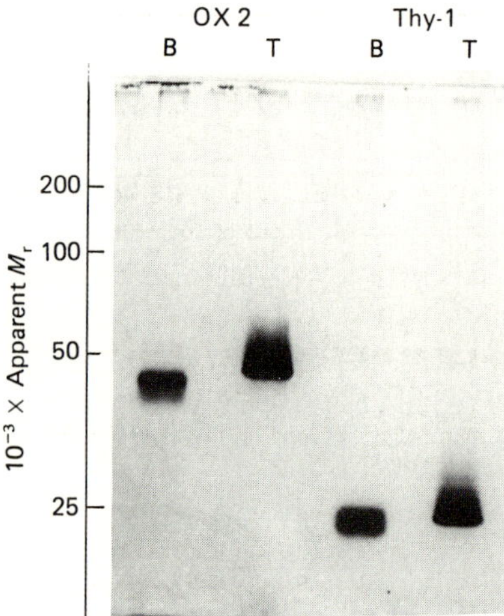


Fig. 1. *Analysis of OX-2 and Thy-1 antigens by polyacrylamide-gel electrophoresis in sodium dodecyl sulphate*

B and T above the lanes denote whether the antigens were purified from brain or thymus. Samples were reduced with dithiothreitol. The gel was stained with a silver stain although similar patterns were observed when stained for protein with Coomassie Blue or carbohydrate with periodic acid/Schiff stain. Reproduced from [3] with permission.

might be related to immunoglobulins in evolution and this was investigated by determining the amino acid sequence of the OX-2 antigen. In this article we summarize the evidence for inclusion of OX-2 antigen in the Ig superfamily and present preliminary evidence on the organization of the gene for OX-2 and its significance in the evolution of the Ig superfamily.

## Sequence of the MRC OX-2 Antigen [6]

The approach used was to obtain sufficient protein sequence to design an oligonucleotide probe for screening cDNA libraries. OX-2 antigen was purified from rat brains by solubilization in sodium deoxycholate, affinity chromato-

Table 2. *Biochemical characteristics of OX-2 and Thy-1 antigens*

Data are from [3,20,24,25].

| Characteristic | OX-2 | Thy-1 |
|---|---|---|
| Number of molecules/thymocyte | $3 \times 10^4$ | $10^6$ |
| Behaviour on polyacrylamide-<br>  gel electrophoresis in SDS | | |
|   Brain | $M_r$ 41 000 | $M_r$ 24 000 |
|   Thymus | $M_r$ 47 000, trails to<br>  higher-$M_r$ forms | $M_r$ 25 000, trails to<br>  higher-$M_r$ forms |
| Carbohydrate composition | | |
|   Brain form | 24% carbohydrate | 29% carbohydrate |
|   Thymocyte form | 33% carbohydrate | 32% carbohydrate |

The carbohydrate compositions of the brain forms of both antigens are very similar, but differ from the thymocyte forms, which contain more galactose and sialic acid but less fucose.

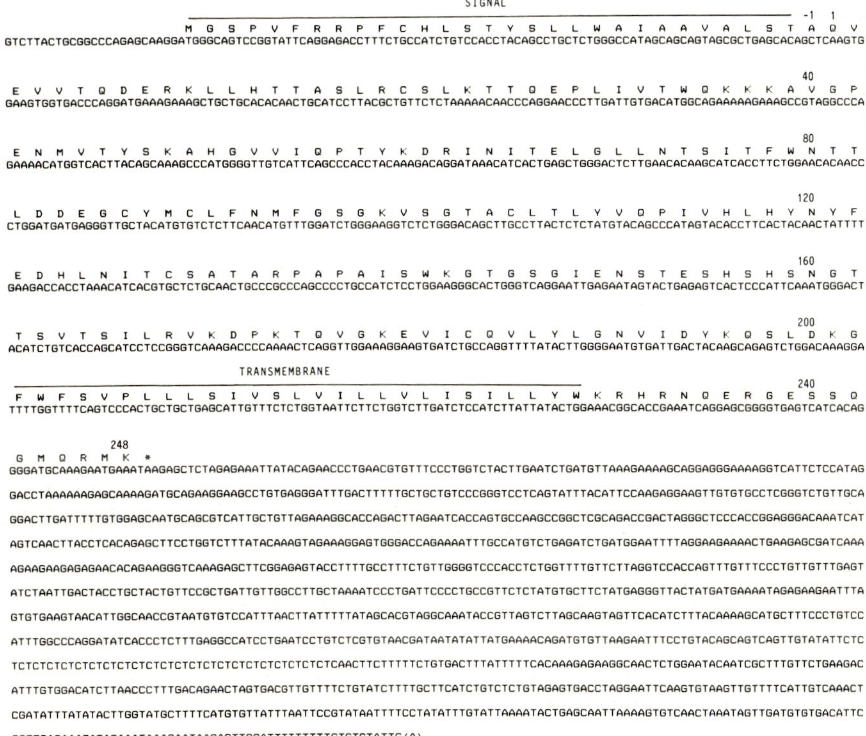


Fig. 2. *Nucleotide sequence and corresponding amino acid sequence of thymocyte OX-2 cDNA*

The possible polyadenylation signal is underlined and the sequence ends in a poly(A) tail. The coding sequence is derived from more than one clone, as described in [6]. The 3′ non-coding sequence was obtained from clone pX 2/13 [6] by restriction enzyme digestion, end-labelling with [32]P and Maxam and Gilbert sequencing (M.J. Clark, unpublished work).

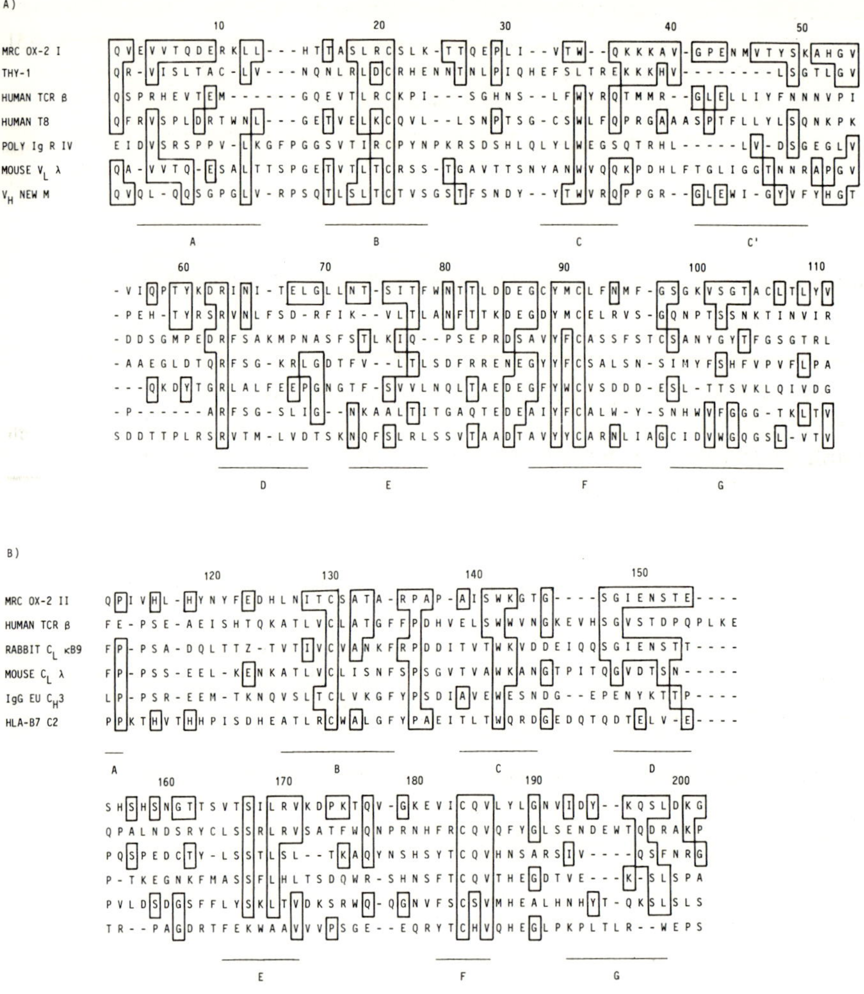

Fig. 3. *Alignment of OX-2 sequence with immunoglobulin domains*

(A) Alignment of OX-2 domain I (residues 1–111) with V-like domains: rat Thy-1 [26], human T cell receptor (TCR) β-chain V-like domain [27], human T8 (CD8) antigen [28], poly Ig receptor [15], mouse $V_\lambda$ MOPC 104E and $V_H$ NEW M [29]. (B) Alignment of OX-2 domain (II) residues (112–202) with C-like domains: human T cell receptor β-chain C-like domain [27], rabbit κ B9, mouse λMOPC 315, IgG EU $C_H3$ and HLA-B7 C2 [29]. Residues identical with OX-2 are boxed. Dashes indicate where gaps have been introduced to maximize homology. The bars with letters below them indicate the residues that are involved in forming the β-strands of immunoglobulin domains (see [6] and [30]).

graphy with the MRC OX-2 monoclonal antibody and gel filtration. After reduction and alkylation, tryptic peptides were isolated by gel filtration and h.p.l.c. and sequenced. Using one of these sequences, an oligonucleotide probe 17 residues long was synthesized containing all 64 possible sequences. This was used to screen a rat thymocyte cDNA library and positive clones were isolated and the cDNA inserts sequenced. The protein sequence deduced from the thymocyte cDNA sequence indicated that the mature protein consisted of 248 amino acids (Fig. 2). Only one segment (residues 203–229) lacks charged amino acids and contains a high proportion of hydrophobic amino acids, and is likely to span the lipid bilayer. At the *C*-terminus of this stretch there are several basic residues, as commonly found at the cytoplasmic side of transmembrane segments. Thus residues 1–202 are presumably outside the membrane and 230–248 inside the membrane. The six potential *N*-linked glycosylation sites (Asn-Xaa-Thr/Ser) at residues 65, 73, 80, 127, 151 and 160 would be consistent with this and the observed percentage of carbohydrate [3].

## Homology of OX-2 Glycoprotein with other Members of the Ig Superfamily

Comparison of the sequence of OX-2 glycoprotein with members of the Ig superfamily shows that the *N*-terminal 111 amino acids (domain I) have most homologies with Ig V-domains and residues 112–202 with Ig C-domains (Fig. 3). In making these alignments the two cysteines that form the conserved disulphide bridge between the two $\beta$-sheets of all Ig folds are aligned, and then the Trp found almost invariably about 15 residues *C*-terminal to the first Cys. The V and C type domains are distinguished by extra sequence in the V-domain (strand C′ in Fig. 3*a*) and the presence of certain highly conserved characteristic residues, e.g. Arg (position 63) Asp (85) and Gly or Ala (87) in V-domains. These characteristic residues are present in the OX-2 sequence and the number of overall identities between the OX-2 domains and other members of the Ig superfamily are comparable with those obtained when the latter are compared amongst themselves (discussed in [6, 7].). Domain I of OX-2 may be slightly more homologous to Thy-1 (29 identities) than to the other members (18–30 identities in those compared in Fig. 3*a*), especially when one considers that only two Thy-1 sequence (mouse and rat) are available for comparison compared with the large numbers of immunoglobulin domains.

## Evolution of the Ig Superfamily

All the members of the Ig superfamily (Fig. 4) contain regions with homologies to either Ig V- or C-domains based on sequence comparisons as above. It seems likely that the Ig superfamily arose in evolution from a single primordial domain which by gene duplication and divergence gave first a light-chain-like structure containing both V and C-domains and then by further gene duplication gave the larger members, as suggested by Hill *et al.* [8]. Williams & Gagnon [5,9] have suggested that Thy-1 may be like the primordial Ig domain, as it is the only Ig domain existing as a single domain ($\beta_2$-microglobulin is associated with MHC Class I heavy chain) and although its sequence shows more homologies with V-domains it also has sequences characteristic of C-domains [7,10]. Thus gene duplication, limited divergence and deletion could give rise to both C and V-domains and hence the remainder of the superfamily. The finding that OX-2 antigen also has homology with Ig's and in particular has a structure like a single Ig light chain suggests that it is like the next stage proposed in the evolution of the Ig superfamily from the primordial domain [8].

The Thy-1 antigen is present in large amounts in brain in all species examined from chicken to man, but its expression in thymocytes varies, being present in

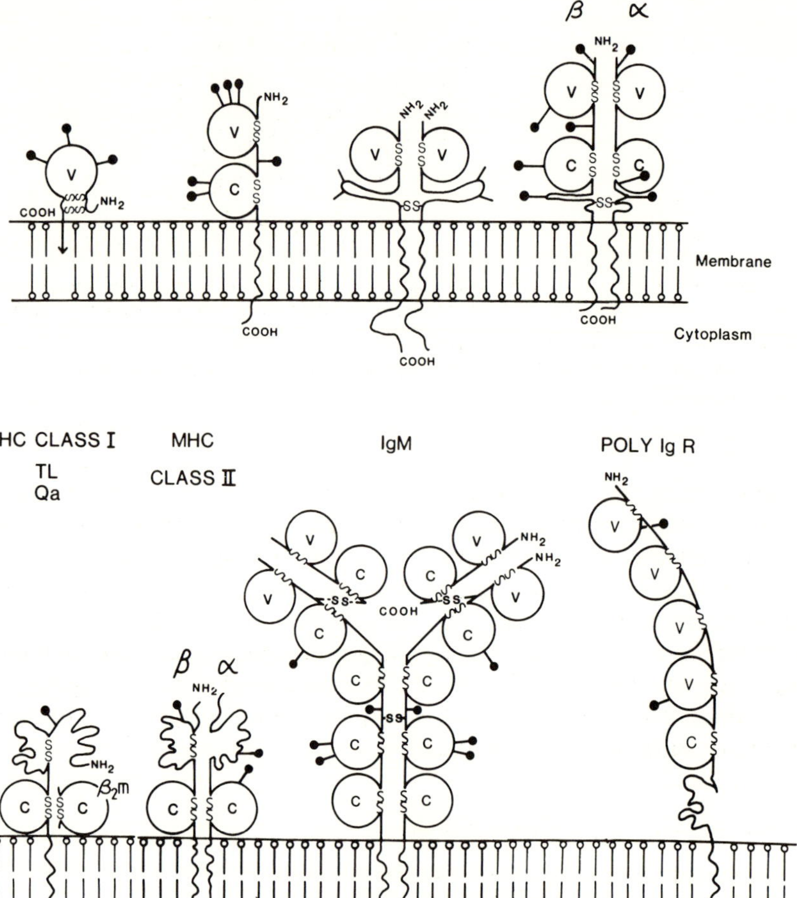


Fig. 4. *Models for molecules in the Ig superfamily*

Circles marked C are like Ig C-domains and those marked V are like Ig V-domains. Intrachain disulphide bonds are shown by symbols (§), interchain disulphide bonds by (S–S). *N*-linked carbohydrate structures by ( ♦ ) and possible location of *O*-linked carbohydrate structures on CD8 by (|). The Figure is modified from [6] and [31]. Data are from: Thy-1 [5], CD8 [28], T cell receptor (Tcr) [32, 33], MHC antigens [34, 35], IgM [36] and poly Ig receptor (poly IgR) [15].

rodents but not in man and only present on peripheral T-cells in mice [5]. Thus if Thy-1 is like the primordial Ig domain then presumably the immunoglobulin superfamily arose in the nervous system to mediate interactions at cell surfaces as discussed in [5], [7] and [9]. The search for Ig-related molecules in invertebrates has yielded a squid nervous system glycoprotein with some but not all the sequence characteristics of the Ig fold [5, 7] and a protein from tunicate haemocytes with antigenic cross-reactivity with Thy-1.1 and biochemical similarities, but which has yet to be sequenced [11]. OX-2 antigen is present in brain

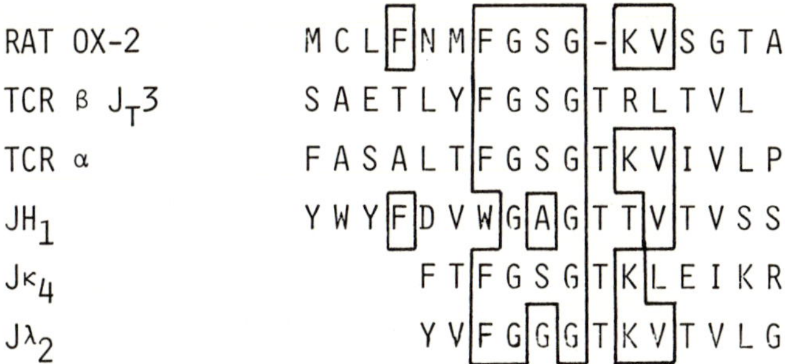


Fig. 5. *Alignment of C-terminal region of OX-2 domain I (residues 90–105) with typical J-regions from the mouse T-cell receptor α and β chains and heavy and light chains of mouse immunoglobulins*

Residues that are identical to OX-2 are boxed. A gap has been introduced in the OX-2 sequence to maximize homology. Data are from: T cell receptor (TCR) $\beta$ [13], TCR $\alpha$ [33], mouse $J_H1$ [37], $J_\kappa4$ [38] and $J_\lambda2$ [39].

and thus would fit with the hypothesis that the Ig superfamily arose to mediate recognition events in the nervous system, than other tissues, and was finally adapted in the immune system to recognize foreign antigens [7,9]. A key point is whether OX-2 is present in the nervous system of other species, but homologues have yet to be identified, although genes for OX-2 homologues are present (see below). Antigens of comparable size to OX-2 are found on human thymocytes, but given the differences in tissue distribution between species for Thy-1 (see above) this is not a reliable way to find the homologues, and of course there are no functional data for OX-2.

## Isolation of the Gene for Human OX-2

One characteristic of light-chain-like structures involved in antigen recognition on B-cells (Ig) and T cells (T cell receptor) is that the genes are organized in segments which are rearranged prior to transcription [12,13]. Thus a V region gene is spliced to a short J region segment in Ig light chains and to both D and J region segments in Ig heavy chains and the T cell receptor. In OX-2 there is a region towards the C-terminus of domain I (V-like domain) that has homology with J-region sequences of both Ig's and T cell receptors and includes the highly conserved F G X G sequence, as shown in Fig. 5. Thus the gene for human OX-2 was isolated to verify if there was a separate exon for the J-region and also to characterise the human homologue for OX-2. A rat cDNA probe was used to isolate a genomic clone for human OX-2 from a human genomic library prepared from peripheral blood DNA in λEMBL3, kindly provided by Colin Sharpe, Department of Chemical Pathology, University of Oxford (A. N. Barclay, M. J. Clark & G. W. McCaughan, unpublished work). Preliminary analysis indicates that a single exon codes for virtually all of the human OX-2 V-domain corresponding to residues 2–110 in the rat OX-2 sequence. There is approximately 80% homology between the rat and human sequences. This exon includes the region of homology with J-regions and ends with a typical intron splicing site, as shown in Fig. 6. Thus the human OX-2 gene contains the

```
Human OX-2   L  F  N  T  F  G  F  G  K  I  S  G  T  A  C  L  T  V  Y
             CTCTTCAATACCTTTGGTTTTGGGAAGATCTCAGGAACGGCCTGCCTCACCGTCTATGGTGAGAATCTC


Rat OX-2              M        S        V                             L
```

Fig. 6. *C-Terminal end of V region exon of human OX-2 gene*

Nucleotide sequence and translation of the human OX-2 region with homology with J-regions of immunoglobulins and T cell receptors. The rat OX-2 sequence is identical to the human sequence except at the four positions indicates. The presumed intron splice site 'GT' is underlined [40].

putative J-like sequence as part of the exon encoding the V-like domain. Analysis by Southern blots of rat thymus and liver DNA probed for OX-2 failed to show any evidence of rearrangement in thymus (results not shown). Thus it is unlikely that the DNA being examined in this genomic clone had already been rearranged (the human DNA had been prepared from peripheral blood).

The data on the human OX-2 gene are consistent with a scheme in which Thy-1 looks like the primordial Ig-domain which by duplication and some divergence and deletion gave rise to both V and C domains in the form of a single light-chain-like structure similar to that found today in OX-2. Further gene duplication could lead to the other members of the Ig superfamily. The final steps in the process being the production of separate exons for the J-regions — possibly by splitting off from V-domains of OX-2 to account for its homology — and then the D-regions and finally the mechanism for generation of somatic mutation for increasing the repertoire of Ig V-domain sequences.

**Functions of the Ig Superfamily**

Immunoglobulins, the T cell receptor and the MHC are all clearly involved in the process of recognition of antigen by B and T cells. The CD8 antigen may be involved in the recognition of MHC class I antigens during T cell recognition [14] and the poly Ig receptor is involved in the transport of Ig's [15], but what of OX-2 and Thy-1? There are no functional data on OX-2 and only limited data on Thy-1, namely that antibodies against Thy-1 can be mitogenic for mouse thymocytes [16,17], that antibodies injected into the hypothalamus can induce behavioural changes by inhibiting a carbamoyl chloride-induced drinking response [18] and when coated on glass enhanced the growth of neural processes [19]. Attempts to find receptors for Thy-1 by binding of radiolabelled Thy-1 to isolated lymphocytes or tissue sections were unsuccessful (A. N. Barclay, unpublished work). Thy-1 is very abundant ($10^6$ molecules/thymocyte) and has been estimated to cover about 25% of the surface of rat thymocytes [20] and about 5% of the surface of axons [21]. Thus low-affinity interactions which are undetectable by binding studies could still give potent effects due to the extreme multivalency. It seems possible that Thy-1, OX-2, and presumably the many Ig-related molecules yet to be characterized, could play a role in the interactions necessary in development in brain and other tissues.

## References


1. McMaster, W. R. & Williams, A. F. (1979) *Eur. J. Immunol.* **9**, 426–433
2. Barclay, A. N. (1981) *Immunology* **44**, 727–736
3. Barclay, A. N. & Ward, H. A. (1982) *Eur. J. Biochem.* **129**, 447–458
4. Cohen, F. E., Novotný, J., Sternberg, M. J. E., Campbell, D. G. & Williams, A. F. (1981) *Biochem. J.* **195**, 31–40
5. Williams, A. F. & Gagnon, J. (1982) *Science* **216**, 696–703
6. Clark, M. J., Gagnon, J., Williams, A. F. & Barclay, A. N. (1985) *EMBO J.* **4**,113–118
7. Williams, A. F., Barclay, A. N., Clark, M. & Gagnon, J. (1985) in *Proceedings of the Sigrid Juselius Symposium: Gene Expression during Normal and Malignant Differentiation* (Andersson, L. C., Gahmberg, C. G. & Ekblom, P., eds.), Academic Press
8. Hill, R. L., Delaney, R., Fellows, R. E. & Lebovitz, H. E. (1966) *Proc. Natl. Acad. Sci. U.S.A.* **56**, 1762–1769
9. Williams, A. F. (1982) *J. Theor. Biol.* **98**, 221–234
10. Campbell, D. G., Williams, A. F., Bayley, P. M. & Reid, K. B. M. (1979) *Nature (London)* **282**, 341–342
11. Mansour, M. H., De Lange, R. & Cooper, E. L. (1985) *J. Biol. Chem.* **260**, 2681–2686
12. Tonegawa, S. (1983) *Nature (London)* **302**, 575–581
13. Chien, Y.-H., Gascoigne, N. R. J., Kavaler, J., Lee, N. E. & Davis, M. M. (1984) *Nature (London)* **309**, 322–326
14. MacDonald, H. R., Glasebrook, A. L., Bron, C., Kelso, A. & Cerottini, J.-C. (1982) *Immunol. Rev.* **68**, 89–115
15. Mostov, K. E., Friedlander, M. & Blobel, G. (1984) *Nature (London)* **308**, 37–43
16. Maino, V. C., Norcross, M. A., Perkins, M. S. & Smith, R. T. (1981) *J. Immunol.* **126**, 1829–1836
17. Gunter, K. C., Malek, T. R. & Shevach, E. M. (1984) *J. Exp. Med.* **159**, 716–730
18. Williams, C. A., Barna, J. & Schupf, N. (1980) *Nature (London)* **283**, 82–84
19. Leifer, D., Lipton, S. A., Barnstable, C. J. & Masland, R. H. (1984) *Science* **224**, 303–306
20. Williams, A. F. & Barclay, A. N. (1985) in *Handbook of Experimental Immunology* (Weir, D. M. & Herzenberg, L. A., eds.), 4th edn., chapter 25, Blackwell Scientific Publications
21. Beech, J. N., Morris, R. J. & Raisman, G. (1983) *J. Neurochem.* **41**, 411–416
22. Booth, C. M., Brown, M. C., Keynes, R. J. & Barclay, A. N. (1984) *Brain Res.* **308**, 380–382
23. Webb, M. & Barclay, A. N. (1984) *J. Neurochem.* **43**, 1061–1067
24. Barclay, A. N., Letarte-Muirhead, M., Williams, A. F. & Faulkes, R. (1976) *Nature (London)* **263**, 563–567
25. Mason, D. W. & Williams, A. F. (1980) *Biochem. J.* **187**, 1–20
26. Campbell, D. G., Gagnon, J., Reid, K. B. M. & Williams, A. F. (1981) *Biochem. J.* **195**, 15–30
27. Yanagi, Y., Yoshikai, Y., Leggett, K., Clark, S. P., Aleksander, I. & Mak, T. W. (1984) *Nature (London)* **308**, 145–149
28. Littman, D. R., Thomas, Y., Maddon, P. J., Chess, L. & Axel, R. (1985) *Cell* **40**, 237–246
29. Kabat, E. A., Wu, T. T., Bilofsky, H., Reid-Miller, M. & Perry, H. (1983) *Sequences of Proteins of Immunological Interest*, U.S. Department of Health and Human Services, National Institutes of Health, Bethesda
30. Amzel, L. M. & Poljak, R. J. (1979) *Annu. Rev. Biochem.* **48**, 961–997
31. Williams, A. F. (1985) *Nature (London)* **314**, 579–580
32. Hedrick, S. M., Nielsen, E. A., Kavaler, J., Cohen, D. I. & Davis, M. M. (1984) *Nature (London)* **308**, 153–158
33. Saito, H., Kranz, D. M., Takagaki, Y., Hayday, A. C., Eisen, H. N. & Tonegawa, S. (1984) *Nature (London)* **312**, 36–40
34. Ploegh, H. L., Orr, H. T. & Strominger, J. L. (1981) *Cell* **24**, 287–299
35. Kaufman, J. F., Auffray, C., Korman, A. J., Shackelford, D. A. & Strominger, J. (1984) *Cell* **36**, 1–13
36. Kehry, M., Ewald, S., Douglas, R., Sibley, C., Raschke, W., Fambrough, D. & Hood, L. (1980) *Cell* **21**, 393–406
37. Sakano, H., Maki, R., Kurosawa, Y., Roeder, W. & Tonegawa, S. (1980) *Nature (London)* **286**, 676–683
38. Sakano, H., Hüppi, K., Heinrich, G. & Tonegawa, S. (1979) *Nature (London)* **280**, 288–294
39. Blomberg, B. & Tonegawa, S. (1982) *Proc. Natl. Acad. Sci. U.S.A.* **79**, 530–533
40. Breathnach, R. & Chambon, P. (1981) *Annu. Rev. Biochem.* **50**, 349–383

*Biochem. Soc. Symp.* **51**, 159–172
*Printed in Great Britain*

# $C_H$ Isotype Switching in B Cell Differentiation

JOHN J. CEBRA, ETHEL R. CEBRA, DEBRA B. KOTLOFF, DEBORAH A. LEBMAN,
PETER A. SCHWEITZER and ROBERTA D. SHAHIN

*Department of Biology, University of Pennsylvania, Philadelphia, PA 19104, U.S.A.*

## Synopsis

We have functionally defined a number of B cell subsets that likely represent
B cells at different stages of development, based on the pattern of $C_H$ isotypes
expressed by their clones in splenic fragment or microcultures and on those
factors necessary in culture to support the growth of a clone displaying a
particular isotype or set of isotypes. Our observations are consistent with isotype
switching being a stochastic process which results in the occurrence of progressive
isotype restriction in members of a diversifying clone. The surface marker best
predictive of the pattern of isotypes a clone may secrete is the sIg isotype of its
B cell precursor. Those B cells that have switched to the expression of non-IgM
isotypes *in vivo* can be stimulated *in vitro* in splenic fragments to give an
antibody-secreting clonal culture but so far cannot be stimulated in a microculture
of dispersed cells that supports clones secreting IgM alone or with other isotypes.

During the past 15 years, a large majority of cellular immunologists have
tended to use cultures *in vitro* of B lymphocytes as indicator systems to 'read-out'
and operationally define antigen-presenting cells, regulatory T lymphocytes, or
growth and differentiation factors in conditioned medium that could modulate
an antibody response (Sterzl & Riha, 1970; Yamamura & Tada, 1984). Our own
work has focused more on the B cell itself and has been directed towards defining
subsets of B cells based on their intrinsic potential to respond to antigen as
expressed by proliferation, maturation, and pattern of secreted antibody
isotypes displayed after clonal outgrowth. For purposes of this discussion we
propose that B cells can differentiate at two distinct levels: (*a*) with respect to
their $C_H$-isotype potential and its restriction, including commitment to expression
of a single isotype, and (*b*) with respect to their response pattern to a given set
of stimuli, including antigen/mitogen and factors responsible for activation,
growth, maturation and antibody secretion. Differentiation of B cells at these
two levels is likely to be mechanistically unrelated but may be temporally and
physiologically correlated. For instance, $V_H/D/J_H$-gene 'switching' to the
various $C_H$-gene segments may change $C_H$-gene content and context during the
genetic rearrangement and hence alter isotype potential, restriction, and
commitment. Response pattern may be regulated by the level of expression of
surface receptors for activation, growth, maturation, and secretion factors. Both
differentiation processes could be dependent on transits through the cell cycle

and thus be somewhat linked both temporally and physiologically. Certainly both differentiation processes act together to determine the full potential of a B cell: the size of the clone it generates, the burst size within the clone of antibody-secreting cells and the proportions of each $C_H$-isotype that contributes to the total antibody secreted.

Since we define B cell potential functionally as expressed over many (five to ten) generations we have relied on clonal cultures *in vitro* to be more certain of precursor–product cell relationships. The two clonal assays *in vitro* for B cells we have used are the splenic fragment culture method developed by Klinman (1972) and microculture (10 $\mu$l) of very small numbers of B cells enriched for a particular antigen specificity (Nossal & Pike, 1978). The first assay involves the adoptive transfer of limiting numbers of B cells into lethally-irradiated, 'carrier' primed recipients. Ideally, B cells of a particular specificity lodge in different regions of the recipient's spleen. At 18–24 h the spleens are removed, diced into small fragments and cultured with the relevant antigen determinant conjugated to the protein 'carrier' used to prime the recipient. This culture system has the disadvantages of complexity and low plating efficiency: a somewhat variable 4–10% of the inoculated B cells lodge in the spleen. However, it is the only system that suggests both clonal outgrowth and antibody secretion by B cells that have previously 'switched' *in vivo* from a productively rearranged $V_H/D/J_H/C\mu$ gene to a productively rearranged $C_H$-gene of another isotype. The second clonal assay, microculture, has the advantage of potentially giving a much higher plating efficiency (Pike & Nossal, 1985) and allowing better characterizations of 'filler', 'feeder', and 'helper' cells and of growth and differentiation factors that are necessary to reveal the potential of a B cell. In the following discussion we will present data from splenic fragment cultures that we have used to define operationally subsets of B cells based on either isotype pattern or response pattern displayed to a given set of stimuli. Then, we will present a progress report on our efforts better to define the requirements of these subsets of B cells in clonal microcultures.

One distinctive clonal phenotype defined in the splenic fragment culture is the exclusive secretion of IgM antibodies. Precursors that give rise to such clones tend to be prevalent in neonatal mice (Klinman & Press, 1975; Teale *et al.*, 1979; Shahin & Cebra, 1981) (see characteristics of B cells from 3-week-old mice, Table 1) and in spleens of formerly germ-free mice shortly (1–2 weeks) after their gut is deliberately colonized with intestinal commensal bacteria (Table 2). A second distinctive phenotype is displayed by clones that secrete IgM along with other isotypes. Intraclonal differentiation with respect to secretion of multiple isotypes along with IgM was originally demonstrated in splenic fragment culture (Gearhart *et al.*, 1975). The clonal precursors were found to bear surface IgM/IgD (sIgM/sIgD) and the intraclonal differentiation was found to involve both single step and successive 'switches' in the expression of IgM followed by IgG and/or IgA (Gearhart & Cebra, 1981; Gearhart *et al.*, 1980). Clonal precursors giving this pattern of isotype expression are the most prevalent sort in spleens of young adult mice that have been raised in a conventional environment (Gearhart & Cebra, 1979) and in formerly germ-free mice within a few weeks after their intestines have been colonized by normal, commensal

Table 1. *Neonatal mice have a low frequency of anti-inulin B cells by 3 weeks of age which is increased either by deliberate antigenic stimulation at birth or by natural priming at 3–5 weeks*

Neonatal mice were given 50–100 µg of inulin–haemocyanin (In-Hy) or bacterial levan (BL) intraperitoneally at 1–3 days of age for deliberate immunization. Inocula of $(20$–$40) \times 10^6$ nucleated cells from spleen (Spl.) or Peyer's patches (PP) were transferred to donors for splenic fragment assay.

| B cells | | Spl. | Spl. | Spl. | Spl. | PP | PP |
|---|---|---|---|---|---|---|---|
| | Source... | | | | | | |
| | Age (weeks)... | 3 | 5 | 3 | 3 | 3 | 5 |
| | Immunization... | – | – | InHy | BL | – | – |
| No. of clones analysed | | 32 | 58 | 52 | 54 | 4 | 19 |
| Frequency/$10^6$ | | 1.5 | 8.9 | 7.1 | 3.5 | 2.6 | 8.8 |
| Percentage of clones expressing: | | | | | | | |
|   IgM only | | 53 | 16 | 27 | 67 | 25 | 21 |
|   Some IgM | | 84 | 38 | 63 | 89 | 100 | 79 |
|   IgG $\pm$ IgA, no IgM | | 3 | 28 | 21 | 6 | 0 | 21 |
|   Some IgG3 | | 6 | 12 | 33 | 0 | 0 | 37 |
|   Some IgA | | 34 | 50 | 25 | 26 | 75 | 47 |
|   IgA only | | 13 | 21 | 10 | 4 | 0 | 0 |

Table 2. *Isotype patterns expressed by clones from anti-inulin specific B cells taken from formerly germ-free mice at varying times after colonization with gut bacteria*

Young adult mice were colonized for short times with the lactobacilli and group N streptococci (Lac./strep.) described by Dubos *et al.* (1965) and then some were given the remaining commensals described by these authors and then exposed to a conventional environment over a longer time (comm./conv.). Their splenic (Spl.) and Peyer's patch (PP) lymphocytes were analysed for inulin-specific clonal precursors in splenic fragment culture.

| Tissue source | Time after colonization (weeks) | Bacterial colonizers | No. of clones | Frequency/$10^6$ | Clones (%) making: IgM only | Some IgM | IgA only |
|---|---|---|---|---|---|---|---|
| Spl. | 1 | Lac./strep. | 10 | 7.5 | 70 | 100 | 0 |
| PP | 1 | Lac./strep. | 4 | 11 | 50 | 100 | 0 |
| Spl. | 3 | Lac./strep. | 44 | 39 | 9 | 100 | 0 |
| PP | 3 | Lac./strep. | 18 | 100 | 0 | 89 | 5 |
| Spl. | 18 | Comm/conv. | 45 | 41 | 0 | 18 | 29 |
| PP | 18 | Comm./conv. | 19 | 92 | 0 | 31 | 42 |
| Spl. | 52 | Comm./conv. | 47 | > 40 | 0 | 49 | 9 |
| PP | 52 | Comm./conv. | 17 | 96 | 0 | 12 | 47 |

bacteria (see Table 2). Generally, the non-IgM isotypes that are coexpressed with IgM by these clones appear to occur in random mixtures of any or all IgG and IgA isotypes, though IgE is rarely among the secreted antibodies (Cebra *et al.*, 1983; see Table 3). Table 3 also indicates that two or three isotypes are commonly expressed by these 'switching' clones in splenic fragment culture and, with the exception of clones exclusively secreting IgA, no obvious hierarchy for the expression of the non-IgM isotypes is apparent. Deliberate parenteral immunization and natural or deliberate infection usually perturbs antigen-specific thymus dependent (TD) B cells in two broad ways. Firstly, the frequencies of

Table 3. *Proportions of B cell clones of different specificities secreting each $C_H$-isotype and different multiplicities of isotypes*

Anti-phosphocholine (PC) and anti-inulin (In) B cells from formerly germ-free mice colonized with a 'cocktail' of normal intestinal bacterial (Dubos *et al.*, 1965) and then exposed to a conventional environment were analysed at 52 weeks in splenic fragment culture. Data from clones derived from all lymphoid tissues are pooled and presented both including and excluding (Excl.) the clones making exclusively IgA that are most prevalent in Peyer's patches.

| Specificity | No. of clones analysed | Proportion (%) of clones secreting: | | | | | | | Proportion (%) of clones secreting following number of isotopes: | | | | |
|---|---|---|---|---|---|---|---|---|---|---|---|---|---|
| | | M only | M | G3 | G1 | G2 | A | A only | 1 | 2 | 3 | 4 | 5 |
| Anti-PC | 190 | 3 | 38 | 27 | 17 | 10 | 90 | 43 | 48 | 31 | 12 | 7 | 2 |
| Anti-PC | 109 | 2 | 66 | 39 | 29 | 17 | 83 | Excl. | 12 | 51 | 21 | 12 | 4 |
| Anti-In | 213 | 5 | 31 | 25 | 39 | 43 | 71 | 23 | 36 | 27 | 23 | 11 | 3 |
| Anti-In | 164 | 3 | 40 | 33 | 51 | 56 | 63 | Excl. | 20 | 35 | 28 | 13 | 4 |

Table 4. *Perturbations in frequency and isotype potential of antigen-specific B cells upon deliberate immunization*

Mice were given three weekly intraperitoneal injections of 1 $\mu$g of *N*-acetyl-glucosamine (GlcNAc)-coupled goat IgG (a single intraduodenal injection of 10 $\mu$g of cholera toxin (Txn) or $10^4$ embryonated *Ascaris* eggs (*Asc.*) intragastrically. The mice receiving the *Asc.* were $F_1$ CBA/N × BALB/c males, which ordinarily do not carry detectable phosphocholine (PC)-specific B cells. These $F_1$ mice were either primed 4 weeks previously with 1 $\mu$g of PC-haemocyanin (PC-Hy) or not prior to receiving Asc. The specific B cells from spleen (Spl) and Peyer's patches (PP) were cultured in splenic fragments to generate clones. [The Table is a composite of data from Fuhrman & Cebra (1981), Clough & Cebra (1983) and Cebra *et al.* (1984).]

| | | | | | | | | | | |
|---|---|---|---|---|---|---|---|---|---|---|
| | | Immunological state of mice unprimed (Un.) or immunized *in vivo* with: | | | | | | | | |
| | Un. | GlcNAc-IgG | Un. | Txn | Un. | Txn | *Asc.* | PC-Hy, *Asc.* | *Asc.* | PC-Hy, *Asc.* |
| Tissue | Spl. | | Spl. | | PP | | Spl. | | PP | |
| Frequency/$10^6$ | 2 | 92 | 1–3 | 96 | 1–3 | 37 | 4 | 60 | 0.7 | 61 |
| No. of clones | 21 | 77 | 19 | 23 | 20 | 57 | 31 | 29 | 7 | 9 |
| Percentage of clones expressing: | | | | | | | | | | |
|   Some IgM | 100 | 64 | 58 | 30 | 75 | 19 | 61 | 48 | 62 | 44 |
|   IgG±IgA, no IgM | 0 | 32 | 36 | 52 | 10 | 42 | 26 | 48 | 43 | 22 |
|   Some IgA | 10 | 10 | 21 | 56 | 20 | 68 | 39 | 41 | 71 | 78 |
|   IgA only | 0 | 1 | 5 | 17 | 10 | 39 | 10 | 7 | 14 | 33 |
| Time of assay (weeks) | – | 4 | – | 2 | – | 2 | 4 | 4+3 | 4 | 4+12 |

TD antigen-specific B cells may arise 10–100-fold following immunization and, secondly, a significant proportion of these B cells develop into precursors of clones with a third phenotype, clones that secrete IgG isotypes±IgA in the absence of detectable IgM (Table 4). Such shifts also occur 'spontaneously' in young mice at times after birth related to a particular specificity of B cells (Sigel *et al.*, 1977; Shahin & Cebra, 1981). For instance, splenic and Peyer's patch B cells specific for the $\beta2\rightarrow1$ fructosyl determinant (inulin or In-determinant) display about a 6-fold 'spontaneous' rise in frequency along with a significant increase in their proportion that give TD clones secreting IgG isotypes±IgA without IgM (Table 1) in splenic fragment cultures. Quite possibly this expansion is driven by naturally colonizing gut bacteria, since it can be accomplished within 1–3 weeks in germ-free, young adult mice by feeding them the lactobacilli and milk streptococcii that ordinarily colonize neonatal mice shortly after their birth (Table 5). It is still not clear why neonatal mice take 3–5 weeks to display a functionally significant rise in In-specific B cells (Shahin & Cebra, 1981), 5–8 weeks to display a detectable, serum anti-In response upon parenteral challenge with bacterial levan (Bona *et al.*, 1978) and 7–8 weeks before they show a marked 'spontaneous' rise in 'natural' anti-In antibodies (Shahin, 1985), since potentially antigenic In-determinants accompany the earliest commensals to colonize newborn mice (Schaedler *et al.*, 1965). Probably these three, late-developing manifestations of responsiveness are interrelated. While trying to probe the bases for their delayed appearance we encountered another possible B cell clonal phenotype that is expressed in the form of a distinct response

Table 5. *Neonatal and young-adult, germ-free mice both show a prompt increase in frequency of inulin (In)-specific B cells after parenteral immunization with inulin-haemocyanin but neonatal mice fail to express a serum antibody response*

Young adult, germ-free mice and 3-day-old neonates were given 100 $\mu$g of inulin–haemocyanin intraperitoneally. Frequencies of splenic anti-In B cells from immunized mice and unprimed controls were assayed in splenic fragment culture and serum anti-In antibodies were determined by radioimmunoassay.

| Donor of B cells | Immune status | Time after priming (weeks) | Frequency of anti-In B cells/$10^6$ | Serum anti-In ($\mu$g/ml) |
|---|---|---|---|---|
| Neonate | Immunized | 1 | $\leqslant 0.3$ | 0.32 |
| Neonate | Control | 1 | 9.4 | 0.41 |
| Neonate | Immunized | 3 | 1.5 | 3.1 |
| Neonate | Control | 3 | 7.1 | 2.3 |
| Germ-free | Immunized | 1 | 8.6 | 2.9 |
| Germ-free | Control | 1 | 48 | 630 |

pattern – an increase in frequency of TD, antigen-specific B cells in the absence of antibody secretion. We wondered whether the late appearance of competence to make an anti-In response was due to the delayed rearrangement of $V_H$-, D-, $J_H$- gene segments necessary to produce antibodies with effective binding sites for the In-determinant. Thus, we deliberately immunized 1–3-day-old mice parenterally with a variety of molecular forms of $\beta 2 \rightarrow 1$ fructosyl: bacterial levan, In-haemocyanin, In-lipopolysaccharide, or *Morganella morganii* vaccine. In all cases the neonates displayed a prompt rise in TD, In-specific B cells within 1–3 weeks (Tables 1 and 5), but in the absence of any rise in serum antibodies (Table 5). In contrast, germ-free, young adult mice responded with *both* a prompt rise in In-specific clonal precursors and with a substantial rise in serum anti-In antibodies within 1–3 weeks after either parenteral immunization with In-haemocyanin (Table 5) or colonization with Gram-positive, natural commensals (Table 2). We have called this response pattern *in vivo* of neonatal mice 'silent priming' – the sharp increase in TD antigen-reactive B cells in the absence of a serum antibody response. The clonal phenotype of B cells responsible for 'silent priming' may not be restricted to B cell populations in neonatal mice. In fact, we have found a similar phenotype expressed by B cells of mice with an x-linked immunodeficiency (*xid* trait) found in the CBA/N substrain (Ohriner & Cebra, 1985). These mice, homozygous females (*xid/xid*) or hemizygous males (*xid/*Y), appear not to respond to a variety of bacterial or synthetic polysaccharides or to immunodominant antigenic determinants of these attached to TD protein carrier, since they express no significant rise in circulating antibody concentrations upon immunization with such antigens (Amsbaugh *et al.*, 1972; Mond *et al.*, 1977). Nevertheless, we have found an antigen-dependent response in mice expressing the *xid* trait upon stimulating them with the apparently non-immunogenic polysaccharides or their immuno-dominant determinants (Clough *et al.*, 1981; Clough & Cebra, 1983; Ohriner & Cebra, 1985). We have detected their response in three forms: (*a*) 'priming' as revealed by an increase in TD, specific clonal precursors which seem to provide the basis for a quantitatively and qualitatively 'normal' secondary response compared with phenotypically normal *xid*/X littermates; (*b*) the development of successful fusion partners for the SP2 cell line to give hybridomas secreting specific antibodies of the sort expressed by normal mice during a primary response to polysaccharide antigens (IgM and IgG3); and (*c*) the appearance of non-secretory, antigen-binding plasmablasts in the spleen. These manifestations of 'silent priming' of mice expressing the *xid* trait may be analogous to 'silent priming' of neonatal mice. Perhaps both sorts of mice display a pre-eminence of a subset of B cells that occurs as an important component of the B cell compartment of all normal mice. These B cells may proliferate extensively upon antigen/growth factor stimulation yet contribute little to a primary antibody response. Instead, such B cells may be important in generating 'memory' cells for secondary responses.

The final clonal phenotype of B cells we would like to propose is that of exclusive secretion of a single isotype other than IgM. By far the most common B cell of this sort is one that produces clones that secrete exclusively IgA upon stimulation with thymus-dependent antigens (Table 3). About 15 years ago, Susan Craig in our laboratory discovered that the Peyer's patches of the small intestines are enriched sources of precursors for IgA-producing plasmablasts (Craig & Cebra, 1971). Afterwards, Patricia Gearhart found a cellular correlate of this potential, that a high proportion of B cells of some specificities obtained from Peyer's patches generate clones in splenic fragment culture that exclusively secrete IgA (Gearhart & Cebra, 1979). We followed upon these observations by demonstrating that both chronic gut mucosal stimulation by either intestinal commensal bacteria (Cebra *et al.*, 1980) or nematode pathogens (Clough & Cebra, 1983) or acute gut mucosal application of certain non-living antigens such as cholera toxin (Fuhrman & Cebra, 1981) both increases the frequencies of specific B cells in all lymphoid tissue and results in the development of high proportions of IgA-committed cells, most prominently in Peyer's patches (see Tables 2 and 4). We do not yet understand the basis for IgA commitment of B cells, though we have hypothesized that these may be cells that have lost all other $C_H$-genes on the chromosome carrying the productive $C_H$-locus except for the 3' terminal $C_\alpha$-gene (Cebra *et al.*, 1982). Mostly, we have found few B cells of any specificity that generate clones exclusively secreting a single non-IgM isotype other than IgA (Gearhart & Cebra 1981; Gearhart *et al.*, 1980; Fuhrman & Cebra, 1981; Clough & Cebra, 1983). A conspicuous exception is our finding that a very high proportion of the TD, antigen-specific B cells that rise in frequency in all lymphoid tissue following chronic respiratory infection with *Mycoplasma pulmonis* give rise to clones *in vitro* that exclusively secrete IgG1 (Rose & Cebra, 1985*a*). These observations suggest there may be a genetic mechanism for 'fixing' isotype expression *in vivo* during the antigen-driven differentiation of B cells without impairing their ability to respond again to antigen with clonal outgrowth and antibody secretion.

So far, the functionally defined clonal phenotypes we have used to identify B cell subsets with different potentials have been correlated with no surface markers except for sIg isotype itself. For instance, use of fluorescence-activated cell sorting (FACS) B cells from hyperimmunized mice and assay of phospho-choline (PC)-specific clonal precursors indicates that the precursors of clones

Table 6. *Surface immunoglobulin (sIg) phenotype correlates with isotype potential of Peyer's patch B cells*

The plasma membranes of viable Peyer's patch cells were exposed to various fluorochrome-labelled antibodies specific for Ig-isotypes and the stained ($+$ve) and unstained ($-$ve) subpopulations were resolved by fluorescence-activated cell sorting. The $+$ve, $-$ve and unstained, unsorted populations were assayed for the isotype potential of either phosphocholine (PC) or inulin (In)-specific clonal precursors in splenic fragment culture. The data for Expt. 1 are from Gearhart & Cebra (1981).

| | Expt. 1 (anti-PC) | | | Expts. 2 and 3 (anti-In) | | |
|---|---|---|---|---|---|---|
| | Unstained, unsorted | sIgM, sIgD $+$ve | sIgM, sIgD $-$ve | Unstained, unsorted | sIgA $+$ve | sIgA $-$ve |
| Total clones | 20 | 20 | 5 | 39 | 13 | 13 |
| Isotype expressed (%): | | | | | | |
|   IgM only | 5 | 20 | 0 | 0 | 0 | 0 |
|   IgM$+$IgG | 18 | 10 | 0 | 0 | 0 | 0 |
|   IgM$+$IgA | 5 | 10 | 20 | 0 | 0 | 0 |
|   IgM, IgG, & IgA | 27 | 40 | 0 | 0 | 0 | 8 |
|   IgG only | 0 | 5 | 20 | 0 | 0 | 0 |
|   IgG$+$IgA | 9 | 10 | 0 | 28 | 0 | 23 |
|   IgA only | 36 | 5 | 60 | 60 | 100 | 61 |

expressing IgG isotypes $\pm$ IgA in the absence of IgM appear to mostly bear sIgG (Gearhart & Cebra, 1981). Similarly, FACS resolution of B cells from Peyer's patches indicates that clones from sIgA-bearing cells secrete exclusively IgA (Table 6). Separate FACS resolution based on surface content of other isotypes indicates that such precursors are mostly sIgM/sIgD-negative while B cells of the same specificity that bear sIgM and sIgD are enriched for those precursors that generate clones expressing IgM alone or IgM along with other isotypes (Gearhart & Cebra, 1981). Recently, we have also used FACS resolution of sIgG1-bearing B cells to enrich dramatically B cells from mice chronically infected with *M. pulmonis* in favor of precursors that generate clones making anti-(*M. pulmonis*) antibodies (Rose & Cebra, 1985*b*). Clones from B cells taken from such mice have previously been found mostly to secrete exclusively IgG1 anti-(*M. pulmonis*) antibodies. Thus, all these observations support a process of *in vivo*, antigen-driven clonal expansion that results in $C_H$-isotype 'switching' and in the generation of memory cells – cells capable of another round of antigen-driven clonal expansion – that are restricted in isotype potential or committed to expression of a particular isotype.

In order to define better those subsets of B cells that we have distinguished on a functional basis, and to establish the stimuli required for their activation, growth, maturation, and antibody secretion, we have been attempting to grow their B cell clones in microculture. The principles of microculturing we sought to apply to study the subsets of B cells defined functionally in splenic fragment culture were learned from Drs. Beverley Pike and Gustav Nossal while we were guests at the Walter and Eliza Hall Institute in Melbourne. The basic approach was to enrich for antigen-specific B cells by panning cells from spleen or Peyer's patches on thin layers of haptenated gelatin in the cold (Nossal & Pike, 1978). The fluorescein (FLU)– or PC–gelatin adherent cells were almost a pure B cell population enriched in specifically responding cells about 200-fold. Ordinarily, about 2–15 of these enriched B cells were cultured for 5–10 days in 10 $\mu$l of medium containing antigen, often with 5–10% by volume of conditioned medium from the EL4 lymphoma line stimulated by concanavalin A (Pike *et al.*, 1982). Cultures were scored for proliferation by microscopic examination and for antibody secretion by radioimmunoassay (RIA) (Hurwitz *et al.*, 1982). Ordinarily, cultures were fed with an additional 5 $\mu$l of medium the day before harvesting supernatant and then 12 $\mu$l of each culture fluid was diluted with 50 or 100 $\mu$l of 1.0% bovine serum albumin in phosphate-buffered saline. As many as ten separate assays for total secreted antibody, isotype, idiotype, and avidity could be carried out by radioimmunoassay on supernatants of clones that made 1.25–250 ng of total antibody. Table 7 shows some typical findings of cultures stimulated by specific hapten conjugated to lipopolysaccharide (LPS) or polymerized flagellin in the presence of EL4 conditioned medium. Proliferation frequencies of about 0.5–5.0% were observed and about 60–80% of these statistically clonal cultures secreted detectable antibodies. Virtually all of the positive cultures made IgM antibodies and only IgM. For PC-specific clones 70–90% made IgM with an avidity $\geq$ that of monoclonal IgM of T15 idiotype as estimated by hapten inhibition, supporting the specificity of enrichment of B cells on PC–gelatin plates. In order to provide a more supportive microculture we tried a variety of 'feeder' cells, including X-irradiated (3000 R) and mitomycin-C-treated spleen and thymus cells, but the most successful 'feeder', tried at the suggestion of Dr. B. Pike, were $10^5$ non-irradiated thymocytes from CBA/N $\times$ BALB/c F$_1$ male (*xid*/Y) mice. Table 7 shows that about 15% of the antibody-secreting clones express non-IgM isotypes along with IgM. This effect on expression of the 'switch' is usually accompanied by a 1.5–3-fold increase in frequency of secreting clones and IgM may be co-expressed with any and all isotypes (IgE was not assayed) although IgG2 expression was relatively rare and the majority of 'switched' clones only made one isotype in addition to IgM.

We wondered whether clones proliferating in the absence of 'feeder' cells and only secreting IgM were able to undergo $V_H/D/J_H$-gene 'switching' during the five to eight generations of observable growth. The effect of the 'feeder' cells to promote clonal secretion of isotypes other than IgM could be on B cells either prior to or after the rearrangement at the $C_H$-locus necessary for the expression of 'switching'. Recovery of all or part of individual clones as adherent balls of cells and fixation followed by staining for cytoplasmic IgM indicated intraclonal diversification: only some cells, usually smaller blast cells, showed detectable intracellular IgM while other, larger cells did not stain more than thymocytes (Schweitzer & Cebra, 1985). Subsequent staining of unfixed B cells, often in clusters, for sIg of non-IgM isotypes indicated significant fractions bore sIgG or sIgA isotypes, usually together with IgM (Cebra *et al.*, 1984*b*). Due to the controversy over whether expression of sIg of isotypes other than IgM/IgD correlates with 'switching' at the DNA level (Katona *et al.*, 1985), we are seeking independent evidence for isotype 'switching' within these clones in the absence of secretion of non-IgM isotypes. Meanwhile we have been seeking culture conditions more supportive of secretion of non-IgM isotypes and which might permit growth and antibody expression by B cell subsets displaying isotype

Table 7. *Summary of characteristics of anti-fluorescein (FLU) and anti-phosphocholine (PC) clones from B cells enriched on haptenated-gelatin and grown in microcultures*

B cells enriched on either FLU–gelatin or PC-gelatin were cultured with either the appropriate haptenated lipopolysaccharide (0.1 $\mu$g/ml) or haptenated polymerized flagellin (0.2 $\mu$g/ml) in medium containing 7.5% EL4 conditioned medium (Pike *et al.*, 1982) and in the presence or absence of $10^5$ thymocytes from $F_1$ male CBA/N × BALB/c mice. The cell input and proliferation frequency were determined microscopically. The culture supernatants were assayed for detectable antibody ($\bar{a}$ Fab+), IgM antibody ($\bar{a}$ $\mu$+), and non-IgM antibodies with a mixture of anti-IgG1, IgG2, IgG3, and IgA ($\bar{a}$ non-$\mu$+) by radioimmunoassay.

| | | | Analysis of microcultures after time indicated: | | | | |
| | | | − Thymocytes | | | + Thymocytes | |
| | | | Days 5–6 | | | Days 5–6 | |
| Specificity | Cell input at day 1 | Proliferation (%) at days 3–4 | $\bar{a}$ Fab+ (no. of clones) | $\bar{a}$ Fab+ / proliferation+ (%) | $\bar{a}$ non-$\mu$+ / $\bar{a}$ $\mu$+ (%) | $\bar{a}$ Fab+ (no. of clones) | $\bar{a}$ non-$\mu$+ / $\bar{a}$ $\mu$+ (%) |
|---|---|---|---|---|---|---|---|
| Anti-FLU | 2–16 | 1.5–1.5 | 244 | 58 | 1.5 | 238 | 16 |
| Anti-PC | 2.5–13 | 0.6–3.4 | 34 | 77 | 0 | 60 | 17 |

Table 8. *Cells from helper T cell lines ($T_H$) enhance both the frequency and 'switching' to expression of non-IgM isotypes of haptenated-gelatin enriched, antigen-stimulated B cells in microculture in a dose-dependent manner*

About 10 B cells enriched on phosphocholine (PC)–gelatin were cultured in 10 $\mu$l with $10^5$ thymocytes (Thy) from $F_1$ male CBA/N × BALB/c mice together with varying doses of $T_H$ cells. The cultures were stimulated with PC-lipopolysaccharide (PC-LPS, 0.1 $\mu$g/ml), PC-haemocyanin (PC-Hy, 0.05 $\mu$g/ml), or without antigen. Antibodies were measured in day 8 culture supernatants by radioimmunoassay.

| 'Feeder' and helper cells... | Frequencies (%) | | | | |
|---|---|---|---|---|---|
| | None | Thy | $0.75 \times 10^3$ $T_H$ + Thy | $1.5 \times 10^3$ $T_H$ + Thy | $3.0 \times 10^3$ $T_H$ + Thy |
| Antigens: | | | | | |
| None | 0.1 | 0.3 | 1.6 | 0.8 | 1.4 |
| PC-Hy | 0.8 | 0.3 | 1.1 | 1.9 | 2.5 |
| PC-LPS | 0.8 | 2.9 | 2.2 | 4.4 | 7.0 |
| | 'Switching' of all clones (%) (no. switched/total) | | | | |
| All clones | – | – | 17 (6/35) | 23 (11/47) | 57 (34/60) |

restriction or non-IgM isotype commitment in the splenic fragment cultures. Since clonal responses by these latter B cell subsets have only been detected in TD cultures, we have generated T-helper ($T_H$) cell lines specifically dependent on haemocyanin for growth, by using established procedures (Kimoto & Fathman, 1980). Such $T_H$- lines were found by ourselves and others (Cebra *et al.*, 1984a; Teale, 1983) to substitute for 'carrier' priming of recipients for the splenic fragment assay. Cotransfer of such $T_H$-cells plus limiting numbers of hapten-specific B cells allowed the growth and antibody secretion of TD clones, including those from Peyer's patch precursors that exclusively secreted IgA (Cebra *et al.* 1984a). Thus, we added these $T_H$-cells ($1.5 \times 10^3$–$4.5 \times 10^3$) to microcultures containing 5–20 hapten–gelatin-enriched B cells. Usually, the effect of the $T_H$-cells was most pronounced when added together with $10^5$ *xid*/Y thymocytes rather than with X-irradiated splenocytes or alone. Generally, 20–70% of the resulting clones expressed isotype switching, most of the 'switched' clones also secreted IgM, and the frequency of responding cultures increased with dose of $T_H$ cells. Though the response was dependent on antigen corresponding to the B cell specificity, it did not need to carry determinants relevant to the $T_H$-cell specificity. Table 8 illustrates many of these points and Table 9 shows the distribution of number and kinds of isotypes among the clones derived from this same experiment. Further experiments indicated that the $T_H$-lines, either 'resting' or undergoing antigen-stimulated blastogenesis continued to produce lymphokines in microculture at sufficient concentration to increase the frequency, antibody output, and expression of non-IgM isotypes of clones from activated B cells. It also became obvious that few 'switched' clones generated in this system expressed IgG and/or IgA isotypes in the absence of IgM. Thus, it was necessary to determine whether the enrichment procedure for B cells that adhere to haptenated gelatin also enriched for those specific B cells that were restricted or committed to the generation of clones in splenic fragment cultures that only secreted IgG and/or IgA isotypes. Table 10 shows

Table 9. *Isotype diversity of antibodies secreted by anti-phosphocholine (PC) clones grown in microculture with 'feeder' cells and cells of helper T cell lines*

About 10 B cells enriched on PC–gelatin were stimulated in culture with relevant antigens in the presence of $10^5$ thymocytes from $F_1$ male CBA/N × BALB/c mice and $(0.75–3.0) × 10^3$ helper T cells. Isotypes of secreted antibodies were determined by radioimmunoassay.

| All clones (172, 100%) | | | 'Switched' clones (63, 100%) | | |
|---|---|---|---|---|---|
| Secreting: | No. | % | Secreting: | No. | % |
| M only | 109 | 63 | G1+M | 28 | 44 |
| Non-M | 63 | 37 | A+M | 13 | 21 |
| One isotype | 119 | 69 | A only | 9 | 14 |
| Two isotypes | 45 | 26 | G1+G3+M | 3 | 5 |
| Three isotypes | 4 | 2 | G3+M | 3 | 5 |
| Four isotypes | 4 | 2 | A+G1+G3+M | 3 | 5 |
| M only, high avidity | 51 | 57 | G1 only | 1 | 1.5 |
| M only, low avidity | 38 | 43 | A+G1+M | 1 | 1.5 |
| Non-M, high avidity | 34 | 100 | G2b+M | 1 | 1.5 |
| Non-M, low avidity | 0 | 0 | A+G1+G2b+G3 | 1 | 1.5 |

Table 10. *Comparison of clones from either Peyer's patch (PP) or splenic (Spl) B cells enriched on phosphocholine (PC)–gelatin and grown in splenic fragment cultures*

The PC-enriched B cells $(6 × 10^3–12 × 10^3)$ were inoculated into haemocyanin-primed, lethally irradiated recipients. The resulting splenic fragments were cultured with PC–haemocyanin.

| No. of B cells inoculated... | | | |
|---|---|---|---|
| B cell source... | $6 × 10^3$ PP | $12 × 10^3$ Spl. | $6 × 10^3$ Spl. |
| +ve/no. of cultures | 52/72 | 27/136 | 17/120 |
| Frequency (%) | 20 | 2.8 | 3.5 |
| Percentage of clones expressing: | | | |
| IgM only | 0 | | 2 |
| 'Switch | 100 | | 98 |
| M+A | 18 | | 77 |
| M+G | 0 | | 0 |
| M+G+A | 18 | | 4 |
| Some G | 18 | | 9 |
| IgA only | 64 | | 11 |

that both Peyer's patch and splenic B cells enriched by binding to PC–gelatin give statistically clonal cultures in the splenic fragment assay when inoculated in very low numbers $(6 × 10^3–12 × 10^3)$ compared with the numbers of unfractionated B cells ordinarily needed $(6 × 10^6–12 × 10^6)$. If a normal lodging frequency in the spleen (4%) is supposed, then frequencies of responding B cells ranged from 2.8–20%. Since the frequency of PC-specific B cells in unfractionated B cells responding in this assay is about $5/10^5$, the enrichment was at least 1000-fold, in agreement with the relative inoculum sizes needed for clonal cultures. The PC-enriched B cells from both spleen and Peyer's patches gave both a higher frequency of responding B cells and a higher proportion of 'switched' clones in splenic fragment cultures than in microcultures. Further, the isotype patterns expressed by the clones grown in splenic fragments from PC-enriched cells from spleen and Peyer's patches were very typical of those shown by unfractionated B cells from these tissues taken from mice not deliberately immunized but raised in a conventional environment. Particularly PC-enriched B cells from Peyer's patches gave a majority of clones that exclusively secreted IgA (Table 10). Thus, haptenated-gelatin fractionation also enriched for specific B cells that respond in the splenic fragment assay and does not seem to distort the isotype potential of the population. Presently, we are attempting to discover the conditions in microculture necessary for stimulating clonal growth and secretion of those B cells restricted or committed to non-IgM isotypes. Thus far, we have established microculture conditions sufficient for vigorous clonal growth and secretion of IgM by some hapten-enriched B cells. At least some of these clones undergo isotype 'switching' without secretion of non-IgM isotypes. Addition of 'feeder' cells together with cells of a $T_H$-line both increases the frequency of responding cells and the proportion of their clones that secrete non-IgM isotypes. These non-IgM isotypes appear to be expressed in random mixtures together with IgM. The stimulation by cells of $T_H$-lines appears not to require linkage of T- and B-cell specific determinants but rather to be via 'bystander' effects of T-cell lymphokines on antigen-stimulated B cells.

Our working postulates for trying to stimulate the subsets of B cells restricted or committed to non-IgM isotypes in microculture are presently: (*a*) that these may require stimulation via linked recognition; (*b*) that $T_H$-clones may vary in their constitutive versus antigen/I-A stimulated production of lymphokines and thus be more informative than $T_H$-lines in deducing T cell requirements of B cell subsets; and (*c*) that B cells of some subsets may require consecutive action of several T cells at different stages of clonal outgrowth in order to reveal their full isotype potential via secreted antibody. This latter possibility may be supported by the successful establishment of FcR$\alpha$-bearing cells with $T_H$-phenotype from Peyer's patches and their effect on sIgA-bearing B cells in bulk culture to increase the burst size of IgA-secreting plaque-forming cells (Kiyono *et al.*, 1982). Presently we are collaboratively testing such T cells in our microculture along with the other cellular components we have so far used. We conclude by suggesting that studies of B cells in clonal culture have been and will continue to be informative about the intrinsic potential of B cells at different developmental stages with respect to their response patterns to various stimuli and their capacity to display the various $C_H$-isotypes of antibody.

## References

Amsbaugh, D. F., Hansen, C. T., Prescott, B., Stashak, P. W., Barthold, D. R. & Baker, P. J. (1972) *J. Exp. Med.* **136**, 931–949

Bona, C., Lieberman, R., Chien, C. C., Mond, J., House, S., Green, I. & Paul, W. E. (1978) *J. Immunol.* **120**, 1436–1442

Cebra, J. J., Gearhart, P. J., Halsey, J. F., Hurwitz, J. L. & Shahin R. D. (1980) *J. Reticuloendothel. Soc.* **28**, 61s–71s

Cebra, J. J., Fuhrman, J. A., Gearhart, P. J., Hurwitz, J. L. & Shahin, R. D. (1982) in *Recent Advances in Mucosal Immunity* (Strober, W., Hanson, L. A. & Sell, K. W., eds.), pp. 155–171, Raven Press, New York

Cebra, J. J., Cebra, E. R., Clough, E. R., Fuhrman, J. A., Komisar, J. K., Schweitzer, P. A. & Shahin, R. D. (1983) *Ann. N.Y. Acad. Sci.* **409**, 25–38

Cebra, J. J., Fuhrman, J. A., Griffin, P., Rose, F. V., Schweitzer, P. A. & Zimmerman, D. (1984*a*) *Ann. Allergy* **53**, 541–549

Cebra, J. J., Komisar, J. L. & Schweitzer, P. A. (1984*b*) *Annu. Rev. Immunol.* **2**, 493–548

Clough, E. R. & Cebra, J. J. (1983) *Mol. Immunol.* **20**, 903–915

Clough, E. R., Levy, D. A. & Cebra, J. J. (1981) *J. Immunol.* **126**, 387–389

Craig, S. W. & Cebra, J. J. (1971) *J. Exp. Med.* **134**, 188–200

Dubos, R., Schaedler, R. W., Costello, R. & Hoet, P. (1965) *J. Exp. Med.* **122**, 67–75

Fuhrman, J. A. & Cebra, J. J. (1981) *J. Exp. Med.* **153**, 534–544

Gearhart, P. J. & Cebra, J. J. (1979) *J. Exp. Med.* **149**, 216–227

Gearhart, P. J. & Cebra, J. J. (1981) *J. Immunol.* **127**, 1030–1034

Gearhart, P. J., Sigel, N. H. & Klinman, N. R. (1975) *Proc. Natl. Acad. Sci. U.S.A.* **72**, 1707–1711

Gearhart, P. J., Hurwitz, J. L. & Cebra, J. J. (1980) *Proc. Natl. Acad. Sci. U.S.A.* **77**, 5424–5428

Hurwitz, J. L., Tagart, V. B., Schweitzer, P. A. & Cebra, J. J. (1982) *Eur. J. Immunol.* **12**, 342–348

Katona, I. L., Urban, J. F., Jr. & Finkelman, F. D. (1985) *Proc. Natl. Acad. Sci. U.S.A.* **82**, 511–515

Kimoto, M. & Fathman, C. G. (1980) *J. Exp. Med.* **152**, 759–770

Kiyono, H., McGhee, J. R., Mosteller, L. M., Eldridge, J. H., Koopman, W. J., Kearney, J. F. & Michalek, S. M. (1982) *J. Exp. Med.* **156**, 1115–1130

Klinman, N. R. (1972) *J. Exp. Med.* **136**, 241–260

Klinman, N. R. & Press, J. L. (1975) *J. Exp. Med.* **141**, 1133–1146

Mond, J. J., Lieberman, R., Inman, J. K., Mosier, D. E. & Paul, W. E. (1977) *J. Exp. Med.* **146**, 1138–1141

Nossal, G. J. V. & Pike, B. L. (1978) *J. Immunol.* **120**, 145–150

Ohriner, W. D. & Cebra, J. J. (1985) *Eur. J. Immunol.*, **15**, 906–915

Pike, B. L. & Nossal, G. J. V. (1985) *Proc. Natl. Acad. Sci. U.S.A.* **82**, 3395–3399

Pike, B. L., Vaux, D. L., Clark-Lewis, I., Schrader, J. W. & Nossal, G. J. V. (1982) *Proc. Natl. Acad. Sci. U.S.A.* **79**, 6350–6354

Rose, F. V. & Cebra, J. J. (1985*a*) *Infect. Immun.* **49**, 428–434

Rose, F. V. & Cebra, J. J. (1985*b*) *Fed. Proc. Fed. Am. Soc. Exp. Biol.* **44**, 978

Schaedler, R. W., Dubos, R. & Costello, R. (1965) *J. Exp. Med.* **122**, 59–66

Schweitzer, P. A. & Cebra, J. J. (1985) *Fed. Proc. Fed. Am. Soc. Exp. Biol.* **44**, 1868

Shahin, R. D. (1985) Thesis, Johns Hopkins University, Baltimore

Shahin, R. D. & Cebra, J. J. (1981) *Infec. Immun.* **32**, 211–21

Sigel, N. H., Pickard, A. R., Metcalf, E. S., Gearhart, P. J. & Klinman, N. R. (1977) *J. Exp. Med.* **146**, 933–948

Sterzl, J. & Riha, I. (eds.) (1970) *Developmental Aspects of Antibody Formation and Structure*, vols. I and II, Academia Publishing House, Prague

Teale, J. M. (1983) *J. Immunol.* **131**, 2170–2177

Teale, J. M., Layton, J. E. & Nossal, G. J. V. (1979) *J. Exp. Med.* **150**, 205–217

Yamamura, Y. & Tada, T. (eds.) (1984) *Progress in Immunology*, vol. 5, Academic Press, New York

*Biochem. Soc. Symp.* **51**, 173–182
*Printed in Great Britain*

# Molecular Events during the Onset and Maturation of the Antibody Response

CESAR MILSTEIN, JOS EVEN and CLAUDIA BEREK

*Medical Research Council Laboratory of Molecular Biology, Hills Road, Cambridge
CB2 2QH, U.K.*

## Introduction

The paper describing the model of antibodies consisting of a pair of light and heavy chains (Porter, 1962) signalled the birth of Molecular Immunology. The stage was set for the next big question, which was a deceptively simple one. Is antibody specificity generated by differences in the amino acid sequences of the antibodies? Since the answer was yes, then what was the genetic mechanism which was capable of such remarkable sequence diversity? The progress since then has been phenomenal (e.g. Leder *et al.*, 1980; Tonegawa, 1983). We are still struggling with important details but basically the diversity of antibody structure is generated at four different levels (Table 1).

To start with, the light and heavy chain variable regions are encoded in two and three segments respectively and the multiplicity of such segments allows a considerable diversity in the germ line. The combinatorial association between the different segments allows a large number of antibody structures to be formed which will be greatly increased by flexibility during the joining of the DNA segments [combinatorial and junctional diversity (Fig. 1)]. How many antibody molecules can these processes alone – in the absence of somatic point mutations – generate?

## The Size of the Potential Antibody Repertoire

The total number of V segment germ line genes is still unknown. In the old days we used to have long and heated arguments about this point and everyone attempted his own calculations.

We have recently analysed the $V_\kappa$-Ox gene family, namely those genes which cross-hybridize with a probe ($V_\kappa$-Ox1) which codes for the light chain of the anti-oxazolone idiotype response. We have isolated and sequenced 16 germ line genes (Even *et al.*, 1985). Among them we found three repeats (i.e. identical sequences present in two isolates coming from separate gene libraries). In addition two of the sequences were the same as sequences previously described while we failed to find one previously described sequence sufficiently similar to cross-hybridize. The DNA and mRNA sequences define a minimum of 20, and a statistical analysis of the data tells us that the total size is most likely below 50, genes. These values agree with previous studies indicating a total of between 100 and 300 mouse $V_\kappa$-genes (Cohn *et al.*, 1980; Cory *et al.*, 1981).

Table 1. *Mechanisms that generate antibody diversity*

---

1. Germline: multipe V-gene segments
2. Combinatorial: (*a*) different combinations of V-(D)-J
   (*b*) different combinations of $V_H$ and $V_L$
3. Junctional: variation at V-J, V-D, and D-J boundaries
4. Somatic point mutation: nucleotide substitutions throughout the V region

---

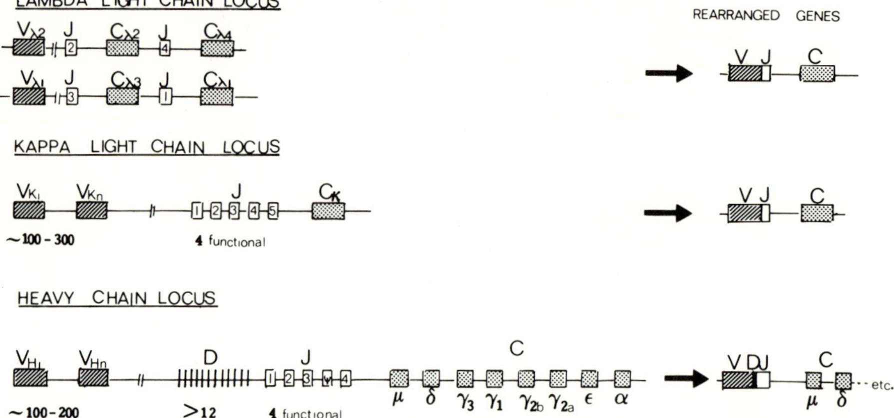


Fig. 1. *The organization of antibody genes in the mouse*

Antibody gene fragments are present in chromosomes 6 ($\kappa$), 16 ($\lambda$) and 12 (H). Germ line gene fragments are assembled in pre-B cells and each committed B cell expresses one specific combination. Considerable diversity is generated by combinatorial assembly of fragments and by flexibility in the joining process.

Let us then take the average value of 200 V genes for each chain. Random assortment with four J segments and 20 D segments gives a value of 800 $\kappa$ and 16000 heavy chains. The number of residues which can be generated by junctional diversity in $V_\kappa$-$J_\kappa$, $V_H$-D and D-$J_H$ will be taken as 2.5 for each pair. In addition there are very common size differences in the use of D and the use of $J_H$ (Kaartinen *et al.*, 1983*b*) and if we assume a factor of 5 (probably an under-estimate) we come to a value of $800 \times 2.5 = 2000$ light chains and $16000 \times 2.5 \times 2.5 \times 5 = 500000$ heavy chains. The potential combination is thus $2000 \times 5000000 = 1 \times 10^9$. To this we must add the diversity introduced by other probable processes during the joining of fragments (e.g. N-region) (Alt & Baltimore, 1982).

## Virgin B Cell Repertoire and the Early Primary Response to Oxazolone

Although there may be some restrictions to the random nature of the processes discussed above, there are factors which are probably underestimated (particularly in the D and N segments). Therefore the potential number of structures which virgin B cells can express as receptor for antigen is larger than the total number of B cell clones. At first sight, one could easily conclude that there is no need for further generation of diversity by somatic mutation.

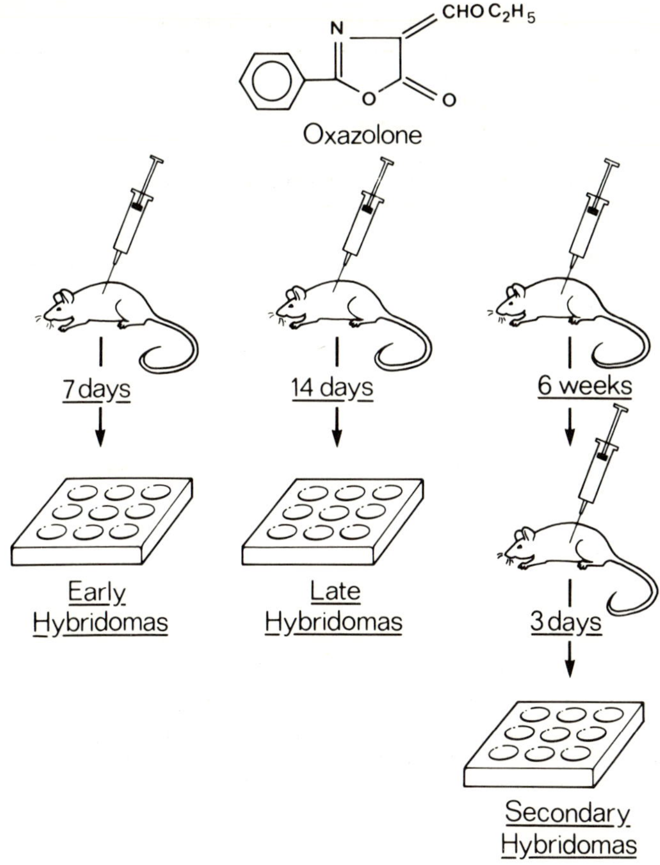



Fig. 2. *Experimental manipulations*

Animals were immunized with an intraperitoneal injection of phenyloxazolone coupled to chicken serum albumin. Hybridomas were prepared at different stages after immunization. The mRNA sequences from which heavy and light chain amino acid sequences were deduced were determined as described by Kaartinen *et al.* (1983*a,b*) and Griffiths & Milstein (1985).

This is far from the truth. Such enormous diversity is likely to be capable of producing structures which somehow can recognize any potential antigenic stimulus. But the whole point of the immune response, what makes the immune system such a unique biological phenomenon, is its exquisite specificity, which resides in high affinity of ligand recognition. Such high affinity goes far beyond what the virgin B cell repertoire can offer. Let me expand the point using as an example the response to the hapten phenyloxazolone, which we have been studying in considerable detail. The experimental set-up is described in Fig. 2.

The primary antibody response to oxazolone is a good example of the dominance of certain idiotypic specificities (Kaartinen *et al.*, 1983*a*). Namely in the early primary responses, a large number of isolated hybridomas give rise to identical or almost identical antibodies (Fig. 3). These express the same $V_\kappa Ox1\text{-}J_\kappa 5$ and $V_H Ox1\text{-}J_H 3$ germline gene combinations, the main exception being the frequency usage of alternative $J_H 4$ and $J_H 2$. Otherwise the diversity

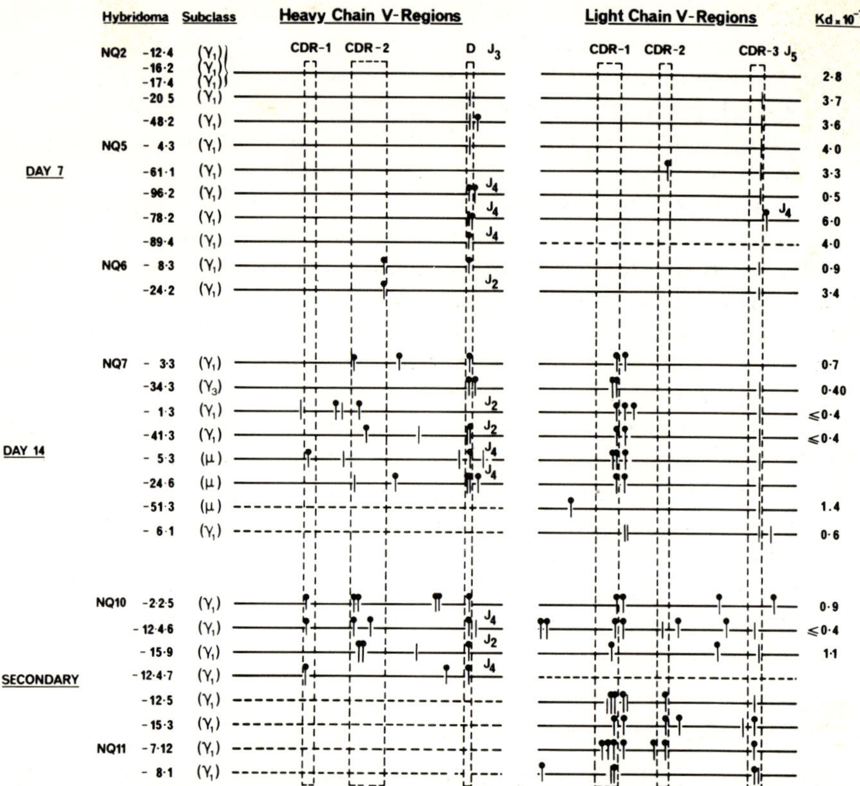

Fig. 3. *Diagrammatic comparison of selected mRNA sequences from anti-phenyloxazolone secreting hybridomas derived at different stages after immunization with phenyloxazolone–albumin*

Only sequences considered to be derived from the $V_\kappa$-Ox1 or $V_H$-Ox1 germ line genes are included. Unbroken horizontal lines denote sequence identity, broken lines represent extensive sequence differences. A vertical bar shows the relative position of codons containing nucleotide differences. A black circle indicates that these changes predict an amino acid difference at this position. Complementarity determining regions (CDR-1, -2, -3) have been marked, as have the D and J regions. Where different J segments are oberved these are indicated. Dissociation constants ($K_d$) were determined by fluorescent quenching. For further analysis using equilibrium dialysis see Fig. 5. NQ5 is from DBA mice. All others are from BALB/c. Taken from Kaartinen *et al.* (1983*a,b*) and Griffiths *et al.* (1984).

is almost totally circumscribed to point mutations in the D segment and the boundaries between segments. Almost all the diversity at this stage seems to be generated during the somatic recombination events which are needed to express the 'virgin' antibody receptor repertoire.

Junctional diversity plays an important role in the overall diversity (Weigert *et al.*, 1980). In the anti-oxazolone response, the most commonly expressed light chain derived from a germ line gene $V_\kappa$-Ox1 joined to a $J_\kappa 5$ fragment. $V_\kappa$-Ox1-J5 boundary residue is either G (germ line of $J_\kappa 5$) or A (germ line of $V_\kappa$-Ox1) in a ratio of 16/9. The joining is therefore ambiguous at this point. This ambiguity is still visible at neighbouring residues, although its frequency seems to be much lower. For instance, NQ5-78.2.2 is one of the few examples where $V_\kappa$OX1 is joined to a $J_\kappa 4$. The first base of codon 96 of NQ5-78.2.6 is a C which differs from the germ line residue of J4 and gives rise to the amino acid leucine instead of the germ line phenylalanine (Kaartinen *et al.*, 1983*b*). C is the corresponding residue of the germ line gene $V_\kappa$-Ox1 recently sequenced. Out of 27 examples where $V_\kappa$-Ox1 germ line gene is expressed (both primary and secondary anti-oxazolone antibodies) all except NQ5-78.2.6 use J5, presumably because it is the only $J_\kappa$ fragment which codes for leucine at residue 96. In the anti-arsonate responses the situation is reversed. The light chain $V_\kappa$-J1 boundary residue 96 of anti-arsonate antibodies is arginine, which most likely arises by junctional diversity. The reason for this strict preference is that the germ line residue of $J_\kappa 1$ is tyrosine which destroys antibody binding, even though it retains the idiotypic specificity (Jeske *et al.*, 1984). The conclusion is that, regardless of whether such junctional diversity is randomized or not (Cohn *et al.*, 1980), it is sufficiently common to be used in primary responses.

Monoclonal antibodies from the early response probably represent good examples of the natural antibody pool. Typically they display affinity constants of about $3 \times 10^{-7}$ mol/litre. Primary responses to other haptens are generally of the same or lower quality. Of course, we know that good antibodies are much better. Otherwise, immunoassays would not have the sensitivity which have made them such powerful analytical tools. Higher affinity antibodies arise by hyperimmunization. This increase in binding affinity is what we refer to as the maturation of the response.


## Somatic Mutation and the Maturation of the Response

Late primary and secondary response antibodies to oxazolone steadily increase in affinity and easily reach dissociation constants of $10^{-8}-10^{-9}$ mol/litre (Fig. 4). Among these higher affinity antibodies there are those which differ from the antibodies of the primary response by a few point mutations. There are good reasons to say that these are somatic mutants generated and selected following antigenic stimulation.

Early response antibodies express primarily unmutated germ line V-genes. I have already argued that the total number of potential antibody molecules in the absence of somatic mutations was larger than the diversity that an individual mouse could express at any given moment. It is therefore unlikely that most of the somatic mutants which appeared on the late primary response (day 14) were present at very low frequency at day 0 and expanded following oxazolone stimulation (Griffiths *et al.*, 1984). This may be so in isolated cases, but is not likely to be general, as forcefully emphasized by the secondary response antibodies, which show a further increase of somatic mutations. (Fig. 4).

The detailed analysis of the sequences of the variants shows that the changes are not randomly distributed. Not surprisingly, they cluster in the complementary regions (also referred to as hypervariable segments). But this is not as general as once thought. Cohn *et al.* (1980) suggested that a single substitution in the framework segments was indicative of a different germ line gene. Our results demonstrate that somatic mutants in framework segments (some even in the J fragment) are not rare in the secondary reponse.

The amino acid substitutions and base changes are sometimes highly concen-

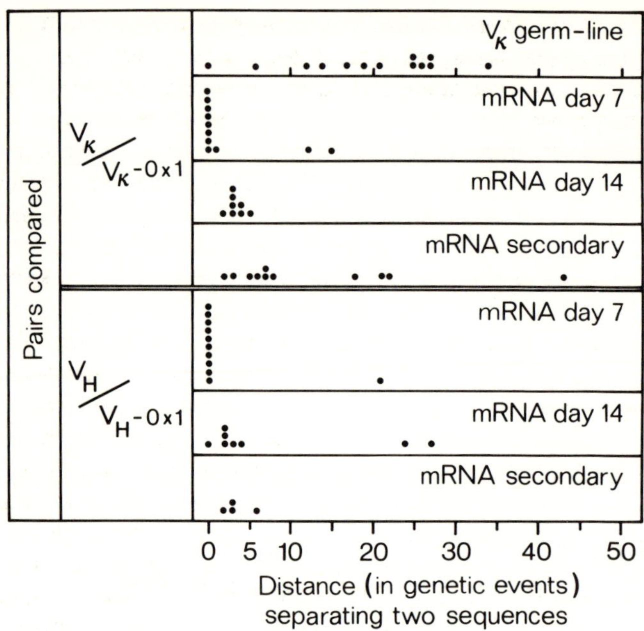


Fig. 4. *Diversity of the V$_\kappa$-Ox family of germ line genes and of mRNAs*

Fourteen germ line genes which cross-hybridize to V$_\kappa$-Ox1 are compared to V$_\kappa$-Ox1, and individual hybridoma heavy and light chain mRNA isolated from day 6, day 14 and secondary response are similarly compared to the germ line sequences. Note that, as the response matures, a group of hybridomas diverges from V$_\kappa$-Ox1 germ line gene sequence. In addition, at all stages of the response there are individual hybridomas which diverge more drastically: these are likely to derive from different germ line genes.

trated at certain positions, particularly residues 34 and 36 of the V$_\kappa$-Ox1. Thus while the germ line gene and all mRNA of the 7-day response code for histidine and tyrosine respectively, in the 15 examples of V$_\kappa$-Ox1 antibodies later in the response, His-34 is generally substituted by glutamine or asparagine (ratio 3:4:8). Tyr-36, retained in five examples (one of which is an altered codon), is substituted by phenylalanine in nine cases and histidine in one case. We have previously discussed why these changes were likely to be somatic mutants (Griffiths *et al.*, 1984). Furthermore, in an analysis of 14 different germ line genes homologous to V$_\kappa$Ox1, none had the sequence His-Trp-Phe, and only one (a pseudogene with a chain termination signal at residue 34) contained a codon for phenylalanine at position 36. Indeed, the pattern of residues 34 and 36 (Table 2) is fully compatible with a high rate of somatic mutation and strong selection at those specific positions.

Based on the crystallographic analysis of other immunoglobulins (Poljak *et al.*, 1975; Clothia *et al.*, 1985) it is very likely that His-34 of the V$_\kappa$-Ox1 chain is at the boundary between the framework and the antigen binding surface area. It probably has direct effects on the folding of the binding site loop. A single base change of the codon for His-34 can give rise to seven different amino acids. Only two of them (glutamine and aspargine) are small and neutral. It is possible that the mutation of histidine to glutamine or asparagine is strongly selected

Table 2. *Residues 34 and 36 in $V_\kappa$-Ox1 variants*

The data are from Griffiths *et al.* (1984).

| Residue number | | Number of antibodies | | |
|---|---|---|---|---|
| 34 | 36 | Day 7 | Day 14 | Secondary |
| His† | Tyr† | 11 | 1+1* | 1 |
| Asn | Tyr | 0 | 1 | 1 |
| Asn | His | 0 | 1 | 0 |
| Gln | Phe | 0 | 2 | 3 |
| Asn | Phe | 0 | 2 | 3 |

* Differ in codon
† Germ line sequence

because it increases the affinity for antigen. Residue 36 is a highly conserved tyrosine or phenylalanine in all light chains (Kabat *et al.*, 1983), buried in the interior of the molecule and thought to be generally essential for H–L chain interaction (Poljak *et al.*, 1975; Clothia *et al.*, 1985). The mutation is therefore not likely to affect the antibody–ligand interaction directly. It is possible however that it has an indirect effect by slightly altering the relative orientations of the combining site of heavy and light chains. These speculations will have to wait for further crystallographic studies at present being carried out in collaboration with Poljak's group (Mariuzza *et al.*, 1985). In any event, it may not be just coincidence that the nucleotide sequence around residues 36 contains a 10-base palindrome:

$$
\begin{array}{cccc}
 & 3\ 5 & 3\ 6 & 3\ 7 \\
 & T\ r\ p & T\ y\ r & G\ l\ n \\
C & T\ G\ G & T\ A\ C & C\ A\ G \\
G & A\ C\ C & A\ T\ G & G\ T\ C
\end{array}
$$

Although the involved amino acids are highly conserved among all light chains, the palindrome itself does not seen so, because of changes in codons (Kabat *et al.*, 1983). It remains nevertheless possible that such a palindrome (and possibly other local DNA sequence peculiarities in other segments) favour mutational errors. A mutation within the palindrome would lead to a decrease in mutational frequency.

That somatic mutants arise as a consequence of antigenic stimulation should not be taken as equivalent to a mutagenic role of antigen. Much more likely is a selective role of antigen on cell proliferation. This selective advantage must be correlated in some way with the energy of interaction between antigen and the antibody which is acting as a receptor.

## Alternative Germ Line Gene Combinations in High-Affinity Antibodies

This takes us to the second, most remarkable property of the maturation of the response. As the response matures, the proportion of highest affinity antibodies begins to shift in favour of gene combinations which differ from the

Table 3. *The variable region segments of high-affinity anti-oxazolone antibodies*

| | Number of antibodies | | | | | |
| | $V_\kappa$-Ox1 | | $V_\kappa$-Ox1 | | Other $V_\kappa$ | |
| Light chain... | $V_H$-Ox1 | | Other $V_H$ | | Other $V_H$ | |
| Heavy chain... | | | | | | |
| $K_d$ ($\times 10^{-8}$ M)... | 20–4 | < 4 | 20–4 | < 4 | 20–4 | < 4 |
|---|---|---|---|---|---|---|
| Day 7 | 2* | 0 | 0 | 0 | 0 | 0 |
| Day 14 | 2 | 2 | 2 | 0 | 1 | 0 |
| Secondary | 2 | 1 | 2 | 2 | 2 | 2 |

* The $K_d$ of most of day 7 antibodies including the $V_\kappa/V_H$-Ox1 prototype was $> 3 \times 10^{-7}$ M.

Table 4. *$V_\kappa/V_H$ combinations found in phenyloxazolone-specific hybridoma lines*

The data are taken from Berek *et al.* (1985)

| | $V\kappa$-Ox | | $V_\kappa$-Ars | $V_\kappa$-45.1 |
| $V_H$‡ | Ox1 | Other | | |
|---|---|---|---|---|
| Group 2<br>($V_H$-Ox1) | *<br>NQ10/2.2.5($\gamma$1)<br><br>12.4.6($\gamma$1)<br>15.9($\gamma$1)<br>NQ11/14.5($\mu$) | NQ2/45.10.4($\mu$)† | NQ5/89.4($\gamma$1)†<br>NQ10/12.4.7($\gamma$1) | |
| Group 1 (J558) | NQ11/7.12($\gamma$1) | NQ2/6.1($\gamma$1)†<br>NQ11/1.18($\gamma$1) | NQ7/7.1($\mu$)† | NQ2/45.1($\gamma$1)† |
| Group 3 | | NQ10/2.12.8($\mu$)<br>/4.6.1($\gamma$1) | | |
| Group 5<br>(MOPC21) | NQ10/12.5($\gamma$1)<br>15.3($\gamma$1)<br>NQ11/8.1($\gamma$1) | NQ10/11.1($\gamma$1) | | NQ10/4.6.2($\gamma$1) |
| Group 6 (J606) | | | | NQ10/2.12.4($\mu$) |
| Group 7 (TEPC15) | | | | NQ7/47.1($\gamma$1)†<br>NQ10/1.12($\gamma$1)<br>2.3($\gamma$1)<br>2.22($\gamma$1)<br>3.8($\mu$)<br>12.22($\gamma$1)<br>NQ11/5.2($\gamma$2a)<br>7.22($\gamma$1)<br>8.4($\mu$) |

* This refers to eleven hybridomas from day 7 and eight from day 14 response having the $V_H$Ox-$V_\kappa$-4Ox1 combination (Kaartinen *et al.*, 1983*a,b*; Griffiths *et al.*, 1984).

† Hybridoma lines NQ2 and NQ5 are derived from the day 7 and hybridoma lines NQ7 from the day 14 response. They are shown in italics. Their sequences are published (Kaartinen *et al.*, 1983*a,b*; Griffiths *et al.*, 1984; Berek *et al.*, 1985).

‡ According to Dildrop (1984).

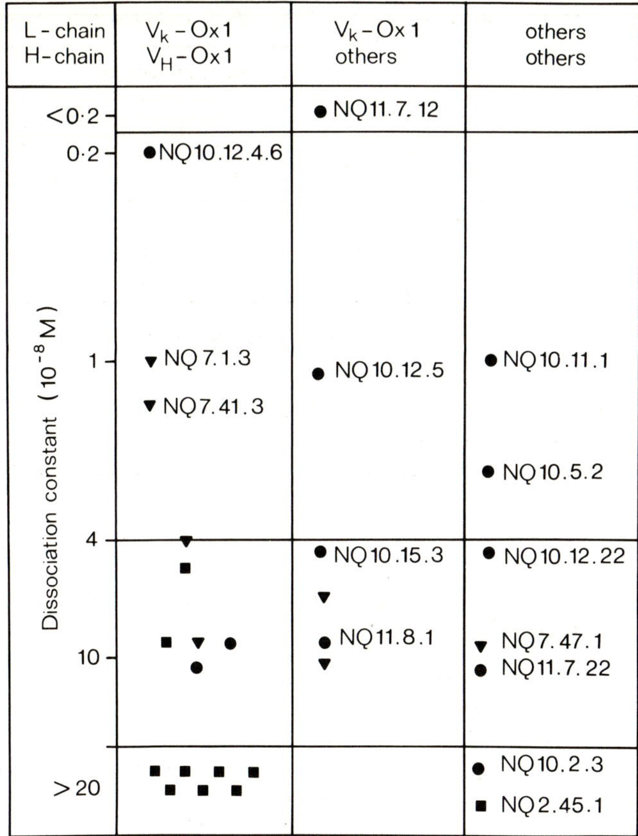


Fig. 5. *Dissociation constants of phenyloxazolone-specific hybridoma proteins*

Dissociation constants of hybridoma lines obtained at three different stages of the immune response to phenyloxazolone [day 7 (■), day 14 (▼), and secondary (●) response] are shown. Data are plotted according to the $V_H/V_\kappa$ combination of the antibody molecules. Symbols with no designation show results from Fig. 3. Affinity measurements were done by the method of fluorescence quenching with the hapten phOx-caproate. Antibodies with $K_d$ lower than $4 \times 10^{-8}$ M were further analysed by the method of equilibrium dialysis, using [³H]phenyloxazolone-aminobutyrate. Taken from Berek *et al.* (1985).

V-Ox1 pair (Tables 3 and 4 and Fig. 5). Even so, the best secondary response antibodies express certain characteristics of the primary response. For instance, NQ11/7.12, although expressing a $V_H$ segment quite different from $V_H$-Ox1 and homologous to J558 (group 1) has a D-J segment with the characteristic length repeatedly associated with the $V_H$-Ox1 of day 7 hybridomas.

The $V_H$-Ox1/$V_\kappa$-Ox1 gene pair, although good to start with, may become increasingly difficult to mutate towards further improvement of binding. Other combinations, which to start with display lower affinity or even do not interact at all with the antigen, may be capable of becoming high-affinity antibodies, perhaps by a single amino acid substitution. Such a somatic mutant is probably not present at the time of antigenic stimulation, but has a certain probability of appearing. To increase that probability, the essential step is expansion of clones from a much larger reservoir than those producing the best affinity antibodies. Indeed, our colleagues from Finland find a larger proportion of non-idiotypic anti-oxazolone antibodies at day 7 when the screening method includes lower affinity antibodies (J. L. T. Pelkonen, M. Kaartinen & O. Mäkelä, personal communication). It is also relevant that during an antibody response there is usually a general increase in immunoglobulin production which is not fully accounted for by specific antibody production.

In conclusion then, it appears that the initial selection of antigen-sensitive clones expressing the initially most suitable germline gene combination is based both on frequency and affinity of the individual combinations. This is followed by two events. One, selection of somatic mutants of the proliferating clones expressing those genes. The other, the active recruitment of other germline gene combinations having a potential to be modified by somatic mutation into the very high affinity antibodies which characterize the mature response. Our experiments point to a very important role of the selective process but do not allow us to say whether the shift to other germ line genes is selected by antigen alone or whether there is some other mechanism which favours it. The first to come to mind is anti-idiotypic response and network interactions, but other factors such as separate compartmentalization of memory and virgin B cells, with a concomitant major role of lymphocyte circulation, deserve careful attention.

## References

Alt, F. W. & Baltimore, D. (1982) *Proc. Natl. Acad. Sci. U.S.A.* **79**, 4118–4122
Berek, C., Griffiths, G. M. & Milstein, C. (1985) *Nature (London)* **316**, 412–418
Clothia, C., Novotny, J., Bruccoleri, R. & Karplus, M. (1985) *J. Mol. Biol.* **185**, 651–663
Cohn, M., Langman, R. & Geckeler, W. (1980) *Progr. Immunol.* **4**, 153–201
Cory, S., Tyler, B. M. & Adams, J. M. (1981) *J. Mol. Appl. Genet.* **1**, 103–116
Dildrop, R. (1984) *Immunol. Today* **5**, 100–101
Even, J., Griffith, G. M., Berek, C. & Milstein, C. (1985) *EMBO J.* **4**, 3439–3445
Griffiths, G. M. & Milstein, C. (1985) in *Hybridoma Technology in Biotechnology and Medicine* (Springer, T., ed.), pp. 103–115, Plenum, New York
Griffiths, G. M., Berek, C., Kaartinen, M. & Milstein, C. (1984) *Nature (London)* **312**, 1090–1092
Jeske, D. J., Jarvis, J. M., Milstein, C. & Capra, J. D. (1984) *J. Immunol.* **133**, 1090–1092
Kaartinen, M., Griffiths, G. M., Hamlyn, P. H., Markham, A. F., Karjalainen, K., Pelkonen, J. L. T., Mäkelä, O. & Milstein, C. (1983a) *J. Immunol.* **130**, 937–945
Kaartinen, M., Griffiths, G. M., Markham, A. F. & Milstein, C. (1983b) *Nature (London)* **304**, 320–324
Kabat, E. A., Wu, T. T., Bilofsky, H., Reid-Miller, M. & Perry, H. (1983) *Sequences of Proteins of Immunological Interest*, NIH Publications, Bethesda, MD
Leder, P., Max, E. E. & Seidman, J. G. (1980) *Progr. Immunol.* **4**, 39–54
Mariuzza, R. A., Boulot, G., Guillon, V., Poljak, R. J., Berek, C., Jarvis, J. M. & Milstein, C. (1985) *J. Biol. Chem.* **260**, 10268–10270
Poljak, R. J., Amzel, L. M., Chen, B. L., Phizackerley, R. P. & Saul, F. (1975) *Immunogenetics* **2**, 393–394
Porter, R. R. (1962) in *Basic Problems of Neoplastic Disease* (Gellhorn, A. & Hirschberg, E., eds.), pp. 177, Columbia University Press, New York
Tonegawa, S. (1983) *Nature (London)* **302**, 571–581
Weigert, M., Perry, R. & Kelley, D. (1980) *Nature (London)* **283**, 497–499

*Biochem. Soc. Symp.* **51**, 183–196
*Printed in Great Britain*

# Activation of Oncogenes by Transposable Elements

DAVID GIVOL

*Department of Chemical Immunology, Weizmann Institute of Science, Rehovot 76100, Israel*

## Synopsis

Mammalian DNA contains several families of highly repeated sequences, some of which have been suggested to be mobile elements. We have screened tumour tissue for the rearrangement of cellular oncogenes and found evidence for the behaviour of repetitive DNA sequences as transposable elements which may activate oncogenes. In the mouse myeloma NSI and XRPC24 we found that intracisternal A particle genome was inserted into the coding region of c-*mos*. In both cases the rearranged c-*mos* was transcriptionally activated and was also able to transform NIH 3T3 cells. In the canine transmissible venereal tumour we found that c-*myc* was rearranged due to the insertion of an 1.8 kilobase pair cellular DNA. Nucleotide sequence analysis demonstrated that the inserted piece is 60% homologous to the monkey *Kpn*I element which is a representative of the LINE group.

## Introduction

During the last several years, a group of cellular genes, termed oncogenes, has been implicated in the spontaneous development of tumours [1–3]. These genes, of which nearly 20 have been identified so far, are evolutionarily conserved [4,5] and are the progenitors for the transforming genes within the genomes of RNA tumour viruses [6]. The oncogenes can be activated and trigger or enhance a malignant process by one of several ways: (*a*) integration of an exogenous retrovirus genome near a cellular oncogene resulting in increased transcription of the latter [7,8]; (*b*) generation of a point mutation and subsequent production of an altered gene product [9,10]; (*c*) amplification of an oncogene and consequent increased expression [11,12]; (*d*) translocation of the oncogene to another chromosome [13–15]; or insertion of another piece of cellular DNA near it [16–18], resulting in both cases in enhanced expression of the oncogene.

The cellular DNA of all high organisms contain repeated elements with the structure of retroviral genomes [18,19]. Those were termed endogenous viruses and may or may not be expressed. The mouse genome contains several such families including endogenous intracisternal A-particle (IAP) genes. These genes, which number about 1000 per haploid mouse genome [20], are expressed abundantly in every mouse plasmacytoma [21] and in other mouse tumours [22]. Transcripts of some of these genes are about 9 kb in size and are encapsidated to form noninfectious retroviral-like entities [23].

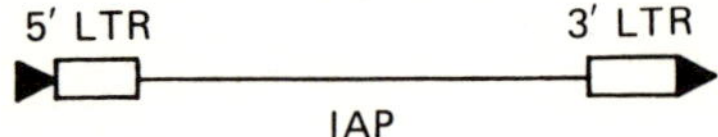

Fig. 1. *Characteristics of two families of mammalian repetitive DNA sequences*
Black triangles represent direct repeats.

Mammalian DNA also contains several families of highly repeated inter-dispersed sequences whose structure is unrelated to retroviruses. They are divided into short (SINE) and long (LINE) elements [24] and it has been suggested that they may behave like mobile elements [25–27]. Part of the evidence for this notion is based on the finding that these elements, such as *Alu*I and *Kpn*I elements, are bounded by direct repeats of 7–20 bp, similar to the integrated provirus of the retroviruses. On the other hand, in contrast to IAP or proviruses, the SINE or LINE elements do not contain symmetrical termini at their ends (long terminal repeats) but contain in one of their ends a poly(A)-rich tail, suggesting their origin in mRNA [28]. For this reason *Alu* and related sequences, as well as processed genes, were recently defined as 'retroposons' [29] or retrotransposons [30] (see Fig. 1).

In this article I review our recent studies concerning transposition of repetitive elements in tumour tissues. We demonstrate that DNA elements such as IAP or LINE can behave like transposable elements and can profoundly alter cellular gene expression. This may be an important source of somatic mutations in tumours as well as in other cases.

### Rearrangement of the c-*mos* Gene in two Plasmacytomas

We screened the DNA of several Balb/c mouse myelomas by Southern blot analysis using v-*mos* specific fragment as a probe. Two of the tumours, XRPC24 and NSI, appeared by initial analysis to contain a rearranged c-*mos* gene and

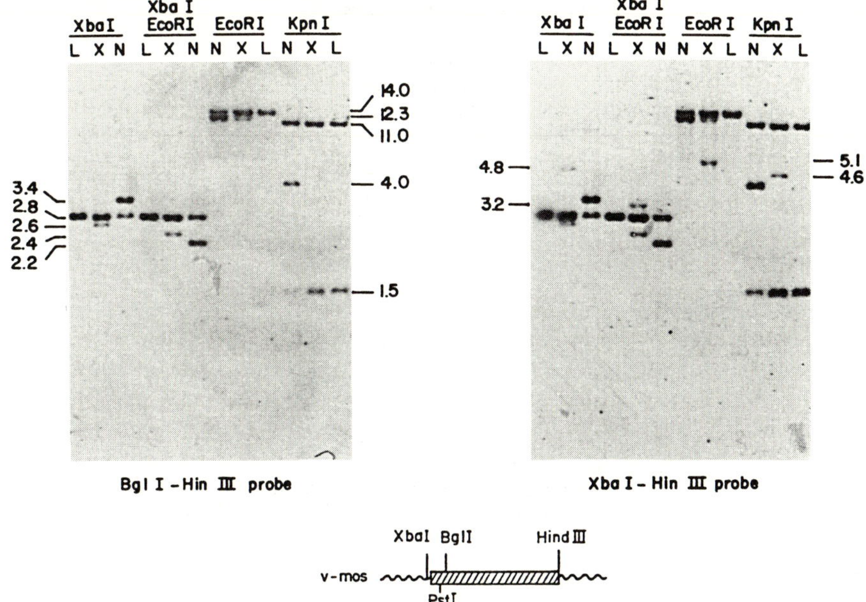



Fig. 2. *Restriction enzyme analysis of the c-mos gene in cellular DNA from Balb/c mouse (L) XRPC24 myeloma (X) and NSI myeloma (N)*

The two probes used are described in the text and schematically shown at the bottom on the partial map of v-*mos* of Moloney murine sarcoma viral DNA. Numbers correspond to kbp and hatched box to v-*mos*.

were studied in detail. The rearranged genes as well as their normal counterpart were compared by hybridization of various DNA digests with two probes. The first one contained sequences bounded by the *BglI* and *Hind*III sites in v-*mos* (Fig. 2, bottom) and detects the 3′ 293 codons of c-*mos* [31]. The second probe was the *XbaI–Hind*III v-*mos* fragment (Fig. 2, bottom) which detects the 3′ 368 codons of c-*mos* (out of 390 codons spanning the entire coding region). Therefore, cellular *mos* DNA fragments detected only by the second probe must originate from the 5′ region of the oncogene. A typical analysis is shown in Fig. 2. It demonstrates, for example, that by probing with the *BglI–Hind*III fragment, a 14 kbp band was detected in *Eco*RI-digested Balb/c DNA as well as in the two tumours. This band represents the normal gene. An additional 12 kbp fragment was apparent in XRPC24 and NSI DNA and corresponded to rearranged c-*mos*. Using the *XbaI–Hind*III fragment as a probe, an additional 5.1 kbp fragment showed only in XRPC24 DNA. This experiment indicates that the two plasmacytomas carry a rearranged c-*mos*, that the rearrangement is not identical in both cases, and that at least in XRPC24 it involves the 5′ terminus of the gene. The rearranged genes in XRPC24 and NSI tumours were termed rc-*mos*[X24] and rc-*mos*[NSI], respectively.

c-*mos* transcription has not been observed in normal and tumour tissues examined so far [32]; therefore it was of interest to determine whether the oncogene's rearrangement is correlated with its transcriptional activation. Fig. 3 shows that *mos* specific RNA of 1.2 kb is synthesized in XRPC24 tumour.

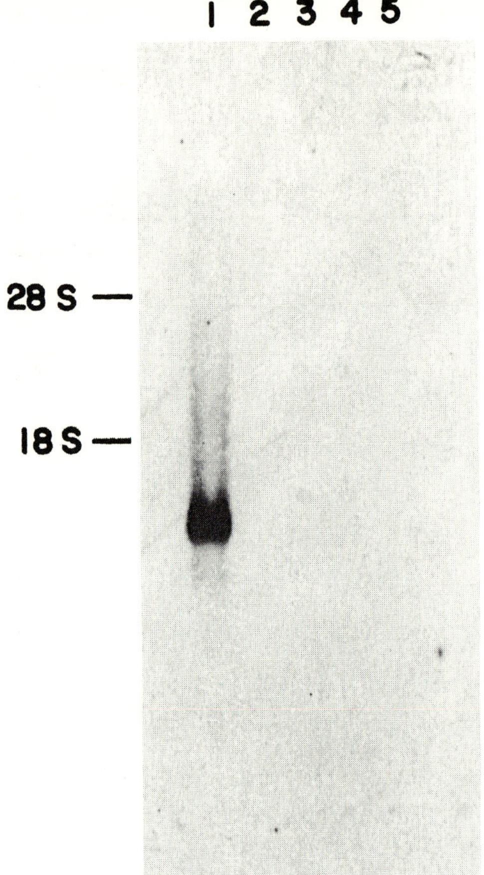



Fig. 3. *Northern blot analysis of c-mos transcripts in myeloma XRPC24 (1), hybridoma 160 (2), hybridoma 103 (3), myeloma NSI (4) and myeloma MPC 11 (5)*

The positions of ribosomal RNA markers are indicated.

These results indicate that in XRPC24 c-*mos* rearrangement resulted in turning on of the oncogene's transcription. Recently a c-*mos* transcript of 1.7 kb mRNA and 1.4 kb mRNA was detected in mouse testes and ovaries respectively [33]. Hence, the 1.2 kb *mos* mRNA in XRPC24 correlates well with the removal of the 5′ region of the gene by the IAP insertion.

## Organization of rc-*mos*$^{X24}$ and rc-*mos*$^{NSI}$

Using the two probes described in the previous section and by performing several double digestions, it became apparent that in both tumours the rearrangements involved the 5′ region of *mos*. The physical map of the rearranged oncogenes as determined by Southern blots is shown in Fig. 4. To compare in detail the normal and rearranged gene we molecularly cloned the normal c-*mos* 14 kbp *Eco*RI DNA fragment from Balb/c tissue, the rearranged

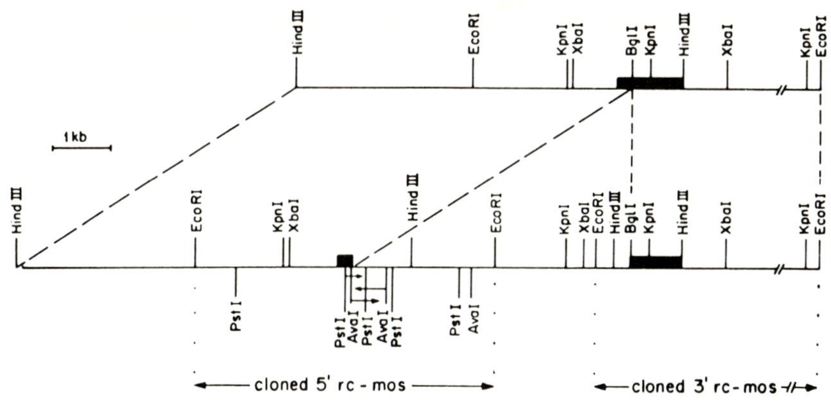



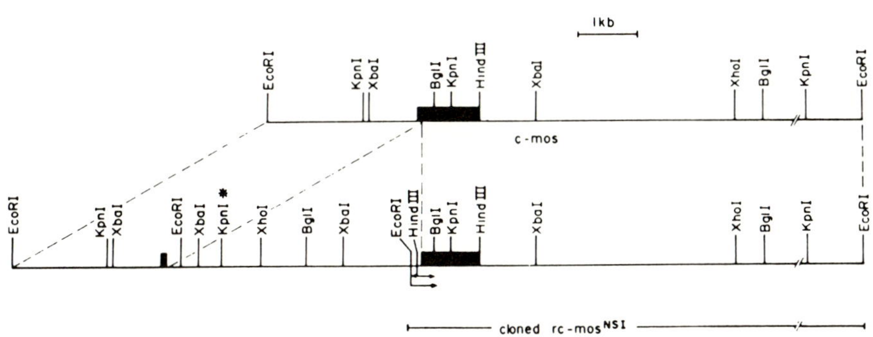



Fig. 4. *Physical maps of normal c-mos, and rc-mos$^{X24}$ (top) and of rc-mos$^{NSI}$ (bottom)*
Black boxes correspond to coding regions. Arrows denote strategy of DNA sequencing.

12 kbp and 5.1 kbp corresponding *Eco*RI fragments from XRPC24 tumour, and
the rearranged 12 kbp rc-*mos Eco*RI DNA of NSI myeloma. The rc-*mos*$^{X24}$
originated by the insertion of a novel cellular DNA element of 4.7 kbp into the
coding region near the *Bgl*I site. This resulted in splitting the gene into two
regions and the shifting upstream of the 5′ flanking *mos* sequences, away from
the main coding body of the gene.

c-*mos* in NSI tumour has undergone a similar but not identical DNA
rearrangement. Here, a 4.5 kbp DNA split the oncogene's coding region very
close to its 5′ terminus (Fig. 4). The two novel DNA elements inserted within
c-*mos* in XRPC24 and NSI tumours show a remarkable similarity in size.
Moreover, a constellation of three restriction enzyme sites (*Kpn*I, *Xba*I, *Eco*RI)
at the 3′ end of the element in rc-*mos*$^{X24}$ appears to resemble closely a similar
constellation at the 5′ end of the element in rc-*mos*$^{NSI}$. This suggests that similar
cellular elements were inserted in opposite orientation into c-*mos* in the two
myelomas.

### Cellular Elements Inserted within c-*mos* are Endogenous Intracisternal A-Particle (IAP) Genes

To define precisely the points of insertion and further characterize the cellular elements inserted within c-*mos* we determined the nucleotide sequence at the junctions between the main bodies of c-*mos* coding regions and the inserted cellular elements (sequences near the *Bgl*I site at c-*mos* coding region of the two rearranged genes). The sequence of rc-*mos*[X24] junction [16] indicated that the element was inserted at nucleotide 513 and codon 88 of c-*mos* presumed coding sequence [31]. The sequence of rc-*mos*[NSI] junction [18] shows that c-*mos* was split at nucleotide 151 or codon 30. The two cellular DNA elements terminate at the junction points with the tetranucleotide AACA which is identical to the terminus of the *Drosophila* transposable element *copia* [34], yeast Tyl mobile element [35,36], and spleen necrosis virus DNA [37]. This last finding immediately suggested that the inserted cellular DNAs are related to transposable elements and indeed it was found that a 349 base pair segment of the cellular DNA immediately adjacent to the retained main body of c-*mos* sequences in rc-*mos*[X24] has close homology with the long terminal repeat (LTR) of a known intracisternal A-particle gene [38]. Comparison of the two sequences (Fig. 5) indicates 88% homology between the sequence of the inserted element and the 'anti sense' strand of the 5' LTR of the IAP gene MIA 14. This means that a 5' IAP LTR was inserted into c-*mos* in a head to head (5' to 5') configuration. Similar comparison of the sequence inserted in rc-*mos*[NSI] [18] showed that in this case a 3' IAP LTR was inserted in a head to tail (3' to 5') orientation (Fig. 6). The sequence data have indicated that in XRPC24 and NSI tumours an IAP LTR integrated within c-*mos*. To study whether a complete IAP gene was inserted we used the 5.1 kbp *Eco*RI fragment which we have molecularly cloned from XRPC24 DNA (termed 'cloned 5' rc-*mos*[X24]' in Fig. 4). This fragment was partially sequenced according to the scheme depicted in Fig. 4. The analysis showed that the 5' end of the element integrated within c-*mos* in XRPC24 myeloma is also an IAP LTR [17]. Therefore, both the 3' and 5' termini of the cellular element in rc-*mos*[X24] are IAP LTRs. To examine if the non-LTR sequences within the 4.7 kbp element are also of IAP origin we tested a non-LTR segment of the 5.1 kbp *Eco*RI fragment for hybridization to endogenous IAP sequences distributed within the mouse genome. Southern blots of *Eco*RI-and *Hind*III-digested Balb/c DNA were probed for IAP-specific fragments by hybridization to non-LTR probes from a typical IAP genome, pMIA1, and from the 5.1 kbp fragment. The pMIA1 probe detected an extensive array of IAP fragments ranging in size from 0.5 to 6 kbp (Fig. 7, lane a). The rc-*mos*[X24] probe cross-hybridized with most of these fragments (Fig. 7, lane b). These results indicate that rc-*mos*[X24] contains a genuine IAP genome. Although a similar experiment was not done with the element inserted into c-*mos* in NSI myeloma, the similarity in size and restriction enzyme pattern of the two elements makes it highly probable that rc-*mos*[NSI] too contains an entire IAP genome.

```
            1        10       20       30       40       50       60       70       80
rc-mos    †TCTTCTTAACAGTCTGCTTTACGGGAACCTTTATTACCGTGACCCGCAGTTCTGGTTCTGGAATGAGCGATCTTCCTTGC


            90       100      110      120      130      140      150      160
rc-mos    GCCGGTCCCGAGTTTTCTTGTATTTTTTCGTCCCGGATTTTTTCTCGTCCTGATTTTCTCGTCCCGGATTTCAGCACCAA
                                                                            :::   :::::::
MIA14                    (Body of IAP gene MIA14)                          †AATTCGGTCACCAA
                                                                            350    340


            170      180      190      200      210      220      230      240
rc-mos    TTGTTATTAGACGCGTTCTCACGACCGGCCAGGAAGAACACCACAGACCAGAATCTTCTGCGGCAAAGCTTTATTTCTTA
          ::::::::::::  ::::::::::: :::::  ::::::::::::::::::::::::::::::::::::::: ::::::: ::::
MIA14     TTGTTATTAGAGCCGTTCTCACGCCCGGC AGGAAGAACACCACAGACCAGAATCTTCTGCGGCAAAACTTTATTGCTTA
           ┌─ 330      320      310      300      290      280      270      260
          ↑ 3'


            250      260      270      280      290      300      310
rc-mos    CATCTTCAGGAGCCAGGGTCGAGGAAGCAAGAGAGCAAGAAGCAAGAGAGCGAGAAAACGAAACCCCGTCCCTCTTA AG
          :::::::::::::: :: :: : ::::::::::::              ::::::::::::::::::::::::: :: ::
MIA14     CATCTTCAGG.GCAAGAGTGTAAGAAGCAAGAGAG              AGAAAACGAAACCCCGTCCCTTTTTTAG
           250      240      230                          220      210      200


                                              PstI
          320      330      340      350      360      370      380      390
rc-mos    GAGCATTCTCCTTCGCCTCGGACCTGTCACTCCCTGATTGGCTGCAGCC ATCGGCC AGTTGACGTCACGGGGAAGGCA
          :::  :: : :::::::: :::::::::::::::::::::::::::::::: ::::::: :::::::::::::::::: ::::
MIA14     GAGAGTTATATTTCGCCTAGGACGTGTCACTCCCTGATTGGCTGCAGCCCATCGGCCGCGAGTTGACGTCACGGGGAGGGCA
           190      180      170      160      150      140      130      120


                                           Met.......►
          400      410      420      430      440      450      460      470
rc-mos    GAGCACAAGTAGTCATAAGATACCCTTGGCACATGCGCAGATTATTTGTTTACCACTTAGAACACACGGATGTCAGCGCC
          ::::::::: : ::: :     :::::  ::::  ::::::  :::::::::::::::::::::::::::::::: ::::::::::
MIA14     TGAGCACATGAAGTAGGGAACCACCCTCGGCATATGCGCGGATTATTTGTTTACCACTTAGAACACAGC TGTCAGCGCC
           110      100      90       80       70       60       50       40


          480      490      500      510      520      530
rc-mos    ATCTTCTACCGGCGAATGTCGGCGCGCCTCCCAACAcactcctcctcgggctcccggactgcc†
          ::::::::: : :     :::::::::::::::::::::::::
MIA14     ATCTTGTAACCGCGATGTCGCGCGCGGCTCCCAACA (Unique flanking sequence)†
           30       20       10        5'↑
```

Fig. 5. *Comparison of the nucleotide sequences of a portion of rc-mos[X24] and the LTR of the IAP genes MIA 14*

Arrows indicate the boundaries of IAP LTR. Short inverted repeats at the LTR termini are underlined. The rc-*mos* sequence is presented in the direction of transcription of c-*mos* and it is aligned with the antisense strand of IAP LTR. 'Met' indicates the beginning of an open reading frame extending into the retained c-*mos* sequence.

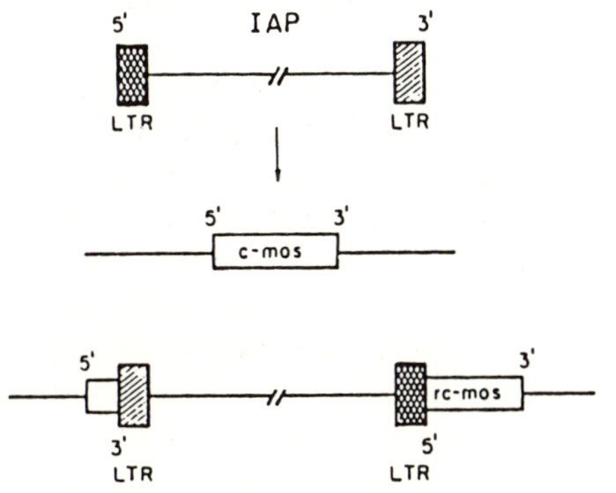


Fig. 6. *Schematic illustration of the insertion of IAP genomes into c-mos in myeloma XRPC24*

In myeloma NSI the IAP was inserted in a 3′ to 5′ orientation.

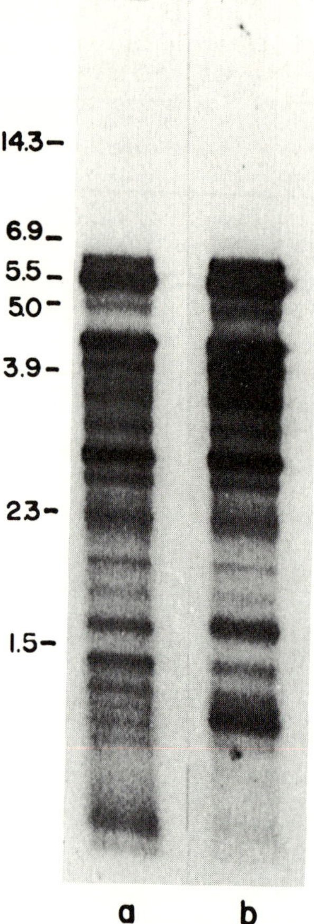



Fig. 7. *Southern blot analysis of HindIII + EcoRI-digested Balb/c mouse DNA, examining hybridization to non-LTR probes derived from pMIA1 IAP genome (lane a) and the IAP genome in rc-mos$^{X24}$ (lane b)*

Sizes of DNA markers are denoted in kbp; Or, origin.


## Biological Activity of the Rearranged c-*mos* Genes

Extensive experiments have shown that the molecularly cloned 14 kbp *Eco*RI fragment corresponding to the normal c-*mos* gene was inactive in transfection experiments, unless a LTR from Moloney sarcoma virus was attached to it *in vitro* [39]. We asked whether the rearranged genes cloned from the two myelomas are active by themelves when assayed by transfection. Two tested clones of the *Eco*RI 12 kbp fragments of rc-*mos*$^{X24}$ induced a large number of foci on NIH/3T3 monolayers with specific activities of 3100 and 4900 focus forming units/$\mu$g of DNA, respectively (Table 1). Molecularly cloned 12 kbp *Eco*RI rc-*mos*$^{NSI}$ DNA induced a small number of foci. No foci were observed when the cloned 14 kbp *Eco*RI fragment of normal c-*mos* was assayed. Two recombinant phages containing integrated Moloney sarcoma viral DNA produced 10 200 and 13 200 focus forming units/$\mu$g of DNA and served as positive control.

Table 1. *Biological activity of molecularly cloned c-mos and rc-mos DNAs*

Intact recombinant phage DNA was transfected into NIH-3T3 fibroblasts by the calcium phosphate method. The cultures were kept in medium containing 5% fetal calf serum and were scored for focus formation 2 weeks after transfection. λ-MSV 26 and λ-MSV 3-3 represent Mo-MSV molecules integrated with normal rat kidney cell DNA and molecularly cloned in Wes B vector. λ-c-*mos* and λ-rc-*mos* are molecular clones of the c-*mos* and rc-*mos* genes, respectively.

| Clone | Insert DNA (ng) | Number of foci | Specific infectivity (foci/pmol of DNA) |
|---|---|---|---|
| λ-MSV 26 | 50 | 150 | 10 200 |
| λ-MSV 3-3 | 100 | 220 | 13 200 |
| λ-c-*mos* IA | 125 | 0 | < 67 |
| λ-rc-*mos*$^{X242}$2B | 100 | 43 | 3100 |
| λ-rc-*mos*$^{X242}$2F | 100 | 68 | 4900 |
| λ-rc-*mos*$^{NSI}$26 | 700 | 8 | 90 |

These results demonstrate that the rearrangement of c-*mos* in XRPC24 tumour resulted in a striking biological activation of the gene. The situation is not clear with regard to rc-*mos*$^{NSI}$, where a significant but very low activity was observed. This may be due to the fact that the cloned rc-*mos*$^{NSI}$ contains only one-half of the 3′ LTR due to the presence of an *Eco*RI site at this position.

## Rearranged c-*myc* in Canine TVT

Canine transmissible venereal tumour (TVT) has been known for over 100 years as a naturally occurring neoplastic disease of uncertain histological origin that affects the external genitalia of both sexes and is transmitted during coitus. In addition to the natural mode of transmission, TVT can also be transplanted experimentally to adult, immunocompetent, allogeneic dogs by subcutaneous inoculation of living tumour cells. TVT is the only naturally occurring tumour which is transmitted by cell transplantation. It is of interest to study the state of oncogenes in this tumour since it may shed some light on its origin.

Since we found that IAP can activate the oncogene c-*mos* by DNA transposition, it seems reasonable to look for transposition of other DNA elements in tumour tissues. A change in the position of the gene on a Southern blot may suggest somatic DNA rearrangement. Using this approach we analysed the organization of c-*myc* in canine TVT and demonstrated the transposition of a retroposon-like element upstream to the c-*mys*.

The analysis of the c-*myc* in TVT and normal canine tissue (spleen) obtained from the same dog (A) is presented in Fig. 8. DNA (10 μg) was digested with *Eco*RI, separated by electrophoresis on 0.7% agarose, transferred to nitrocellulose, and hybridized to the *myc* probe. Both tumour and spleen DNA showed two *myc* bands, 7.5 and 15 kbp. The tumour DNA contains an additional band of approx. 16.8 kbp, which we designate rearranged c-*myc* (rc-*myc*). To compare c-*myc* (the 15-kbp band) with rc-*myc* (the 16.8-kbp band), we cloned the two genes in Charon-4A after enrichment of the respective bands by preparative agarose-gel electrophoresis.

Fig. 9 depicts the physical map of canine *myc* and rc-*myc* as determined by

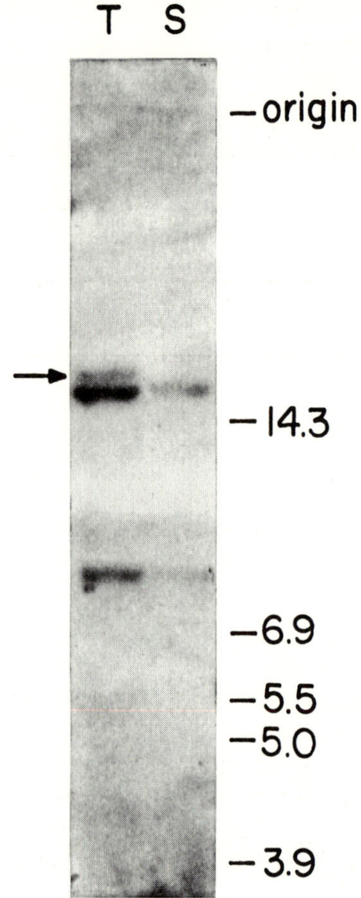


Fig. 8. *Southern blot analysis of c-myc sequences in canine TVT and spleen DNA*

Aliquots (10 $\mu$g) of high-$M_r$ DNA were digested with *Eco*RI, electrophoresed on 0.7% agarose gels, and, after blotting onto nitrocellulose filters, hybridized to the radiolabelled *myc* probe pM104BH. Numbers at the right side are in kbp according to DNA markers of Charon 4A digested with *Eco*RI/*Bam*HI. T, tumour DNA; S, spleen DNA. Both tumour and spleen were from the same dog (dog A), but Southern blots of tumour DNA from two other dogs showed the same results.

restriction enzyme analysis and hybridization with the two *myc* probes, which hybridize either to exon 1 or to exons 2 and 3 of c-*myc*.

The map demonstrates that c-*myc* and rc-*myc* share identical restriction sites downstream and upstream to the two *Pst*I sites. However, rc-*myc* differs from c-*myc* by the insertion of approx 1.8 kbp of DNA between the two *Pst*I sites.

**rc-*myc* in TVT Contains a Retroposon-like Element**

To locate repetitive DNA in the cloned c-*myc* and rc-*myc* we hybridized the cloned phages or a subclonal region of rc-*mys* (pRCM1, containing the *Hin*dIII fragment of rc-*myc*; Fig. 9) with nick-translated canine DNA (1 $\mu$g). The results (not shown) demonstrated that rc-*myc* contains a highly repetitive DNA

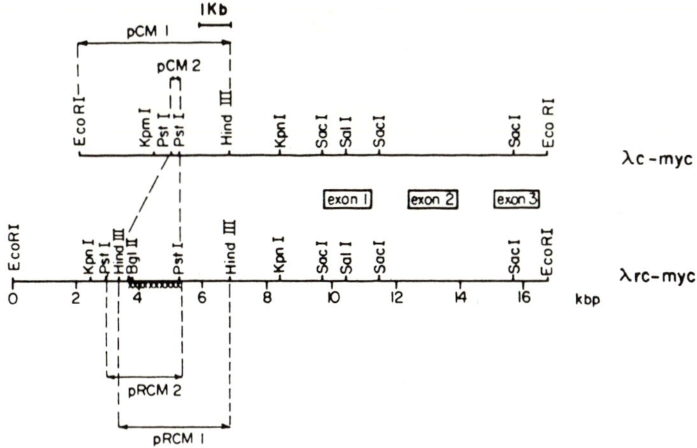


Fig. 9. *Physical maps of c-myc and rc-myc inserts in cloned phages*

Phage DNA was digested with several restriction enzymes in various combinations, separated on agarose gel, and hybridized to various *myc* probes to map *myc* exons, XXXX, Repetitive DNA. Subcloned fragments are named in the corresponding regions.

fragment located in the 1.5-kbp *Bgl*II/*Pst*I fragment that is contained in the inserted DNA of rc-*myc*.

We wished to know the nature of the inserted DNA in rc-*myc* and in particular to analyse the junctions between the inserted DNA and the c-*myc* gene. To do this we compared the sequence of the 270-bp *Pst*I fragment of c-*myc* [subcloned and designated pCM2 (Fig. 9)] with the sequence of the termini of the 2.0 kbp *Pst*I fragment [subcloned as pRCM2 (Fig. 9)] of rc-*myc*.

The results (Fig. 10) showed identity between the sequence of 258 bp and 60 bp of c-*myc* and rc-*myc* at the 5′ and 3′ ends, respectively, of the two *Pst*I fragments. It is also clearly shown that the *Pst*I fragment of rc-*myc* contains an approx. 1.8-kbp insert of DNA that splits the *Pst*I fragment of c-*myc*. The squence data showed that this insert is bounded by a 10-bp repeat (A-T-T-C-T-C-T-G-G-C) of the c-*myc* sequence and the insert itself has a tail of approx. 60 bp of A-rich sequence just before the junction with the cellular DNA. These features of the insert may classify it as a retroposon [29] and suggests that the insertion was in a tail-to-head orientation with respect to c-*myc*, approx. 5 kbp upstream to the first exon of c-*myc* (Fig. 9). Fig. 10 also shows the homology between the partial sequence of the inserted sequence in rc-*myc* and that of the monkey *Kpn*I element. The comparison was done by Maxine Singer, using a consensus sequence obtained from various *Kpn*I fragments (M. Singer, personal communication). The inserted sequence shows 62% homology to a region of the *Kpn*I element beginning approx. 900 bp from the 3′ end of the *Kpn*I element. This is similar to the homology between mouse MIF repetitive element and monkey *Kpn*I element. It is of interest that the homology is in the 3′ to 5′ orientation of the *Kpn*I element, suggesting a possibility of inversion of the inserted element.

```
          PstI                                                                          70
rc-myc    CTGCAGCATA GGGGTGGGGC AGGGAGACAA CATTTTAAAA ACAATGCATT TTTAAAAATG CACCAAGATT
c-myc     ---------- ---------- ---------- ---------- ---------- ---------- ----------


                                                                                       140
rc-myc    TTCTTCACTG CCTTTTTTTT TTATCATTCC TACGAATGAA TGATTGGCCA GATTTCGTCT GCTCGTCTGC
c-myc     ---------- ---------- ---------- ---------- ---------- ---------- ----------


                                                                                       210
rc-myc    TGAAGAGCTT CCCAGTGTTC CTCTCACTGG GACACATGGT TAGCACAGGA AATACTGGTG AGGCTTTCCC
c-myc     ---------- ---------- ---------- ---------- ---------- ---------- ----------


                                                                             HinIII    280
rc-myc    ATCCTTTAAC [ATTCTCTGGG] TGTCTTCGAC CCCCGCCGAC TGTATCCTTT GCTGTGCAAA AGCTTCTTAT
c-myc     ----------
Kpn I                            --C--GT-CA -T-T-AT-GT A--T--T--- --------G- ----CT---G


                                                                                       350
rc-myc    CTTGATGAAG TCCCAATAGT TCATTTTTGC TTTTGTTTCT TTTGCCTTCG TGGATGTATC TTGCAAGAAG
c-myc
Kpn I     T--A--T-GA -----T-T-- CA-----G-- -------C-C A----T--T- GA-T-T--GA CATG---TCT


                                                                             417BglII
rc-myc    TTACTATGGC CGAGTTCAAA AAGGGTGTTG CCTGTGTTCT TCTCTAGGAT TTTGATGGAA CCTTGCC...
c-myc
Kpn I     --G-CCAT-- -T-TG--CTG --T---A--- ---AG---T- CT------G- ---T----T- TTAG-T-


                                                                                       510
rc-myc    TGAAGTGTAT GGTGAAAGAG AGTGGTCTAG TTTCATTCTT TCTGCATGTG GATGTCCAAT TTCCCAGCAC
c-myc
Kpn I     -TTT--A--A ----T---GA --G-A--C-- -----G---- ---A-T--A-- -C--C-A-T- ----------


                                                                                       580
rc-myc    CATTTATTGA GAGCAATGTC TTTCTTCCAA TGGATAGTCT TTCCTCCTTT ATCGAATATT AGTTGCCCAT
c-myc
Kpn I     --------A- ATAGGGAA-- C--TCC---T -TCT-GT--C -CT-AGG--- G--A--G--C --A--GTTG-


                                                                  635
rc-myc    AAAGTTCAGG TCCACTTCTG GATTCTCTAT TCTGTTGGAC TGATCTATGT GTCTG.....~1.4 kbp..
c-myc
Kpn I     -G-TG-GT-- -TT-T----- GAAG----G- ------CC-T --G-----A- -----


                                                                                       2070
rc-myc    TGTTCATCTG TATGTTAGTA AATTGAACAC CAATAAAAAT AAAATAAAAT AAAATAAAAT AAAATAAAAT
c-myc


                                                                                       2140
rc-myc    AAAAAAAAAA AATAAA[ATTC TCTGGC]CGCC TTGGTTTTGA ATCTCATGGC CCTTCTTTCA AAATGATCTT
c-myc                    ---- ---------- ---------- ---------- ---------- ----------
          PstI
rc-myc    TCTGCAG
c-myc     -------
```

Fig. 10. *Partial sequence of the DNA insert in rc-myc*

The sequence shows the beginning and end of the insert and compares the junction between c-*myc* and the insert in rc-*myc*. The insert of pCM2 (Fig. 9) derived from c-*myc* was subcloned in M13mp8 and sequenced in both orientations. The *Pst*I fragment of pRCM2 (Fig. 9) was subcloned in M13mp8 and the two ends were sequenced by the dideoxy method. The sequence of the left-hand *Pst*I/*Bgl*II fragment of pRCM2 was determined by the Maxam and Gilbert method after labelling of the *Bgl*II site. The sequence between the two *Bgl*II sites (downstream from nucleotide 417) was determined for both strands by the Maxam and Gilbert method. A small region (approx. 25 bp) overlapping the *Bgl*II site has not yet been sequenced and is marked by dots after nucleotide 417. Direct repeats at both ends of the insert are boxed. Identical nucleotides are marked by dashes. The sequence of the insert is compared with that of monkey *Kpn*I element (M. Singer, personal communication). The *Kpn*I sequence is of the complementary strand beginning at nucleotide 5079 and ending at nucleotide 4663 of the *Kpn*I element.

## Conclusion

This work presents two examples for the transposition of endogenous repetitive DNA elements in tumour tissues. In the first example the repetitive element was the IAP genome which is an endogenous retrovirus-like element. In the second example the element was one of the LINE family which has the features of a processed gene. In both cases the transposition and relocation of the elements showed duplication of the target sequence of the insertion site: 6 bp in the case of IAP and 10 bp in the case of the *Kpn*I element. In the latter case the insert also contains a 3' tail of dA-rich sequence, suggesting its origin from mRNA. These data suggest that the mechanism of transposition may involve successive steps of transcription, reverse transcription into cDNA and integration. This process of retrotransposition [30] is similar to that involved in the integration of retroviruses. However, here the transposed elements [29], or retrotransposons [30], are endogenous and the process is intracellular. This behaviour of repetitive DNA elements in the mammalian genome qualifies them as transposable elements and indicates that the movement of transposable elements in the somatic cell is a process which may occur regularly. Such a process may generate somatic mutations which still affect the expression of neighbouring genes.

We were particularly interested in the effect of these insertional mutations on oncogenes, and therefore searched for it in tumour tissues. The activation of c-*mos* in XRPC24 was probably due to promoter insertion mechanism [7] since we have demonstrated that the mRNA of c-*mos* starts within the LTR of IAP [40]. This implies that the 5' LTR of IAP provides some promoter activity in the opposite orientation to its normal transcription.

As was shown by others [32], c-*mos* is transcribed also in NS1, implying that the 3' LTR provides some promoter activity in this case. In the case of rc-*myc* in the TVT tumour we have shown an elevated c-*myc* mRNA in the tumour tissue, but we do not yet know what is the normal tissue counterpart of this tumour [41]. Further studies are necessary to understand the effect of the inserted LINE element upstream to c-*myc* in this tumour. Finally, it should be emphasized that human DNA contains representatives of the two repetitive DNA families described here, i.e. endogenous defective retrovirus and the LINE family. It is not impossible that transposition of these elements play a role in human cancer.

I acknowledge the collaboration of Drs. G. Rechavi, J. B. Cohen, O. Dreazen, T. Unger, M. Horowitz and E. Canaani in various aspects of this work.

## Reference

1. Weinberg, R. A. (1982) *Adv. Cancer Res.* **36**, 149–163
2. Cooper, G. M. (1982) *Science* **217**, 801–806
3. Duesberg, P. H. (1983) *Nature (London)* **304**, 219–226
4. Stehelin, D., Varmus, H. E., Bishop, J. M. & Vogt, P. K. (1976) *Nature (London)* **260**, 170–173
5. Shilo, B. Z. & Weinberg, R. A. (1981) *Proc. Natl. Acad. Sci. U.S.A.* **78**, 6789–6792
6. Hayward, W. S., Neel, B. G. & Astrin, S. M. (1981) *Nature (London)* **290**, 475–480
7. Payne, G. S., Bishop, J. M. & Varmus, H. E. (1982) *Nature (London)* **295**, 209–214

 8. Tabin, C. J., Bradley, S. M., Bergmann, C. I., Weinberg, R. A., Papageorge, A. G., Scolnick, E. M., Dhar, R., Lowy, D. R. & Chang, E. H. (1982) *Nature* (*London*) **300**, 143–149
 9. Reddy, E. P., Reynolds, R. K., Santos, E. & Barbacid, M. (1982) *Nature* (*London*) **300**, 149–152
10. Collins, S. & Groudine, M. (1982) *Nature* (*London*) **298**, 679–681
11. Dalla-Favera, R., Wong-Staal, F. & Gallo, R. G. (1982) *Nature* (*London*) **299**, 61–63
12. Shen-Ong, G. L. C., Keath, E. J., Picolli, S. P. & Cole, M. D. (1982) *Cell* **31**, 443–452
13. Dalla-Favera, R., Bregni, M., Erikson, J., Paterson, D., Galo, R. & Groce, C. M. (1982) *Proc. Natl. Acad. Sci. U.S.A.* **79**, 7824–7829
14. Taub, R., Kirsch, I., Morton, C., Lenoir, G., Swan, S., Tronick, S., Aaronson, S. & Leder, P. (1982) *Proc. Natl. Acad. Sci. U.S.A.* **79**, 7837–7841
15. Rechavi, G., Givol, D. & Canaani, E. (1982) *Nature* (*London*) **300**, 607–611
16. Canaani, E., Dreazan, O., Klar, A., Rechavi, G., Ram, D., Cohen, J. B. & Givol, D. (1983) *Proc. Natl. Acad. Sci. U.S.A.* **80**, 7118–7122
17. Cohen, J. B., Unger, T., Rechavi, G., Canaani, E. & Givol, D. (1983) *Nature* (*London*) **306**, 797–799
18. Chattopadhyay, S. K., Lowry, D. R., Teich, N. M., Levine, A. S. & Rowe, W. P. (1975) *Cold Spring Harbor Symp. Quant. Biol.* **24**, 1085–1101
19. Todaro, G. J., Benveniste, R. E., Callahan, R., Lieber, M. M. & Sherr, C. J. (1975) *Cold Spring Harbor Symp. Quant. Biol.* **24**, 1159–1168
20. Lueders, K. K. & Kuff, E. L. (1977) *Cell* **12**, 963–972
21. Dalton, A. J., Potter, M. & Merwin, R. M. (1961) *J. Natl. Cancer Inst.* **26**, 1221
22. Kuff, E. L., Lueders, K. K. Ozer, K. L. & Wivcel, N. A. (1972) *Proc. Natl. Acad. Sci. U.S.A.* **69**, 218–222
23. Paterson, B. M., Segal, S., Lueders, K. K. & Kuff, E. L. (1978) *J. Virol.* **27**, 118–126
24. Singer, M. F. (1982) *Int. Rev. Cytol.* **76**, 67–112
25. Jelinek, W. R. & Schmid, C. W. (1982) *Annu. Rev. Biochem.* **51**, 813–844
26. Lerman, I. M., Thayer, R. E. & Singer, M. F. (1983) *Proc. Natl. Acad. Sci. U.S.A* **80**, 3966–3970
27. Singer, M. F. (1982) *Cell* **28**, 433–434
28. Digovanni, L., Haynes, S. R., Misra, R. & Jelinek, W. R. (1983) *Proc. Natl. Acad. Sci. U.S.A.* **80**, 6533–6537
29. Rogers, J. (1983) *Nature* (*London*) **30**, 460
30. Baltimore, D. (1985) *Cell* **40**, 481–482
31. Van Beveren, C., Van Staten, F., Galleshaw, J. A. & Verma, I. M. (1981) *Cell* **27**, 97
32. Gattoni-Celli, S., Hsiao, W. W. Z. & Weinstein, I. B. (1983) *Nature* (*London*) **306**, 780–799
33. Propst, F. & Vanda Woude, G. F. (1985) *Nature* (*London*) **315**, 516–518
34. Dunsmuir, P., Broein, W. J., Jr., Simon, M. A. & Rubin, G. M. (1980) *Cell* **21**, 575
35. Farabough, D. J. & Fink, G. R. (1980) *Nature* (*London*) **286**, 352–356
36. Gafner, J. & Philippsen, P. (1980) *Nature* (*London*) **286**, 414–418
37. Shimotohno, K., Mizutani, S. & Temin, H. M. (1980) *Nature* (*London*) **285**, 550–554
38. Kuff, E. L., Feenstra, A., Lueders, K., Rechavi, G., Givol, D. & Canaani, E. (1983) *Nature* (*London*) **302**, 547–548
39. Oskarsson, M., McClements, W. L., Blair, D. G., Maizel, F. V. & Vande Wonde, G. F. (1980) *Science* **207**, 1222–1224
40. Horowitz, M., Luria, S., Rechavi, G. & Givol, D. (1984) *EMBO J.* **3**, 2937–2941
41. Katzir, N., Rechavi, G., Cohen, J. B., Unger, T., Simoni, F., Segal, S., Cohen, D. & Givol, D. (1985) *Proc. Natl. Acad. Sci. U.S.A.* **82**, 1054–1058

*Biochem. Soc. Symp.* **51**, 197–204
*Printed in Great Britain*

# T Cell Receptors in the Thymus

PHILIPPA MARRACK,*†§ WILLI BORN,* NEAL ROEHM,* ANDREW FARR‖ and
JOHN KAPPLER*‡§

**Department of Medicine, National Jewish Hospital and Departments of †Biochemistry,
Biophysics and Genetics, ‡Microbiology and Immunology and §Medicine, University of Colorado
Health Sciences Center, Denver, Colorado, and ‖Department of Biological Structure, University of
Washington, Seattle, Washington, U.S.A.*

## Introduction

The protein and genes of the T cell receptor for antigen in association with
major histocompatibility complex (MHC) products have recently been identified.
The receptor on peripheral T cells is a disulphide-bonded heterodimer composed
of two glycopolypeptide chains, $\alpha$ and $\beta$, both around 43 kDa in mouse, and
of about 49 kDa and 39 kDa respectively in man (Allison *et al.*, 1982; Haskins
*et al.*, 1983; Meuer *et al.*, 1983). The polypeptide backbones of each chain are
similar in size, each composed of about 250–270 amino acids (McIntyre &
Allison, 1984; Hedrick *et al.*, 1984; Yanagi *et al.*, 1984; Chien *et al.*, 1984*a,b*;
Saito *et al.*, 1984; Sim *et al.*, 1984). Like immunoglobulins, $\alpha$ and $\beta$ chains each
contain both variable and constant region sequences, J regions, intra-domain
disulphide bonds and many other features conserved in light and heavy chains
and other members of the immunoglobulin family (Hedrick *et al.*, 1984; Yanagi
*et al.*, 1984; Malissen *et al.*, 1984; Chien *et al.*, 1984*a,b*; Kavaler *et al.*, 1984;
Sim *et al.*, 1984). Both $\alpha$ and $\beta$ chains have sequences of amino acids which
probably span the plasma membrane of the T cell bearing them, and may also
bind the receptor-associated polypeptides of the T3 complex, and both chains
have short cytoplasmic tails.

Molecular biological studies have shown that T cell receptor genes use
strategies very similar to those of immunoglobulin genes to construct large
numbers of related molecules with specificities for many different ligands. Thus,
each $\beta$ chain is constructed from a $V\beta$, $J\beta$ and $C\beta$ gene, with a $D\beta$ insert optional.
$\alpha$ chains seem to have similar component parts.

Although much is now known about the structure of the receptor protein and
its genes, we still have very little understanding of how the molecule functions.
For example, it is apparent that this single receptor is responsible for the
specificity of the T cell bearing it for both antigen and MHC (Kappler *et al.*,
1981; Hünig & Bevan, 1982; Heber-Katz *et al.*, 1982; Marrack *et al.*, 1983), but
we do not know whether one chain contributes more to MHC binding than the
other, nor do we understand in any detail the nature of the trimolecular complex
between receptor, antigen and MHC protein which presumably must form on
the surface of the T cell and antigen-presenting cell.

Another phenomenon which will probably be illuminated by these recent discoveries is that of the maturation of T cells in the thymus. The pioneering experiments of Zinkernagel *et al.* (1978) and Bevan (1977) showed that T cells acquire their restriction for recognizing antigen in association with self-MHC products in the thymus of that individual. It is also clear that tolerance to MHC can be induced by thymic cells (Jordan *et al.*, 1985). The fact that T cells can be selected to recognize antigen plus self MHC, in the absence of antigen, and also selected *not* to recognize self MHC alone, in the thymus, is one of the most mysterious problems in modern immunology.

Our own approach to this has been to study the rearrangement of T cell receptor genes and expression of receptor protein in the thymus, since presumably the processes outlined above depend upon the appearance of T cell receptors on the surface of the maturing thymocyte. As described in this paper, our results, and those of others, show that receptor genes are not rearranged and receptor protein is not expressed by very immature thymocytes. As cells mature they rearrange their receptor genes in an ordered fashion, and later express surface protein. In the cortex, these surface receptors collect at the surface of contact between the thymocyte bearing them and an epithelial cell, where selection for self-MHC recognition and/or tolerance induction may occur. There is evidence that cortical thymocytes may bear low levels of receptor/cell because of this capping and patching property. Mature, medullary thymocytes and peripheral T cells express relatively high levels of receptor/cell and this is not associated (in the absence of an immune response) with any other cell.

## Experimental Methods, Results and Discussion

*The rearrangement of receptor and immunoglobulin genes in thymocytes and T cells*

There are difficulties in studying the rearrangement of receptor genes in bulk populations of cells since, unless the genes can rearrange to only a few configurations as in the case of $\gamma$ chain genes (Hayday *et al.*, 1985), Southern blotting techniques are not sensitive enough to detect the positions of new rearrangements in uncloned populations of cells, although they can certainly be used in a quantitative way to detect the disappearance of unrearranged germ line bands.

In order to study receptor $\beta$ chain rearrangements we therefore constructed collections of hybridomas from thymocytes and T cells at different stages of mouse development (Born *et al.*, 1985). If the cells are properly handled, each hybridoma represents the product of a single fusion partner and a thymocyte or T cell, and therefore by studying the DNA from each hybridoma we could analyse the status of the DNA in the normal cell from which the hybrid was derived.

Hybridomas were prepared from the thymocytes and peripheral T cells of mice of various ages by fusing these cells to the AKR thymoma line, BW5147. As a control, cells from 14-day fetal liver were also fused. The DNA of these hybridomas was digested with three restriction enzymes which in mouse allow detection of rearrangements at the $\beta$ chain locus using a $C_\beta$ probe, *Hin*dIII, *Pvu*II

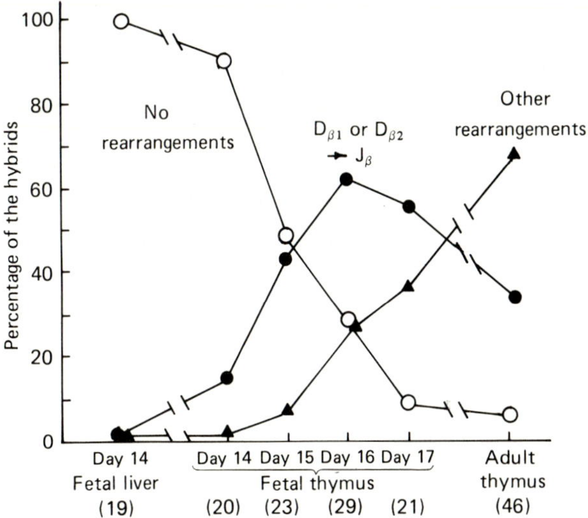


Fig. 1. *Kinetics of B complex rearrangement in fetal thymus*

Hybridomas were produced by fusion of fetal liver and fetal and adult thymocytes to the AKR thymoma BW5147. DNA from individual hybridomas was digested with *Hind*III, *Hpa*I or *Pvu*II, electrophased and blotted. Filters were probed with DNA coding $J_{\beta 2-6}$ and $C_{\beta 2}$. The rearrangements at $\beta$ which had occurred in each hybridoma were deduced from the band patterns so obtained.

and *Hpa*I. Blots were analysed with a cDNA probe containing $J_{\beta 2-6}$ and $C_{\beta 2}$. Comparison of the bands obtained from the same hybridoma with the different enzymes and knowledge of the restriction enzyme map of $C_\beta$ (Malissen *et al.*, 1984) allowed us to deduce what type of rearrangement had taken place in each T cell prior to fusion.

As shown in Fig. 1, rearrangements were absent from hybridomas produced by fusion to fetal liver cells, indicating both that fusion to BW5147 did not induce rearrangement in the DNA of normal cells and also that fetal liver cells (which include thymocyte precursors) did not contain cells with receptor gene rearrangements at any appreciable frequency.

Hybridomas constructed from early fetal thymocytes possessed $\beta$ chain rearrangements only rarely. Those that were found were consistent with D to J movements with no upstream Vs. As the thymuses matured, however, hybridomas prepared from their cells contained $\beta$ chain rearrangements with increasing frequency and complexity with bands consistent with complete $V_\beta D_\beta J_\beta$ genes first occurring at day 16. By birth, most mouse thymocytes contained the complex rearrangements characteristic of $V_\beta D_\beta J_\beta$ rearrangements and appeared to be able to synthesize mature receptor $\beta$ chains.

These results are consistent with the results of mRNA expression studies by Snodgrass *et al.* (1985) and Raulet *et al.* (1985) and Southern blotting of hybridomas produced from very immature thymocytes in adult mice (Samelson *et al.*, 1985). Overall, we can conclude that the earliest thymocytes in ontogeny, and very immature cells in the adult, contain no rearranged $\alpha$ and $\beta$ genes and do not express $\alpha$ and $\beta$ chain mRNA; however $\gamma$ chains may be rearranged and

Table 1. *Progressive rearrangements of the Igh complex in Balb/c T cells*

Cells were fused to the AKR thymoma BW5147. DNA of the indicated number of individual hybridomas was digested with restriction enzymes and probed with a J4 clone. Blanks indicate no rearrangements of this type.

| T cell surface | No. of hybridomas analysed | No. of rearrangements compatible with: | | |
|---|---|---|---|---|
| | | DQ52 to $J_H$ | Others to $J_H$ | Non-$J_H$ |
| Day 14 fetal liver | 27 | | | |
| Day 14 fetal thymus | 23 | | | |
| Day 15 fetal thymus | 25 | | | |
| Day 16 fetal thymus | 29 | | | 1 |
| Day 17 fetal thymus | 26 | 2 | | 1 |
| New-born thymus | 30 | 4 | | |
| 7-day-old thymus | 30 | 3 | 3 | |
| 10-week-old thymus | 30 | 8 | 16' | |

$\gamma$ mRNA may be present. At about day 16 of mouse fetal life, functional $\beta$ chain genes start to appear, together with 1.3 kb $\beta$ chain mRNA; a day or two later mature $\alpha$ chain genes seem to be constructed and start producing $\alpha$ mRNA.

## Rearrangement of immunoglobulin genes in thymocytes

Rearrangement and expression of immunoglobulin genes in thymocytes has been reported in the past, although the function of the proteins produced by these events has yet to be understood. We decided to study the relationship between heavy chain and $\beta$ gene rearrangements in our collection of thymocyte hybridomas.

To do this we used a probe coding for the most 3′ $J_H$ gene, $J_{H4}$, and DNA surrounding it. Such a probe should detect all rearrangements at the Igh locus if the right restriction enzymes are used. By measuring the sizes of rearranged restriction enzyme fragments detected with this probe we could distinguish rearrangements involving the most 3′ D, DQ52, to any $J_H$, from other rearrangements.

A summary of our data is shown in Table 1. Early in thymus ontogeny nearly all rearrangements involved DQ52; later, other types of movement, involving other D genes and perhaps upstream V region genes, occurred. Two other points were clear; Igh rearrangements occurred much more slowly than those for $\beta$ genes, and at no time did all T cells or thymocytes contain Igh rearrangements. These results suggest that if, as has been suggested, the same enzyme catalyses T cell receptor and immunoglobulin gene rearrangements, then this enzyme may persist in T cells long after its probable function has been completed. Also, the contents of adult thymuses, which include quite frequent cells with Igh rearrangements, do not seem to be the same as those of neonatal or young thymi, where cells with Igh rearrangements were found only rarely.

**Expression of thymocyte receptor protein**

Given these results described above for $\beta$ chain genes, we would not expect receptor protein to be present on very immature thymocytes. In fact, that is exactly what has been observed (Roehm *et al.*, 1984). We have studied the appearance of receptors on thymocytes by using a rat monoclonal antibody, KJ16-133, which recognizes a determinant common to the $V_{\beta 8}$ family expressed by about 20% of T cells in most mouse strains but absent in SJL, SWR, C57BR and C57L (Haskins *et al.*, 1984; Epstein *et al.*, 1985; Barth *et al.*, 1985; Roehm *et al.*, 1985; Sim & Augustin, 1985).

When KJ16-133 was used to stain thymocytes a very heterogenous pattern was observed. Mature thymocytes bound about as much antibody/cell as did mature peripheral T cells. KJ16-133 immature thymocytes, on the other hand, bound very variable amounts/cell ranging from practically none to levels equivalent to those of normal T cells. The frequency of KJ16-133-reactive cells in immature thymocytes, at 10%, was about two-thirds that of mature thymocytes and peripheral T cells. During thymocyte development receptors did not appear on the surfaces of these cells until day 17 of fetal life, in nice agreement with the data concerning $\alpha$ and $\beta$ chain mRNA expression. Immunoprecipitation with KJ16-133 or a polyclonal rabbit anti-receptor antibody showed that the receptor on thymocyte surfaces is indistinguishable biochemically from that of mature T cells.

Most of these results are not at all unexpected. In confirmation of molecular biological studies they indicate that early thymocytes bear no receptors, and that these proteins are expressed as these cells mature in the thymus. The fact that at no time does the level of receptor protein on immature cells exceed that on mature T cells suggests that selection of the T cell repertoire in the thymus cannot depend on super-expression of these proteins. As predicted by functional studies, expression of receptors on mature thymocytes seemed equivalent to that of their peripheral T cell relatives.

Perhaps the only unexpected result is the fact that the levels of receptor on immature thymocytes are very heterogenous, averaging about one-fifth that of peripheral T cells. Several explanations for this phenomenon come to mind. It is possible that some cells have only just begun to express $\alpha$, $\beta$ and T3 proteins, and the low amounts of receptor on their surfaces reflect this fact. This seems an unlikely explanation for all our results, since receptor levels are still low on virtually all thymocytes in newborn animals, 4 days after the protein had started to appear on cell surfaces. We therefore sought, and perhaps found, other causes as described below.


*Distribution of receptor on thymocytes*

We have used KJ16-133 in electron microscopic studies on the distribution of receptor proteins on thymocytes. Frozen sections showed that KJ16-133-bearing cells could be found as clusters of eight to ten cells in the medulla and two or three cells in the cortex. The existence and size of such clusters depends presumably on both clonal expansion and cell migration within a restricted region. It has long been known that dividing cells can be found in the cortex; however, the fact that only small clusters of KJ16-133 cells can be found in this part of the organ suggests either that most of this division occurs before receptors have appeared on the surfaces of these cells, or that cell migration is so rapid that the static picture afforded by electron microscopy cannot capture an image of lengthy clonal expansion. If the former explanation is true, this indicates that the process of clonal selection and expansion based on interaction with the T cell receptor may not be the basis for selection of the T cell repertoire for self-MHC.

We were surprised to find clusters of KJ16-133-bearing cells in the medulla since the literature on thymocytes has not suggested that medullary cells divide.

To examine the cellular location of receptor protein, thin sections of thymuses were stained with KJ16-133. Cells in different parts of the thymus had different locations for their receptors, as described below.

Three different kinds of KJ16-133-staining cells were found in the cortex. The most frequent of these contained reactive material in the perinuclear region, and no surface receptor. We assume that these cells have synthesized receptor $\beta$ chains, but that these cannot be expressed on the cell surfaces until $\alpha$ chains and/or T3 proteins are synthesized. A second type of cell, found only rarely, had both intracellular and surface receptor; this cell was presumably in transition. The third, and most interesting, type of cell bore surface receptor. These cells were always found in association with epithelial cells, which bear both Class I and Class II MHC molecules in the thymus. Receptors in these thymocytes were always capped on that part of the cell in contact with the epithelial cell. In the medulla, KJ16-133-reactive cells bore their receptors distributed evenly over their surfaces, as do peripheral T cells (A. Farr, S. Anderson, P. Marrack & J. Kappler, unpublished work).

Several points should be made about the distribution of surface receptors on cortical thymocytes. Since these cells bear low concentrations of these proteins, it is possible that the receptors cannot be detected by our methods unless they are capped and therefore easier to visualize by electron microscopy. Nevertheless, it is clear that a sizeable number of cortical thymocytes do bear receptors capped and in association with epithelial cells. This result suggests that some cortical thymocytes may bear low numbers of receptors because they are constantly being stripped of these molecules after binding to epithelial cells. In support of this idea we cultured suspensions of newborn thymocytes overnight, under conditions where most of the cells were no longer tightly packed and in contact with other cells. The percentage of cells bearing mature levels of receptor rose dramatically. We could not account for this result by selective death of cells *in vitro*, since 70% of cells survived our procedure. We therefore concluded that the receptors on a sizeable proportion of thymocytes were indeed being removed constantly in the thymus, but not when the cells were cultured at low density, as predicted by our epithelial cell–thymocyte contact stripping hypothesis.

The fact that receptor capping was seen only in association with epithelial cells, and not dendritic cells or macrophages, may offer an explanation for cellular immunological studies which have shown that both self-restriction and self-tolerance are imposed by the MHC antigens of epithelial cells. We do not yet know which of these processes is recurring during the capping we observed. However, the fact that capped cortical cells are quite frequent, and that receptor levels are low on nearly all cortical cells, may indicate that selection of epithelial cells with receptors may occur for practically all immature thymocytes.

The authors would like to thank Drs. E. Palmer and J. Yagüe, Dr. J. Cambier and D. DiGiusto and S. Anderson for their help in molecular biological experiments, fluorescent cell analysis and electron microscopic experiments respectively. They are also grateful to James Leibson, Elenora Kushnir and Janice White for technical support and to Kelly Bakke and Donna Thompson for patient secretarial help. This work was supported by USPHS grants AI-18785 and AG-04360 and ACS research grants IM-49 and IM-319. J.W.K. was supported in part by a Faculty Research Award from the American Cancer Society.

# References

Allison, J., McIntyre, B. & Bloch, D. (1982) *J. Immunol.* **129**, 2293–2300

Barth, R. K., Kim, B. S., Lau, N. D., Hunkapiller, T., Sobieck, N., Winoto, A., Gershenfeld, H., Okada, C., Hansburg, D., Weissman, I. L. & Hood, L. (1985) *Nature (London)* **316**, 517–523

Bevan, M. (1977) *Nature (London)* **269**, 417–419

Born, W., Yagüe, J., Palmer, E., Kappler, J. & Marrack, P. (1985) *Proc. Natl. Acad. Sci. U.S.A.* **82**, 2925–2929

Chien, Y., Gascoigne, N., Kavaler, J., Lee, N. & Davis, M. (1984a) *Nature (London)* **309**, 322–326

Chien, Y., Becker, D., Lindsten, T., Okamura, M., Cohen, D. & Davis, M. (1984b) *Nature (London)* **312**, 31–35

Epstein, R., Roehm, N., Marrack, P., Kappler, J., Davis, M., Hedrick, S. & Cohn, M. (1985) *J. Exp. Med.* **161**, 1219–1224

Haskins, K., Kubo, R., White, J., Pigeon, M., Kappler, J. & Marrack, P. (1983) *J. Exp. Med.* **157**, 1149–1169

Haskins, K., Hannum, C., White, J., Roehm, N., Kubo, R., Kappler, J. & Marrack, P. (1984) *J. Exp. Med.* **160**, 452–471

Hayday, A. C., Saito, H., Gilles, S. D., Kranz, D. M., Tonegawa, G., Eisen, H. M. & Tonegawa, S. (1985) *Cell* **40**, 259–269

Heber-Katz, E., Schwartz, R., Matis, L., Hannum, C., Fairwell, T., Appella, E. & Hansburg, D. (1982) *J. Exp. Med.* **155**, 1086–1099

Hedrick, S., Cohen, D., Nielsen, E., Kavaler, J., Cohen, D. & Davis, M. (1984) *Nature (London)* **308**, 153–158

Hünig, T. & Bevan, M. (1980) *J. Exp. Med.* **151**, 1288–1298

Jordan, R. K., Robinson, J. H., Hapkinson, N. A., House, K. C. & Bentley, A. L. (1985) *Nature (London)* **314**, 454–457

Kappler, J., Skidmore, B., White, J. & Marrack, P. (1981) *J. Exp. Med.* **153**, 1198–1214

Kavaler, J., Davis, M. & Chien, Y. (1984) *Nature (London)* **310**, 421–423

Malissen, M., Minard, K., Mjolsness, S., Kronenberg, M., Goverman, J., Hunkapillar, T., Prystowsky, M., Yoshikai, Y., Fitch, F., Mak, T. & Hood, L. (1984) *Cell* **37**, 1101–1110

Marrack, P., Shimonkevitz, R., Hannum, C., Haskins, K. & Kappler, J. (1983) *J. Exp. Med.* **158**, 1635–1646

McIntyre, B. S. & Allison, J. P. (1984) *Cell* **38**, 659–665

Meuer, S., Acuto, D., Hussey, R., Hodgdon, J., Fitzgerald, K., Schlossman, S. & Reinherz, E. (1983) *Nature (London)* **303**, 808–810

Raulet, D. H., Gasman, R. D., Saito, H. & Tonegawa, S. (1985) *Nature (London)* **314**, 103–107

Roehm, N., Herron, L, Cambier, J., DiGiusto, D., Haskins, K., Kappler, J. & Marrack, P. (1984) *Cell* **38**, 577–584

Roehm, N., Epstein, R., Cohn, M., Taylor, B., Kushnir, E., Marrack, P. & Kappler, J. (1985) *J. Immunol.* **135**, 2176–2182

Saito, H., Kranz, D., Takagake, Y., Hayday, A., Eisen, H. & Tonegawa, S. (1984) *Nature (London)* **312**, 36–40

Samelson, L. E., Lindsten, T., Fowlkes, B. J., van der Elsen, P., Terhorst, C., Davis, M. M., Germain, R. N. & Schwartz, R. H. (1985) *Nature (London)* **315**, 765–767

Sim, G. K. & Augustin, A. (1985) *Cell* **42**, 89–92

Sim, G., Yagüe, J., Nelson, J., Marrack, P., Palmer, E., Augustin, A. & Kappler, J. (1984) *Nature (London)* **312**, 771–775

Snodgrass, H. R., Kisielow, P., Kiefer, M., Steinmetz, M. & van Sochmer, H. (1985) *Nature (London)* **313**, 592–595

Sui, G., Kronenberg, M., Strauss, E., Haars, R., Mak, T. & Hood, L. (1984) *Nature (London)* **311**, 344–350

Yanagi, Y., Yoshikai, Y., Leggett, K., Clark, S., Aleksander, I. & Mak, T. (1984) *Nature (London)* **308**, 145–149

Zinkernagel, R., Callahan, G., Althage, A., Cooper, S., Klein, P. & Klein, J. (1978) *J. Exp. Med.* **147**, 882–896

*Biochem. Soc. Symp.* **51**, 205–209
*Printed in Great Britain*

# Paradigm Regained: Similarities and Differences between T Cell Receptor and Immunoglobulin Genes

MARK M. DAVIS, TULLIA LINDSTEN, NICHOLAS R. J. GASCOIGNE, CHRISTOPHER GOODNOW and YUEH-HSIU CHIEN

*Department of Medical Microbiology, Stanford University School of Medicine, Stanford, CA 94305, U.S.A.*

## Introduction

With the clarity of hindsight, one might say that progress was only made in isolating and characterizing T cell receptors when significant portions of the prevailing notions about their use of immunoglobulin genes were dispensed with [1] and replaced with experiments designed to deal with related yet distinct molecules. The first successes in this regard were the discovery of clonotypic structures on the surface of T cell tumours [2] and then on functional lines and hybrids [3–6], and of the biochemical definition of the $\alpha:\beta$ heterodimer [7,8] on the surface of T cells which seems responsible for antigen MHC specificity. The isolation and characterization of the cDNA clones for the $\alpha$ and $\beta$ chain [9–14] and the genes encoding them [15–18] have somewhat paradoxically indicated the highly immunoglobulin-like character of T cell receptor molecules and yet may also provide the means by which the peculiarities of T cell recognition can be more properly explored.

## T Cell Receptor Gene Structure

Analyses of the $\alpha$, $\beta$ and the anomalous $\gamma$ chain of the mouse [15–20] all indicate the strongly immunoglobulin-like character of the T cell receptor, with distinct J, V and C region elements; D in the case (at least) of $\beta$ chains. The C regions are divided into exons with plausible functional properties, as are immunoglobulins [21]. Rearrangement of variable region elements (V, D, and J) occur in a manner indistinguishable from that of antibody genes [12]. This similarity extends to the seven- and nine-nucleotide sequences thought to mediate rearrangement as well. Recent findings have even indicated that T cell receptor D-J rearrangements can be induced to occur in B cell lines (K. A. Blackburn, personal communication). These data all indicate, with minor variations, the commonality of T cell and B cell antigen receptors and how the intuition of earlier workers was essentially correct, although the route by which it has now been demonstrated is circuitous to an extreme. One challenge now becomes the identification of in what ways T cell receptors differ from immunoglobulin and how such differences might explain the unique aspects of T cell recognition.

### Differences from Immunoglobulin

Several possibly significant differences between immunoglobulins and T cell receptor genes have been noted.

1. The spacing of the conserved seven- and nine-nucleotide elements separated by 12 and 23 nucleotides (on average) is different in $\beta$ chain genes than IgH genes, allowing potential D-D rearrangements (to form a VDDJ exon presumably) and V-J rearrangements to form directly [12,21]. The latter rearrangement has been observed [22,23], but does not appear common.

2. D region elements may be read in different frames [21,24]. This is very frequent in T cell receptors and rarely, if ever, seen in immunoglobulins [24], but may be compensated for by a smaller number of D regions ([21], M. M. Davis, unpublished work).

3. There are many more J regions in the T cell receptors $\alpha$ and $\beta$ chain loci; 11–12 functional $J_\beta$ regions [16,21], and perhaps 40 or more $J_\alpha$ regions [18,25,26]. This compares with four $J_H$ regions and four $J_\kappa$ regions in the mouse [27]. Thus, the number of possible $J_\alpha J_\beta$ combinations ($11 \times 40 = 440$) in T cell receptors greatly exceed the number of possible $J_H J_\kappa$ combinations ($4 \times 4 = 16$) in immunoglobulins, assuming an immunoglobulin-like three-dimensional structure.

4. Smaller $V_\beta$ and $V_\alpha$ repertoires. In contrast with the much larger number of T cell receptor J regions, there appears to be a significantly smaller repertoire of $V_\beta$ sequences utilized, perhaps as few as 15 that occur repeatedly [24,28]. Although there appear to be more $V_\alpha$ sequences [24,26], the considerable overlap in the available data indicates that it is unlikely that there will be many more than 50 $V_\alpha$ genes in a genome. This compares with estimates of 100–300 for the $V_\kappa$ and $V_H$ repertoires in the mouse [29,30]. Thus, in this case, we can estimate the approximate number of possible $V_\alpha V_\beta$ pairings at ($15 \times 50 = 750$), whereas current figures for $V_H V_\kappa$ ($100 \times 300 = 30\,000$) indicate a 40 or so fold difference, in exactly the opposite direction as that of the J region comparison above. Therefore, the overall combinatorial diversity available to T cell receptor genes is approximately equal to that of immunoglobulins, the difference being that in the former there would appear to be many more V region combinations and in the latter many more V region pairings.

5. Somatic hypermutation may not contribute significantly to diversity in T cell receptors. A number of reports [15,18,24] have indicated an apparent lack of somatic mutation in T cell receptor $V_\beta$ genes relative to what one might expect for the variety of instances documented in immunoglobulin genes [27,31]. The one exception to these reports is the data of Augustin & Sim ([32], A. A. Augustin & G. K. Sim, personal communication) where mutations are observed in the T cell receptors of hybridoma cultured *in vitro*. The relevance of the latter case to the situation *in vivo* is not yet clear, however; at least preliminarily, it appears that somatic mutation may be confined to immunoglobulin genes at least as a major mechanism of diversity.

6. Hypervariable and framework regions may be different in T cell receptor variable regions. Comparisons of $V_\beta$ sequences have suggested [26,28] that additional hypervariable regions in the protein may form a second site of

interaction on the outside of the T cell receptor heterodimer. In other collections of $V_\beta$ sequences [24], these additional peaks of variability are not so prominent and it becomes a matter of speculation whether or not these regions are functionally more variable than immunoglobulin. The $V_\alpha$ and $V_\beta$ sequences are, however, significantly more diverse than those of immunoglobulins in their overall amino acid homologies (30–36% average homology versus 48% for a comparable number of the $V_H$ sequences [26]), but this may be a product of a longer evolutionary history.

7. The $\gamma$ chain enigma. Potentially the most significant, and certainly the least expected, finding in T cell receptor gene isolation has been the isolation of the so-called $\gamma$ chain [18]. This locus is as immunoglobulin-like as either $\alpha$ or $\beta$ and has been postulated to play a role in early thymocyte selection [34], possibly in association with the $\beta$ chain. Recent characterizations of a putative $\gamma$ chain protein have cast some doubt on this model (H. Eisen *et al.*, personal communication), however but until its role is more precisely defined it remains a 'wild card' in the mechanics of T cell recognition.

## Expression of Soluble T Cell Receptor Molecules

Understanding how the above-mentioned differences between T cell receptors and immunoglobulins might be significant in terms of how T cells recognize antigen/MHC determinants simultaneously will be possible only when we are able to study the chemical structure of the receptor and its binding properties directly. In pursuing this objective, members of my laboratory (N. R. J. Gascoigne & C. Goodnow), in collaboration with V. Oi at Beckton-Dickinson, have been constructing immunoglobulin–T cell receptor hybrid genes and transfecting them into plasma cell lines. Initial experiments with a $V_\beta$–$C_\kappa$ hybrid are promising in that large amounts of protein reactable with anti-$\kappa$ sera are produced in cells which have stably incorporated the hybrid genes. We hope that we would be able to have plasma cell lines which can secrete the $V_\beta C_\kappa : V_\alpha C_H$ hybrid molecule that could be used in binding studies and eventually in X ray structural work.

## T Cell Receptor Ontogeny

Another of the mysteries surrounding T cell receptors has been the question of where receptors are first expressed and what precise role the thymus has in the repertoire selection. In collaboration with L. Samelson and his coworkers at NIH, we have determined that $\alpha$ chain mRNA is not expressed in very early thymocytes, but $\beta$ chain and $\gamma$ chain mRNA is ([33] and unpublished work; see also [34]). T cell hybrids made from these early ('double negative') cells also indicate that there is a subpopulation (one-third of the cells) within this subset which has not yet rearranged its $\beta$ chain genes [33]. Detection of $\alpha$ chain rearrangement is much more difficult and has required the extreme measures described below. It seems clear, however, from these (and other studies in this volume) that $\alpha$ and $\beta$ chain genes must (in most cases) be rearranged and expressed first in the thymus and that only there could they be acted upon by

MHC determinants. The corollary to this is that perhaps a great deal, approx. 50%, of thymic cell death might be the result of improper rearrangement of the $\alpha$ or $\beta$ chain genes. Another interesting finding with the early T cell hybridomas is that many of them have rearranged their $\gamma$ chain genes from the germ line pattern (T. Lindsten *et al.*, unpublished work). Details of this work indicate that the $\gamma$ chain gene is rearranging at approximately the same time in ontogeny as the $\beta$ chain gene.

### Detection of $\alpha$ Chain Rearrangements

Because of the extreme size of the $J_\alpha$–$C_\alpha$ locus ($> 70$ kb) [19], it becomes a difficult problem to detect rearrangements by conventional Southern analysis requiring four or more separate probes from different portions of the locus. Even then, it is hard to know whether all of the $J_\alpha$ regions would be included in the analysis since new sequences are continually being found. In an attempt to deal with this problem, we (T. Lindsten *et al.*, unpublished work) have adapted the pulse-field gel electrophoresis technique of Schwartz & Cantor [35] using the apparatus of Carle & Olson [36] that enables one to electrophorese very large fragments of DNA through ordinary agarose gels. By restriction digesting DNA with relatively infrequent enzymes, Southern blotting and hybridizing the $C_\alpha$ and $J_\alpha$ probes, we have been able to visualize very large fragments containing these sequences and to detect rearrangements in a number of T cell tumours and hybridomas. This method of analysis is generally useful for mapping complex loci and may enable us to link up $V_\alpha$ with $C_\alpha$ (for example) and to address other questions that might otherwise be difficult to answer with existing technologies.

### References


1. Jensenius, J. C. & Williams, A. F. (1982) *Nature (London)* **300**, 583–588
2. Allison, J. P., McIntyre, B. W. & Bloch, D. (1982) *J. Immunol.* **129**, 2293–2300
3. Haskins, K., Kubo, R., White, J., Pigeon, M., Kappler, J. & Marrack, P. (1983) *J. Exp. Med.* **157**, 1149–1169
4. Meuer, S. C., Fitzgerald, K. A., Hussey, R. E., Hodgdon, J. C., Schlossman, S. F. & Reinherz, E. L. (1983) *J. Exp. Med.* **157**, 705–715
5. Samelson, L. E., Germain, R. N. & Schwartz, R. H. (1983) *Proc. Natl. Acad. Sci. U.S.A.* **80**, 6972–6976
6. Kaye, J., Porcelli, S., Tite, J., Jones, B. & Janeway, C. A., Jr. (1983) *J. Exp. Med.* **158**, 836–856
7. Acuto, O., Hussey, R. E., Fitzgerald, K. A., Protentis, J. P., Meuer, S. C., Schlossman, S. F. & Reinherz, E. L. (1983) *Cell* **34**, 717–726
8. Kappler, J., Kubo, R., Haskins, K., Hannum, C., Marrack, P., Piegon, M., McIntyre, B., Allison, J. & Trowbridge, I. (1983) *Cell* **35**, 295–302
9. Hedrick, S. M., Cohen, D. I., Nielsen, E. A. & Davis, M. M. (1984) *Nature (London)* **308**, 149–153
10. Hedrick, S. M., Nielsen, E. A., Kavaler, J., Cohen, D. I. & Davis, M. M. (1984) *Nature (London)* **308**, 153–158
11. Yanagi, Y., Yoshikai, Y., Leggett, K., Clark, S. P., Aleksander, I. & Mak, T. W. (1984) *Nature (London)* **308**, 145–149
12. Chien, Y., Becker, D. M., Lindsten, T., Okamura, M., Cohen, D. I. & Davis, M. M. (1984) *Nature (London)* **312**, 31–35
13. Saito, H., Kranz, D. M., Takagaki, Y., Hayday, A. C., Eisen, H. N. & Tonegawa, S. (1984) *Nature (London)* **312**, 36–40
14. Sim, G. K., Yague, J., Nelson, J., Marrack, P., Palmer, E., Augustin, A. A. & Kappler, J. (1984) *Nature (London)* **312**, 771–775
15. Chien, Y., Gascoigne, N. R. J., Kavaler, J., Lee, N. E. & Davis, M. M. (1984) *Nature (London)* **309**, 322–326

16. Gascoigne, N. R. J., Chien, Y., Becker, D. M., Kavaler, J. & Davis, M. M. (1984) *Nature (London)* **310**, 387–391
17. Siu, G., Kronenberg, M., Strauss, E., Haars, R., Mak, T. W. & Hood, L. (1984) *Nature (London)* **311**, 344–350
18. Hayday, A. C., Diamond, D. J., Tanigawa, G., Heilig, J. S., Saito, H. & Tonegawa, S. (1985) *Nature (London)* **316**, 828
19. Winoto, A., Mjolsness, S. & Hood, L. (1985) *Nature (London)* **316**, 832–836
20. Hayday, A. C., Saito, H., Gillies, S. D., Kranz, D. M., Tanigawa, G., Eisen, H. N. & Tonegawa, S. (1985) *Cell* **40**, 259–269
21. Davis, M. M. (1985) *Annu. Rev. Immunol.* **3**, 537–560
22. Yoshikai, Y., Anatoniou, D., Clark, S. P., Yanagi, Y., Sangster, R., Van den Elsen, P., Terhorst, C. & Mak, T. W. (1984) *Nature (London)* **312**, 521–524
23. Jones, N., Leiden, J., Dialynas, D., Fraser, J., Clabby, M., Kishimoto, T., Strominger, J., Andrews, D., Lane, W. & Woody, J. (1984) *Science* **227**, 311–314
24. Barth, R. K., Kim, B. S., Lan, N. C., Hunkapiller, T., Sobieck, N., Winoto, A., Gershenfeld, H., Okada, C., Hansburg, D., Weissman, I. L. & Hood, L. (1985) *Nature (London)* **316**, 517–523
25. Arden, B., Klotz, J. L., Siu, G. & Hood, L. (1985) *Nature (London)* **316**, 783–787
26. Becker, D. M., Patten, P., Chien, Y., Yokota, T., Eshhar, Z., Giedlin, M., Gascoigne, N. R. J., Goodnow, C., Wolf, R., Arai, K. & Davis, M. M. (1985) *Nature (London)* **316**, 430–434
27. Tonegawa, S. (1983) *Nature (London)* **302**, 575–581
28. Patten, P., Yokota, T., Rothbard, J., Chien, Y., Arai, K. & Davis, M. M. (1984) *Nature (London)* **312**, 40–46
29. Seidman, J. G., Leder, A., Nau, M., Narman, P. & Leder, P. (1978) *Science* **202**, 11–15
30. Brodeur, P. H. & Riblet, R. (1984) *Eur. J. Immunol.* **14**, 922–930
31. Griffiths, G. M., Berek, C., Kaartinen, M. & Milstein, C. (1984) *Nature (London)* **312**, 271–275
32. Augustin, A. A. & Sim, G. K. (1984) *Cell* **39**, 5–12
33. Samelson, L. E., Lindsten, T., Fowlkes, B. J., Van den Elsen, P., Terhorst, C. & Davis, M. M. (1985) *Nature (London)* **315**, 765–768
34. Raulet, D. H., Garman, R. D., Saito, H. & Tonegawa S. (1985) *Nature (London)* **314**, 103–107
35. Schwartz, D. C. & Cantor, C. R. (1984) *Cell* **37**, 67–75
36. Carle, G. F. & Olson, M. V. (1984) *Nucleic Acids Res.* **12**, 5647–5664

*Biochem. Soc. Symp.* **51**, 211–232
*Printed in Great Britain*

# The Ontogeny, Structure and Function of the Human T Lymphocyte Receptor for Antigen and Major Histocompatibility Complex

ELLIS L. REINHERZ, HANS DIETER ROYER, THOMAS J. CAMPEN, DUNIA RAMARLI, MARINA FABBI and ORESTE ACUTO

*Division of Tumor Immunology, Dana-Farber Cancer Institute, and the Departments of Medicine and Pathology, Harvard Medical School, Boston, MA 02115, U.S.A.*

## Synopsis

Recent studies using cloned antigen-specific T lymphocytes and monoclonal antibodies directed at their various surface glycoprotein components have led to identification of the human T cell antigen receptor as a surface complex comprised of a clonotypic 90 kDa Ti heterodimer and the invariant 20 and 25 kDa T3 molecules. Approximately 30000–40000 Ti and T3 molecules exist on the surface of human T lymphocytes. These glycoproteins are acquired and expressed during late thymic ontogeny, thus providing the structural basis for immunological competence. The $\alpha$ and $\beta$ subunits of Ti bear no precursor–product relationship to one another and are encoded by separate germline V, D, J and C segments which rearrange during intrathymic differentiation to form an active gene set. Triggering of the T3–Ti receptor complex induces a rapid increase in free cytoplasmic $Ca^{2+}$ and gives rise to specific antigen-induced proliferation through an autocrine pathway involving endogenous interleukin-2 production, release and subsequent binding to interleukin-2 receptors. The implications of these findings for understanding of human T cell growth and its regulation in disease states are discussed.

## Introduction

T lymphocytes, unlike B lymphocytes, predominantly recognize antigen in the context of membrane-bound products of the major histocompatibility complex (MHC) [1–5]. This 'dual' recognition is important for activation of T cells with cytotoxic effect function as well as immunoregulatory activities. With regard to the former, T cells can lyse specific target cells, including those that have been infected with viruses and carry viral antigens [7–10]. Moreover, they regulate the activity of cells within the immune system such as T cells, B cells and macrophages, as well as haematopoeitic stem cells, fibroblasts, osteoclasts and other cell types extrinsic to the lymphoid system [11–19]. Characterization of T cell receptors encoding various specificities could be particularly valuable for a molecular understanding of the cellular interactions underlying these activities.

Given that T lymphocytes recognize antigens in a precise fashion, discriminative surface structures restricted in their expression to individual T cell clones (i.e. clonotypic) must exist. The recent development of technologies to propagate continually clonal populations of human T lymphocytes *in vitro* [20–24] has provided a new basis for identification of such clonotypic recognition determinants. Thus, clonal human T cell populations of predefined specificities have been used as immunogens to produce a series of clone-specific murine monoclonal antibodies. These anticlonotypic antibodies identify a novel class of 90 kDa heterodimers, termed Ti, that are membrane-associated with the previously described 20 and 25 kDa T3 glycoproteins present on all mature human T lymphocytes [25,26].

With this approach it is possible to provide compelling evidence for the notion that each T lymphocyte, regardless of subset derivation, specificity or regulatory activity, uses an analogous T3-associated Ti heterodimer (one $\alpha$ and one $\beta$ subunit) as its receptor for antigen and MHC. The subunits of the Ti heterodimer are products of gene segments which rearrange in T lineage cells to give rise to an active set of Ti $\alpha$ and Ti $\beta$ genes. The mechanism by which triggering of the T3–Ti molecular complex results in clonal T cell population and its linkage to an interleukin-2 (IL-2) dependent autocrine pathway will also be discussed.

## T4+ and T8+ Cytotoxic T Lymphocytes Recognize Different MHC Gene Products

A number of human T cell lineage restricted surface glycoproteins have been defined by monoclonal antibodies. Thus, for example, each mature T lymphocyte expresses the 20 and 25 kDa T3 glycoproteins. The latter appear in late intrathymic ontogeny at the time of acquisition of immunological competence and play a central role in T cell function [27–33]. Moreover, among T3+ T lymphocytes, two subsets were found that exhibit unique regulatory and effector activities. On the basis of their unique 62 and 76 kDa membrane markers, these were termed T4+ and T8+, respectively [17–19,34]. The T4+ subset (representing approximately two-thirds of peripheral T cells) was shown to provide inducer/helper activities for T–T, T–B and T–macrophage interactions, whereas the T8+ subset (one-third of peripheral T cells, reciprocal to T4+ T cells) principally functioned in a suppressive mode [35,36]. Although both subsets of cells proliferated to alloantigen (foreign cell surface determinants) in mixed lymphocyte culture (MLC), the vast majority of cytotoxic effector function was detected in the T8+ population. Moreover, development of cytotoxicity by T8+ cells in general required interactions with T4+ cells or their soluble products. In contrast, only a minor component of cytotoxic effector or function resided within the T4+ subset, and this was maximal when T4+ cells alone were sensitized in MLC [34].

To characterize individual cytotoxic effector lymphocytes (CTL) in humans, we developed a strategy to clone and propagate antigen-specific T cell populations *in vitro* [24]. This involved expansion *in vitro* with IL-2, a T cell specific growth factor, and frequent restimulation with specific alloantigen. As expected from the above data with purified T cells, the majority of cytotoxic T cell clones was derived from the T8+ subset.

The specificity of T4+ and T8+ clones and subclones was also characterized. T8+ clones were found to kill targets that shared class I MHC antigens (HLA-A,B,C) with the original stimulator cells whereas cytotoxic T4+ clones were directed at class II MHC antigens (Ia-related). These results indicated that the vast majority of T4+ and T8+ T lymphocytes had receptors for different classes of MHC antigens (T8–class I and T4–class II correlation) [24,37–39]. Analogous findings have been reported from studies with autoreactive T cell clones as well.

## Effects of Monoclonal Antibodies Directed at Monomorphic Structures on Antigen-Specific Function

The association between the surface phenotype (that is, surface glycoproteins) of CTL and the class of MHC molecules recognized implied that the subset restricted structures, T4 and T8, might be required to facilitate selective lysis of different target antigens. To determine if individual anti-T4 and anti-T8 antibodies influenced killing function, cytotoxic T cell clones were preincubated with T cell specific monoclonal antibody or medium alone prior to the CML assay. It was found that anti-T8 did not diminish the level of killing by MHC class II restricted T4+ clones but markedly decreased cytotoxicity mediated by the MHC class I restricted T8+ clones [24,40]. In contrast, anti-T4 preincubation resulted in > 80% reduction of cytotoxicity by the T4+ clones but had no effect on killing by the T8+ clones. In addition, anti-T3, unlike anti-T4 and anti-T8, inhibited the killing by both T4+ and T8+ clones. Moreover, anti-T3 was unique among antibodies that define other mature T cell surface structures since other antibodies did not inhibit the cytotoxic capacity of all clones even when used in saturating concentrations. Thus, the observed inhibition was not simply a function of antibody binding to a clonal effector population.

These results implied that at least several surface molecules were important in CML: T3 and T4 molecules on T4+ clones, and T3 and T8 molecules on T8+ clones [41]. Since other studies showed that CML was restored by lectin approximation of killer and target cells even in the presence of monoclonal antibodies to these molecules, it appeared that T3, T4 and T8 were involved in recognition events rather than the lytic mechanism. In further support of this notion was the observation that the target cell specificity of CTL clones was abrogated by lectin approximation.

Antigen-specific CTL clones recognizing autologous B lymphoblastoid lines transformed by Epstein–Barr virus (EBV) are also governed by a series of recognition elements identical to the clones directed at allogeneic targets [9,10,42]. Specifically, CTL expressing the T8 phenotype recognize the autologous B lymphoblastoid line in the context of class I MHC molecules. Anti-HLA antibodies block CTL effector function at the target level, and anti-T8 antibodies abrogate the ability of these class I specific killers to kill. In contrast, T4+ CTL recognize Ia (class II) determinants on the autologous lymphoblastoid cell and are inhibited by anti-T4 but not by anti-T8 antibodies. As with the allogeneic

CTL, anti-T3 antibody preincubation inhibits the killing ability of both T4+ and T8+ effector T cells. Because these clones fail to lyse autologous B cells not infected with Epstein-Barr virus or pokeweed mitogen-stimulated B cell blasts, such T4+ and T8+ effectors seem to recognize virally encoded surface glycoproteins in association with class II or class I molecules, respectively. Thus, they are said to show either a class II or class I MHC restriction. At present, however, these viral proteins have not been identified.

**The T3 Molecular Complex Modulates from the T Cell surface**

In the case of the 20 and 25 kDa T3 surface molecule, their appearance in late intrathymic ontogeny at the time immunological competence was acquired and their critical role in T lymphocyte function suggested that T3 was closely linked to an important recognition receptor or cell–cell interaction molecule [28–33]. Antibodies directed against T3 were unique in the ability to block the induction phase (mixed lymphocyte culture) as well as the effector phase (lysis) of CML, to inhibit T lymphocyte proliferative responses to soluble antigen, and to be mitogenic for resting T cells. In this latter case, activation was accompanied by release of various T cell lymphokines including IL-2 and $\gamma$-interferon [43,44].

Given the central role of the T3 molecules in human T cell function and the known rapid surface receptor loss (modulation) that occurs with a variety of hormone and growth factor receptors after ligand binding, we examined the capacity of anti-T3 to induce modulation of T3. Preincubation of peripheral blood T cells or T cell clones with anti-T3 for 18 h at 37 °C resulted in loss of cell surface T3 antigen [31,41]. In contrast, anti-T3 modulated cells were unaffected in their expression of cell surface T4 or T8 antigen. Modulation of T3 antigen with anti-T3 antibody had no effect on the viability of T cell clones. Furthermore, once anti-T3 was removed from cell culture supernatants, T3 antigen was re-expressed within 48 h [31].

**Functional Effects Resulting from Anti-T3 Antibody Induced Modulation of the T3 Surface Complex on Antigen-Specific Human T Cell Clones**

One could argue that the inhibitory effects of anti-T3 monoclonal antibodies [29–31,41] on CTL function were indirect and occurred as a consequence of antibody-induced agglutination of cells or steric blockade of still undefined but functionally important surface determinants. However, since anti-T3 led to T3 modulation by selective shedding of both the T3 antigen and the anti-T3 directed towards it without alteration of cell viability or change in the density of other T cell antigens (including T1, T4, T8, T11, T12 and Ia), it was possible to examine the lytic activity of T3 modulated cells in the absence of surface-bound monoclonal antibody that could affect steric blockade [31,41].

Anti-T3 modulation of T4 and T8 clones markedly reduced cytotoxic effector function to less than 25% of maximal. Perhaps more importantly, cytotoxic function increased to approx. 60% of maximal when T4 and T8 clones were tested 1 day after modulation and had achieved maximal levels 2 days after modulation. Given the fact that 48 h were required to complete T3 antigen

Table 1. *Monoclonal antibodies influencing T cell function*

| Antibody | Molecular mass (kDa) | Functional effect on: | | |
| --- | --- | --- | --- | --- |
| | | Antigen-induced proliferation | CTL Function | IL-2 response |
| Anti-T3 | 20/25 | ↓ | ↓ | ↑ |
| Anti-T4 | 58 | 0 or ↓ | ↓ | 0 |
| Anti-T8 | 72 | 0 or ↓ | ↓ | 0 |

re-expression after anti-T3 modulation of T4 and T8 clones, this re-establishment of cytotoxic T lymphocyte effector function appeared temporally related.

Since many cytotoxic T lymphocyte clones display specific proliferative responses when confronted by alloantigen, we also determined whether anti-T3 modulation influenced antigen recognition [31]. Anti-T3 modulated clones had a significantly reduced proliferative response to the alloantigen in comparison with unmodulated clones. In contrast, they mounted a greater proliferative response to IL-2 than the unmodulated clones. Taken together, these studies suggested that modulated loss of T3 surface molecules interferes with T cell antigen recognition and that this is a specific effect since IL-2 responsiveness is enhanced and not diminished. Thus, the functional consequences of anti-T3 induced modulation could not be explained on the basis of a nonspecific diminution in cell responsiveness. The relevance of enhanced IL-2 responsiveness after anti-T3 modulation will become apparent later (see below).

### Clonotypic Surface Structures Involved in Antigen-Specific Human T Cell Function

Given that T lymphocytes recognize antigen in a precise fashion, there had to exist, in addition, discriminative surface recognition structures unique to individual T cell clones. To delineate such 'clonotypic' molecules, we produced monoclonal antibodies against several human cytotoxic T cell clones, and developed a screening strategy that selected for such anticlonotypic antibodies.

Unlike anti-T3, which blocked the cytotoxic effector functions of all clones, anticlonotypic antibodies (termed anti-Ti) selectively blocked killing of that individual clone to which they were directed. Also in contrast to anti-T3, anticlonotypic antibodies only blocked the antigen-specific proliferative capacity of that individual clone. Pretreatment of a clone with the appropriate anti-Ti antibody did, however, augment the clones proliferative response to human IL-2. These functional effects are summarized in Table 1, and are compared with those of other antibodies. Note that pretreatment of a given clone with an irrelevant anticlonotypic antibody had no effect on T cell function, even when clones were derived from the same individual. Hence, the anti-Ti antibodies have been termed non-crossreactive.

## Ti and T3 are Associated in the Cell Membrane

The observation that anti-T3 as well as anti-Ti antibody specific for a given clone could (a) inhibit both antigen-specific proliferation and CTL effector function of this clone and (b) enhance its IL-2 responsiveness suggested a relationship between the cell surface structures defined by these antibodies. This was born out in immunofluorescence studies where it was found that anti-T3 induced modulation of T3 also resulted in loss of anti-Ti reactivity. This was not a nonspecific effect since the T8 or T4 antigen density was uninfluenced by this process [26].

Given the strong evidence that the clonotypic structures were involved in antigen recognition and membrane associated with T3, it was important to quantify the number of surface Ti and T3 molecules on representative T cell populations. Three points emerged from such analysis. First, the number of T3 and Ti molecules on each of the clonal populations was similar despite some interclonal variation in the absolute number of T3 and Ti molecules (30000–40000 molecules). This finding suggested a stoichiometric relationship in which one T3 molecule is linked to one Ti molecule in the cell membrane. Second, on the clonal populations, the density of the associative recognition structures T4 and T8 was far greater than T3/Ti (118000 and 175000 binding sites per cells, respectively). Third, in contrast with T3 (and presumably Ti) which is expressed to a similar degree on the resting T lymphocytes and T cell clones, there were three to four times fewer T4 and T8 molecules on resting lymphocytes of the appropriate subset derivation (30000 and 60000, respectively) than on the respective activated clonal populations [25]. Thus, T cell activation may result in expression of additional T4 and T8 molecules. This probably offers one explanation for their critical involvement in clonal effector function. In contrast, the number of antigen receptors defined by anti-T3/Ti remains comparable and hence fully expressed in the resting state before antigen binding.

## Ti: a 90 kDa Disulphide-Linked Heterodimer

Since all the above studies demonstrated that there existed a close functional and phenotypic relationship between 20/25 kDa T3 glycoproteins and the Ti clonotype, it was important to define biochemically the surface molecules detected by anti-Ti monoclonal antibodies. Thus, solubilized membrane preparations were obtained from two representative externally $^{125}$I-labelled clones, $CT4_{II}$ and $CT8_{III}$; antigens defined by anti-Ti antibodies were precipitated and electrophoresed on SDS/polyacrylamide gels [25]. As shown in Fig. 1, the molecule precipitated by anti-$Ti_1$ (specific anticlonotype) and $^{125}$I-labelled $CT8_{III}$ cells appears as two bands and consists of a 49 kDa $\alpha$ chain and a 43 kDa $\beta$ chain in reducing conditions (lane a). Moreover, in nonreducing conditions, this structure appears as a single band at approx. 90 kDa (lane c). In contrast, anti-$Ti_1$ does not immunoprecipitate material from the $^{125}$I-labelled $CT4_{II}$ clone (lanes f and h). This is not surprising since anti-$Ti_1$ reacts with $CT8_{III}$ but not with $CT4_{II}$ by indirect immunofluorescence. In a reciprocal fashion, anti-$Ti_2$ precipitates material from $^{125}$I-labelled $CT4_{II}$ (lanes e and g) but not $CT8_{III}$

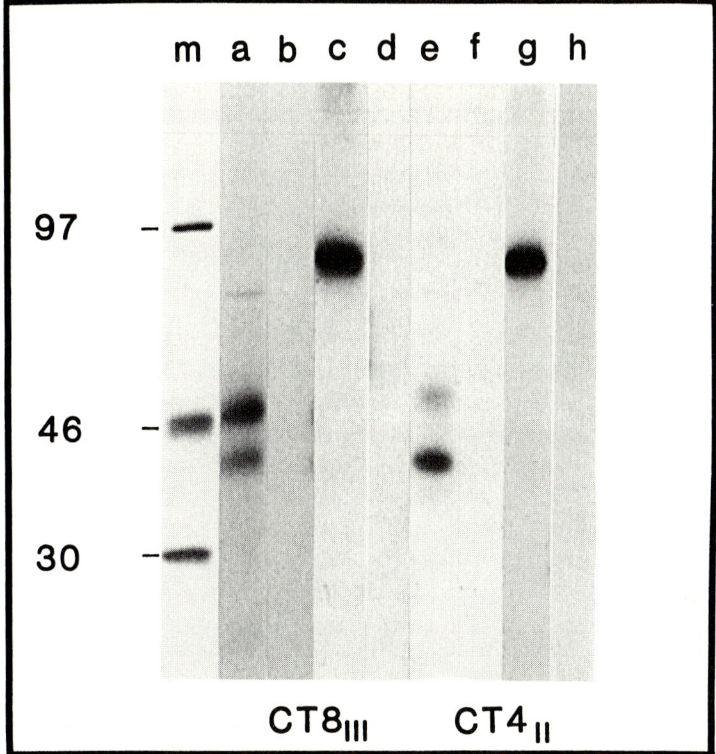



Fig. 1. *Isolation and characterization of Ti$_1$ and Ti$_3$ molecules*

Immunoprecipitations were performed using the clonotypic antibodies anti-Ti$_{1B}$ or anti-Ti$_{2A}$ (both IgM) covalently linked to CnBr-activated Sepharose 4B. Before electrophoresis, immunoprecipitates were washed three times with RIPA solution. SDS/polyacrylamide-gel electrophoresis was performed in reducing (R) and non-reducing (NR) conditions in a 12.5% polyacrylamide gel according to a modification of the Laemmli procedure following by autoradiography. The following [$^{14}$C]methylated molecular mass markers (New England Nuclear) were used (m): carbonic anhydrase (30 kDa); ovalbumin (46 kDa); phosphorylase *b* (97 kDa). Lanes (a–d), CT8$_{III}$; lanes (e–h), CT4$_{II}$. a, anti-Ti$_{1B}$ (R); b, anti-Ti$_{2A}$ (R); c, anti-Ti$_{1B}$ (NR); d, anti-Ti$_{2A}$ (NR); e, anti-Ti$_{2A}$ (R); f, anti-Ti$_{1B}$ (R); g, anti-Ti$_{2A}$ (NR); h, anti-Ti$_{1B}$ (NR).

(lanes b and d). The former appears as two bands of apparent size 51 kDa and 43 kDa on SDS/polyacrylamide-gel electrophoresis in reducing conditions (lane e) and 90 kDa in nonreducing conditions (lane g). Thus, although the Ti$_2$ and Ti$_1$ antigens are comparable in molecular characteristics and are derived from T cell clones of the same individual, they express unique structures that can be defined by non-crossreactive monoclonal antibodies. This, together with the fact that both antibodies did not react with peripheral T cells or with a total of 80 clones derived from the same donor, further suggested that they detected clone-specific rather than allotypic or isotypic determinants on Ti molecules.

More recently, we have described three additional clone-specific monoclonal antibodies which detect similar 90 kDa disulphide-linked heterodimers. Slight differences in molecular mass were noted among all the different α chains. Whether these variations reflect different post-translational modifications or

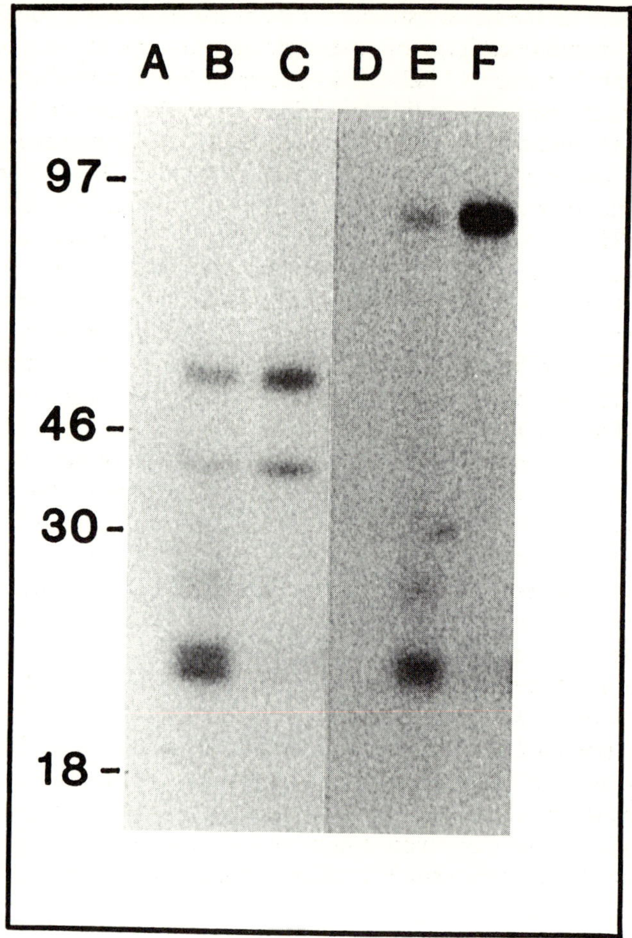



Fig. 2. *Characterization of Ti$_{4A}$ and T3$_C$ surface structures*

RW17C cells ($30 \times 10^6$) were surface-labelled with 1 mCi of Na$^{125}$I (New England Nuclear) and immunoprecipitated with the monoclonal antibodies anti-T6, anti-Ti$_{4A}$, and anti-T3$_C$ covalently linked to CnBr-activated Sepharose 4B. SDS/polyacrylamide-gel electrophoresis was performed in reducing and non-reducing conditions in a 10% polyacrylamide gel according to a modification of the Laemmli procedure. The following [$^{14}$C]methylated molecular mass markers were used: phosphorylase *b* (97 kDa); ovalbumin (46 kDa); carbonic anhydrase (30 kDa); lactoglobulin A (18 kDa). Lanes A–C show reducing conditions: A, anti-T6; B, anti-T3$_C$; C, anti-Ti$_{4A}$; lanes D–F show non-reducing conditions; D, anti-T6; E, anti-T3$_C$; F, anti-Ti$_{4A}$.

protein size differences is still unclear. Biosynthetic labelling of Ti molecules also indicated that they are synthesized by T cells rather than merely absorbed from culture media.

To determine whether the T3-associated disulphide-linked Ti heterodimer was also the receptor for antigen on inducer clones, an anticlonotypic monoclonal antibody, termed anti-Ti$_{4A}$, was produced against the class II MHC restricted ragweed antigen-specific clone RW17C. Not surprisingly, the surface structure defined by anti-Ti$_{4A}$ comodulated with T3 following incubation of RW17C cells with either anti-Ti$_{4A}$ or anti-T3 monoclonal antibodies [41].

To characterize the surface structure recognized by anti-Ti$_{4A}$ on RW17C, we performed immunoprecipitation experiments with membrane preparations externally labelled with $^{125}$I [42,43]. As shown in Fig. 2, anti-Ti$_4$A precipitates two bands under reducing conditions in SDS/polyacrylamide-gel electrophoresis, one at 52 kDa and a second at 41 kDa (lane C), and appears as a single band at about 90 kDa under non-reducing conditions (lane F). These species of these molecular masses were not present in control immunoprecipitates (lanes A and D) or anti-Ti$_{4A}$ from $^{125}$I-labelled clones of differing specificities than RW17C, for example, T15A (results not shown). In contrast with anti-Ti$_{4A}$, anti-T3 precipitated four bands on SDS/polyacrylamide gel electrophoresis under reducing conditions from RW17C (lane B). The major protein band had a molecular mass of 20 kDa with three additional bands at 25, 41 and 52 kDa. Moreover, the two bands of higher molecular mass were identical in size with those found in anti-Ti$_{4A}$ immunoprecipitates from RW17C (lane C), suggesting that they may be related. Further support for this notion came from the observation that, like Ti$_{4A}$ under nonreducing conditions, the 52 kDa and 41 kDa upper bands detected by anti-T3$_C$ appeared as a band at about 90 kDa (lane E). Note that the 20 kDa and 25 kDa species within the anti-T3 precipitate were similar under both reduced and nonreduced conditions. These findings are analogous to those previously obtained with the above series of human cytotoxic T cell clones. That anti-Ti$_{4A}$ (lanes C and F) as well as other anti-Ti antibodies did not precipitate detectable material in the range of the 20–25 kDa T3 molecule implied that binding of the monoclonal antibody anti-Ti$_{4A}$ may dissociate the clonotypic structure from the 20–25 kDa molecules. Given the above T3 association and the subunit composition of Ti$_{4A}$, it appears that the clonotype on an MHC restricted antigen-specific inducer clone is very similar in molecular characteristics to the clonotypes found on CTL clones. Although not shown, clonotypes on T8 + suppressor T lymphocytes are also T3-associated disulphide-linked heterodimers.

## Peptide Variability Exists with Ti Molecules of Different T Cell Clones

Comparison by two-dimensional (isoelectric focusing SDS-gel) electrophoresis of Ti molecules immunoprecipitated from two representative CTL clones, CT4$_{II}$ and CT8$_{III}$, showed that the $\alpha$ subunits of Ti$_1$ and Ti$_2$ had distinct pI values (4.4 versus 4.7) as did the two $\beta$ subunits (6.0 versus 6.2). Both $\alpha$ and $\beta$ subunits were resolved into a series of closely migrating spots [44]. Prolonged treatment of immunoprecipitated Ti molecules with neuraminidase, which cleaves off sialic acid residues from 'complex oligosaccharides' reduced both $\alpha$ and $\beta$ subunits to virtually a single spot at unique pI values, thus indicating that the micro-heterogeneity observed was due to glycosylation (O. Acuto & E. L. Reinherz, unpublished work). Moreover, the latter results demonstrated that the poly-morphism detected by anti-Ti$_1$ and anti-Ti$_2$ antibodies resided in the peptide moiety of the Ti molecules. Note that similar analysis of T3, T4 and T8 molecules derived from clones of differing specificities revealed no such evidence of molecular variability.

Given the above results, we speculated that, similar to the immunoglobulin receptor, Ti molecules may be composed of constant and variable domains. To directly test this hypothesis, comparative two-dimensional peptide maps were performed on isolated $^{125}$I-labelled Ti subunits following digestion with proteolytic enzymes [44]. The tryptic peptide maps of the $\alpha$ chains from clones CT4$_{II}$ and CT8$_{III}$ were similar yet distinct. At least one major and one minor peptide was clearly identical in both clones. Moreover, the major one also was shown to be shared by the Ti $\alpha$ subunit isolated from a thymic tumour (REX), implying that constant domains are preserved among Ti molecules even when derived from genetically unrelated individuals [45].

### Partial *N*-Terminal Amino Acid Sequencing of the Ti $\beta$ Subunit

Peptide map and two-dimensional analysis showed that a 94 kDa disulphide-linked heterodimer existed on the surface of the thymus-derived tumour line (REX). This structure was isolated by monoclonal antibodies and shown to be homologous to both $\alpha$ and $\beta$ subunits of Ti molecules from other IL-2 responsive T cell clones. The growth advantage of the T3+ REX tumour line over clones prompted us to exploit the former for large scale purification of Ti receptor molecules. To this end, microgram amounts were isolated by immunoaffinity chromatography from REX cell membrane and $\alpha$ and $\beta$ were subunits separated by preparative SDS/gel electrophoresis. Given the large variability which is observed within the $\beta$ subunit on peptide map analysis, it was of interest to obtain data on the primary structure of this molecule [46]. Unambiguous amino acid analysis was possible for 17 residues out of 20 cycles in a gas/liquid/solid-phase protein sequenator. A rabbit antiserum raised against synthetic peptide containing the amino acid sequence from residues 2–11 conjugated to bovine serum albumin as a carrier, immunoprecipitated the isolated denatured $\beta$ chain from REX Ti. This result confirmed that the protein sequence was indeed that of the Ti $\beta$ subunit. In order to determine whether there were homologies between this *N*-terminal sequence (residues 2–11 as well as residues 2–20) and known proteins, a computer search employing the Dayhoff protein data bank was utilized. Homologies with the first framework of the variable region of the Ig $\lambda$ and $\kappa$ light chains were found. $\lambda$ homology was evident in a search employing residues 2–11 whereas $\kappa$ homology was obvious when residues 2–20 were used. In fact, 44 of 66 matches were made with human and mouse light chain framework I V-region.

### Genes Encoding the T Cell Receptor

By using a novel technique of cDNA cloning known as subtractive hybridization, two groups very recently described the isolation of human and mouse cDNA clones which encoded putative membrane proteins uniquely expressed in T cells and identified genes that rearranged in these same cells [47–49]. Comparison of the Ti $\beta$ subunit *N*-terminal sequence obtained by ourselves and the predicted gene product from the human cDNA clone demonstrated an identity of all residues that could be analysed. Specifically, it was found that the *N*-terminal amino acid sequence began at residue 22 of the predicted protein indicating that the first 21 residues belong to the leader signal peptide. The predicted gene protein has a remarkable homology to the human Ig λ light chain in both *N*-terminal and *C*-terminal regions and particularly in the area of the cysteine residues. This too is in agreement with the finding from the *N*-terminal acid sequence of the Ti β subunit.

The identity of 17/17 amino acids in these sequences and the fact that the molecular masses for the predicted polypeptide encoded by the human cDNA protein (35 kDa) and the membrane form of the Ti β subunit (43 kDa) are very close, strongly supports the notion that this cDNA is the gene for the β subunit of the human Ti molecule (Fig. 3a). Any differences in molecular mass between the predicted protein and the observed protein are likely to be due to glycosylation. In this regard, it is worthy of note that the human Ti β subunit is a glycoprotein and at least two potential glycosylation sites for complex oligosaccharides have been identified on the human cDNA gene product. Analysis of the murine T cell specific cDNA clone revealed, in addition, that the putative protein encoded by the gene contained an *N*-terminal variable and a *C*-terminal constant region, in agreement with our own peptide map data. The high degree of homology between the *C*-terminal regions in the human and mouse clones (80% between residues 134–306) and a lack of homology of the α and β subunits at the protein level indicates that the mouse equivalent of the Ti β subunit has also been cloned.

To characterize the structure of the predicted protein encoded by one such Ti cDNA clone, p REX, a Kyte–Doolittle plot was generated [50]. This analysis is based on a water-vapour transfer free energy values exterior-interior distribution of amino acid side chains. Several features are worthy of comment. First, *N*-terminal residues −21 to −1 constitute an extremely hydrophobic stretch of sequence, consistent with the fact that it is the leader sequence. Second, while the *C*-terminal residues 257–286 are hydrophobic, the residues 287–291 (Lys-Arg-Lys-Asp-Phe) are hydrophilic and heavily charged. This finding predicts that residues 257–286 define a transmembrane region of the molecule and that amino acids 287–291 comprise a characteristic cytoplasmic tail similarly found on a large number of transmembrane proteins. Third, the alternating hydrophobic–hydrophilic regions throughout the molecule are characteristic of globular membrane proteins, including immunoglobulin [51]. Fourth, the positions of cysteine residues within the Ti β molecules are analogous to those found in immunoglobulin. Thus, the pairs of cysteines in positions 21 and 90 as well as 145 and 210 are separated by approximately the same number of residues as those cysteines which form intrachain disulphide bonds in immunoglobulin molecules (i.e. 69 and 65 residue spans, respectively) and are likely to form disulphide loops defining two distinct Ti β domains. Moreover, the cysteine located at position 245, just *N*-terminal to the transmembrane region, is analogous in position to the cysteine of immunoglobulin light chain which forms an interchain disulphide bond to immunoglobulin heavy chain [51]. Whether the cysteine in position 189 also might participate in interchain disulphide bonding is not known, but appears likely given the fact that rabbit immunoglobulin heavy chains are known to be covalently associated by more than one cysteine residue.

Analysis of a number of Ti β cDNA nucleotide sequences and this predicted

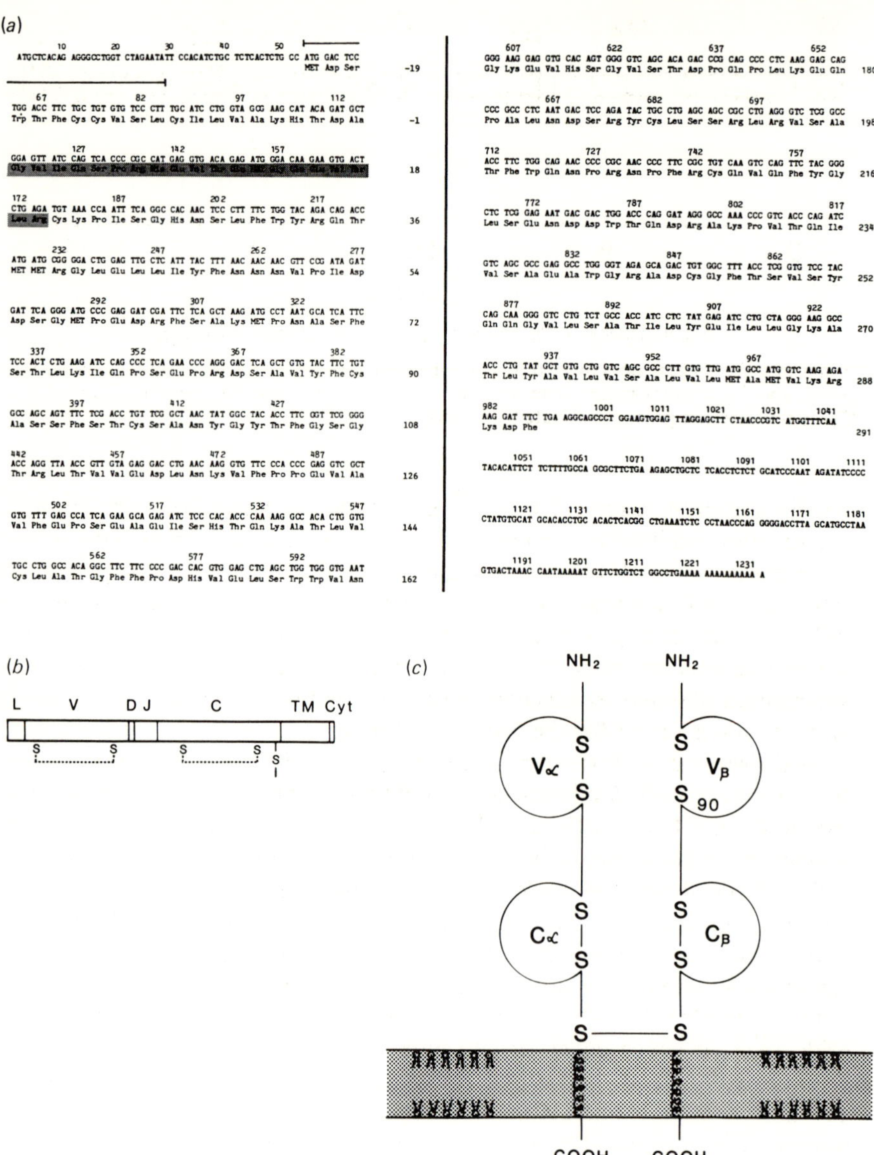

Fig. 3. *Structure of T cell receptor subunits*

(a) DNA sequence and predicted protein sequences of a full-length REX cDNA Ti β probe. (b) Predicted primary structure of the β chain subunit after translation from cDNA sequence. The variable region leader (L), variable domain (V), diversity (D) and joining (J) elements are shown. As part of the constant region (C) a hydrophobic transmembrane segment (TM) and the cytoplasmic part (Cyt) are indicated. Potential intrachain disulphide bonds (S–S) are shown as well as the single thiol group (S) which can form a disulphide bond with the α subunit. (c) Proposed secondary structure of the human T cell receptor α and β chains. Starting at the *N*-terminus (NH₂) the α and β chains have variable regions (V_ρ and V_β) which form a domain stabilized by an intrachain disulphide bond between cysteines. In a similar way, the constant region domains are formed by intrachain disulphide bonds. The subunits are held together by an interchain disulphide bond close to the cell membrane. Both chains have a hydrophobic transmembrane segment and a short cytoplasmic tail at their *C*-terminus.

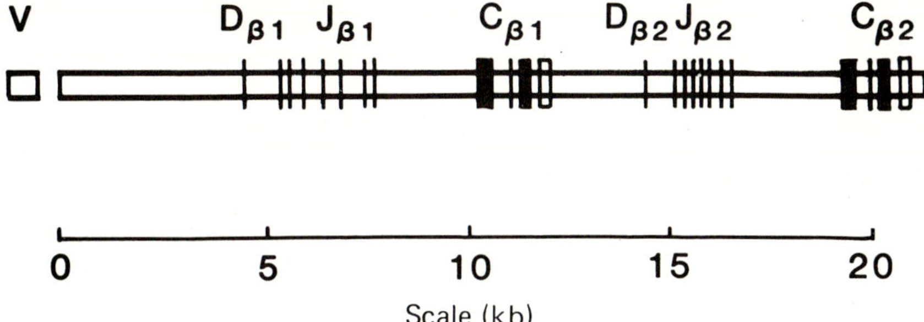


Fig. 4. *Scheme of the genomic organization of human β chain genes*

V indicates the $V_\beta$ gene pool located 5′ at an unknown distance of the $D_{\beta 1}$ (diversity) element, $J_{\beta 1}$ cluster and $C_\beta$ constant region gene. Further downstream a second $D_{\beta 2}$ element, $J_{\beta 2}$ cluster and $C_{\beta 2}$ constant region gene are indicated. Distances are noted in kb.

protein sequence suggested that the gene products were strikingly similar to immunoglobulin insofar as containing variable, constant, joining and diversity-like elements (Fig. 3*b*). That rearrangement of germline Ti β segments was detected in all functional mature T cell populations further supported the view [56]. Comparison of genomic clones from murine liver DNA and functional T cell populations has led to an understanding of the germline arrangement of Ti β elements and signal sequences resulting in somatic recombination [57]. Similar organization is found for the human Ti β locus.

As shown in Fig. 4, the genomic organization of the constant region locus, which encodes the Ti β subunit, consists of two tandemly arrayed set of segments termed $D_{\beta 1}$-$J_{\beta 1}$-$C_{\beta 1}$ and $D_{\beta 2}$-$J_{\beta 2}$-$C_{\beta 2}$. The two constant regions are remarkably similar and differ from each other by only six amino acids in the translated region. Note should be made of the fact that they are dissimilar in this 3′ untranslated region [58]. Each constant region in the mouse in turn is composed of four distinct exons: a first exon encoding an external domain of approx. 125 amino acids; a mini exon encoding six amino acids, including a cysteine thought to form a disulphide bridge to the Ti α subunit; a third exon encoding 36 amino acids, most of which constitute the hydrophobic transmembrane region; and a fourth exon which encodes the cytoplasmic tail and 3′ untranslated region. The transmembrane and cytoplasmic exons are the only coding regions of the molecule where amino acid differences are noted. In contrast, the 3′ untranslated region of the $C_{\beta 1}$ and $C_{\beta 2}$ loci are < 50% homologous [58].

5′ to each $C_\beta$ region is a cluster of J segments. In the murine system seven J elements have been located 5′ to the C regions, of which only six appear to be functional. Moreover, a D element is noted to be 5′ of each of the $J_\beta$ clusters [59]. The presence of such D elements was anticipated from the nucleotide sequencing data, wherein it was found that eight nucleotides between V and J segments could not be accounted for. Fig. 4 also indicates that V gene segments are located upstream of $D_{\beta 1}$ at an unknown distance. As will be discussed below, rearrangements in T cell receptor genes occur during intrathymic ontogeny. In this process the D gene segment is brought into a location adjacent to a given J element with deletion of intervening J components. Subsequently, an upstream

V segment is repositioned to a site immediately 5′ of the D-J segment. Thus, a contiguous V-D-J sequence is formed which is separated from the C exons by an intervening sequence [57].

In the immunoglobulin system, germline $V_H$, $D_H$ and $J_H$ elements are flanked by recombination sequences separated in space by one or two turns of the DNA helix [60]. A feature of the recognition sequences for Ig $V_H$ and $V_L$ gene rearrangement is that a one-turn element is always joined to a two-turn recognition sequence. This rule of one- and two-turn joining is found in the Ti $\beta$ gene family as well. Thus, 3′ of $V_\beta$ and 5′ of $J_\beta$ genes are segments with two-turn and one-turn recognition signals, respectively, whereas the $D_\beta$ gene segment has a one-turn recognition signal on its 5′ side and a two-turn recognition signal on its 3′ side [59]. This organization would permit several types of $V_\beta$ gene formations consistent with the one- to two-turn joining rule: $V_\beta$ gene segments could join directly to $J_\beta$ gene segments, $V_\beta$-$D_\beta$-$J_\beta$ joining may occur involving a single D gene segment, or alternatively, several $D_\beta$ gene segments. Thus, additional combinatorial joining possibilities exist within the Ti $\beta$ gene family.

Similar analysis based on biochemistry and molecular cloning of the human and murine Ti $\alpha$ subunit indicated that these molecules contain an $N$-terminal V domain probably created by joining of dispersed germline genes. Moreover, the Ti $\alpha$ subunit, like the Ti $\beta$ subunit, bears homology to immunoglobulin heavy and light chains [52–55]. Thus, it is likely that variable regions of the Ti $\alpha$ and $\beta$ chains will interact to form a single combining site (Fig. 3c). To date, all evidence has indicated that functional isotypy (i.e. help versus suppression) is not encoded in the T cell receptor subunits themselves [56].


**The Ontogeny of the T Cell Receptor**

Earlier studies delineated three discrete stages of human intrathymic differentiation based upon surface expression of multiple T lineage specific glycoproteins (stage I: T11+T3+T4−T6−T8−; stage II: T11+T3−T4+T6+T8+; and stage III: T11+T3+T4±T6−T8±) [17,28,63]. Of note was the finding that T3 appeared on a minor population of cells and only during late intrathymic ontogeny (stage III). To determine whether clonotypic structures were expressed on T lineage cells during thymic differentiation, we produced monoclonal antibodies to a thymic-derived tumour termed REX (T11+T3+T4+T6−T8+)[64]. The results of SDS/polyacrylamide-gel electrophoresis and peptide map analyses indicate that an homologous T3-associated heterodimer is synthesized and expressed by REX. In addition, similar Ti-related molecules appear during intrathymic ontogeny in parallel with surface T3 expression. The latter could be documented by the presence of disulphide-linked heterodimers on Goding map analysis of $^{125}$I-labelled membranes from T3+, but not T3−, thymocytes.

The presence of Ti-related, disulphide-linked heterodimers as defined by two-dimensional gel analysis on T3+ thymocytes and T3+ and T cell tumours is not surprising given the unique association of Ti with the T3 molecule and the fact that the above cells are derived from late stages of intrathymic and

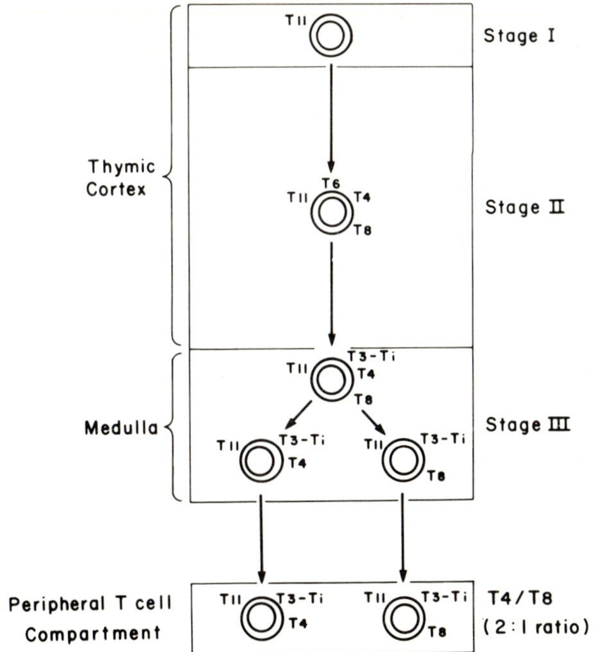


Fig. 5 *Stages of differentiation in the human thymus*

Three discrete stages of thymic differentiation are defined based on the sequential expression of T cell differentiation antigens and the T3–Ti molecular complex.

post-thymic differentiation. The fact that T3- T-ALL cells or thymocytes derived from stage I or early stage II compartments lack Ti strongly further implies that the Ti antigen receptor is intimately linked to T3 expression during ontogeny. This result provides a structural basis for the observation that immunological competence (such as responsiveness to alloantigens in MHC) is acquired only among the population of thymocytes that expressed surface T3 (Fig. 5) [32].

It seemed likely that surface Ti expression on thymocytes was preceded by DNA rearrangements analogous to those occurring in immunoglobulin heavy chain genes prior to surface immunoglobulin expression during B cell differentiation [65]. To obtain further information about the ontogeny of the T cell antigen/MHC receptor, a Ti $\beta$ subunit cDNA probe and heteroantisera specific for the Ti $\alpha$ and Ti $\beta$ subunits were utilized to characterize human T lineage cells. Analysis of thymic tumours and normal thymocytes at both the DNA and protein levels demonstrates that Ti $\beta$ gene rearrangement is evident in stage II (T11 + T6 + T3 −) and stage III (T11 + T6 − T3 +), but not stage I (T11 + T6 − T3 −) thymocytes. In contrast, surface expression of Ti $\alpha$ and $\beta$ molecules is exclusively restricted to stage III thymocytes. Thus human T lineage ontogeny is characterized by an orderly series of differentiation steps wherein Ti $\beta$ gene rearrangement precedes surface expression of the T3–Ti molecular complex (Table 2) [63]. Recent data also indicate that Ti $\alpha$ gene rearrangement occurs during intrathymic ontogeny, probably after Ti $\beta$ gene activation. Transcriptional analysis also indicates that Ti $\beta$ gene activation precedes Ti $\alpha$ gene

Table 2. *Relationship of Ti β gene rearrangement to surface Ti expression*

| Stage | Phenotype | Surface Ti | Ti gene rearrangement |
|---|---|---|---|
| I | T11+ T3− T6− | − | − |
| II | T11+ T3− T6+ | − | + |
| III | T11+ T3+ T6− | + | + |

activation. Although the T3 molecule has not been defined in the mouse, there are disulphide-linked dimers analogous to Ti that are expressed on murine thymocytes and T–T hybridomas [66,67]. Future analysis of these structures will indicate whether their ontogeny is identical to that of Ti in humans.

## Antigen-Like Effects of Monoclonal Antibodies Directed at T Cell Receptor Structures

If anti-Ti monoclonal antibodies defined variable regions of the T cell receptor, then under the appropriate conditions, anti-Ti antibodies might induce clonal T cell activation in a fashion analogous to that of antigen itself. Because the alloantigens that serve as receptor ligands are membrane-bound and are likely to interact via multi-point surface attachment, and because anticlonotypic monoclonal antibodies, by themselves, were not mitogenic, we investigated the functional effects of purified monoclonal antibodies bound to a solid surface support (Sepharose beads) [43].

A number of important points emerge from these experiments: (*a*) antigen, Sepharose-linked anti-Ti and Sepharose-linked anti-T3 caused very similar functional effects, initiating clonal proliferation and secretion of lymphokines including IL-2; (*b*) triggering of a single clonally unique epitope appears to be sufficient to induce antigen-specific functions and to substitute for antigen plus MHC determinant; (*c*) multimeric interaction between ligand and antigen receptor was an essential requirement for the initiation of clonal T cell responses because non-surface-linked monoclonal antibodies do not mediate these effects; (*d*) the T4 and T8 surface structures, although critical for MHC restricted CTL effector function, could not be triggered by monoclonal antibodies to induce clonal proliferation or lymphokine secretion; and (*e*) these clones produced and responded to IL-2 [61].

## A Model of Antigen Recognition by T Cells

From these studies it is possible to construct a unifying model of antigen recognition by T lymphocytes. The structure responsible for dual recognition (i.e. antigen and MHC) consists of the clonally unique Ti molecule, which is associated with the T3 glycoproteins. In addition, depending on the subset derivation of the individual T lymphocyte, T4 or T8 serves as an 'ancillary' binding structure for an invariant portion of class II or I MHC gene production, respectively. These latter glycoproteins do not appear to be critical for T cell activation and may therefore rather be considered to serve as stabilizing elements that facilitate the cell–cell contact necessary to allow for efficient target-cell lysis by CTL. This might be particularly important for killer–target cell interactions as well as triggering of primary immune responses prior to clonal selection of high-affinity antigen-responsive cells. However, it seems likely that at the clonal level, some effector cells exist that display high-affinity Ti/T3 receptors for specific antigen and thus do not require T4 or T8 in order to interact with stimulator/target cells. Indeed, such T cell clones have been reported recently [62]. Given the extent of T cell diversity, one might in addition expect to find an occasional clone viewing antigen plus MHC gene product only with its high affinity Ti/T3 antigen receptor and in apparent contradiction to the T4 class-II and T8 class-I correlation.

### Triggering of the T3–Ti Antigen Receptor Complex Results in Clonal T Cell Proliferation through an Interleukin-2 Dependent Autocrine Pathway

As noted above, upon binding to the surface of a clone, monoclonal antibodies to either T3 or its associated unique Ti heterodimer induced rapid loss of the T3–Ti complex and inhibited all antigen specific functions of a given clone. However, at the same time these antibodies produced a markedly enhanced clonal proliferative response to IL-2. Moreover, when coupled to Sepharose, the appropriate surface-linked anti-Ti and anti-T3 antibodies were able to induce IL-2 secretion and clonal proliferation, analogous to the physiological ligand (i.e. antigen) itself [43]. These findings suggested that clonal proliferation resulting from triggering of the T3–Ti antigen receptor might be mediated through the growth factor IL-2. The latter is a 15 kDa sialoglycoprotein that interacts with activated T cells through specific membrane receptors distinct from the T3–Ti complex and is known to induce T cell growth upon surface binding [68–73].

As expected, the IL-2-driven proliferative responses of clones in either an unmodulated or anti-Ti modulated state could be abrogated by anti-(IL-2 receptor) antibodies or anti-IL-2 antibodies ($\geq$ 90% inhibition). Moreover, these same studies showed that the enhanced proliferative response *in vitro* to IL-2 by anti-Ti or anti-T3 modulated T cell clones was a consequence of rapid induction (hours) of IL-2 receptors (6–10-fold increase). Perhaps more importantly, similar increases in IL-2 receptor expression and T3–Ti modulation were triggered by the appropriate combination of antigen and MHC.

That triggering of the T structure should result in endogenous IL-2 production, enhanced IL-2 receptor expression and clonal proliferation appeared more than coincidental and suggested that stimulation of clonal proliferation through antigen receptor triggering was due to release and subsequent binding of endogenous IL-2, i.e. an autocrine mechanism. Thus, one would anticipate that clonal proliferation after antigen receptor triggering and crosslinking could be inhibited by anti-IL-2 or anti-(IL-2 receptor) antibodies. This was found to be precisely the case [61].

Therefore, it appears that T cell proliferation occurs as a series of complex and precisely orchestrated events, as outlined in Fig. 6. Resting T cell clones or

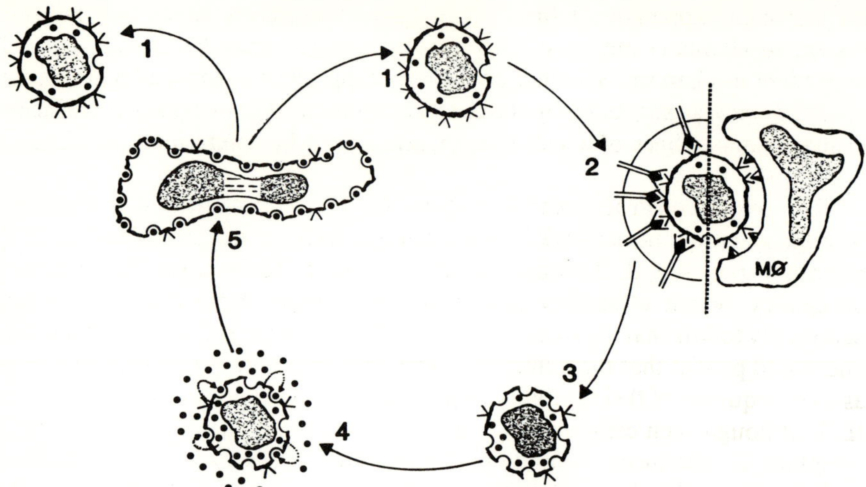


Fig. 6. *Schematic model of T cell proliferation mediated by an IL-2 dependent autocrine mechanism*

Stage 1: resting T cells or T cell clones express few or no IL-2 receptors while reciprocally displaying a maximal number of antigen receptors. Stage 2: T3–Ti triggering by antigen/MHC restricting element or surface bound anti-(T cell receptor) antibodies results in T3–Ti modulation, thus reducing the number of surface antigen receptors and rapidly inducing surface IL-2 receptor expression. Stage 3: this event appears to occur prior to IL-2 secretion itself because IL-2 receptor expression can be observed 2–4 h after T3–Ti triggering, whereas IL-2 is not detectable in culture supernatants until at least 10–12 h later. Stage 3 also can be achieved with anti-Ti or anti-T3 monoclonal antibodies in soluble form. Stage 4: activation via T3–Ti leads to production and secretion of endogenous IL-2 and subsequent binding to its own IL-2 receptors. Stage 5: once a critical density of occupied IL-2 is achieved, DNA synthesis and mitosis occur. Finally, in the absence of additional antigenic stimulation, there is re-expression of the surface T3–Ti receptor complex (stage 1).

T cells express few or no IL-2 receptors but display a maximal number of surface antigen receptors (stage 1). However, T3–Ti receptor triggering by the appropriate antigen plus MHC gene product or Sepharose-bound anticlonotypic monoclonal antibodies results in modulation of the T3–Ti complex (stage 2), thus diminishing the number of surface antigen receptors and rapidly inducing surface IL-2 receptor expression (stage 3). Furthermore, such activation also leads to endogenous IL-2 production, secretion, and subsequent binding to IL-2 receptors on the same clones (stage 4). Once a critical number of IL-2 receptors have bound IL-2, DNA synthesis and cell mitosis occurs (stage 5). Finally, in the absence of continued antigenic stimulation, there is re-expression of surface T3–Ti antigen receptor complex and, reciprocally, a reduction in the number of IL-2 receptors (stage 1).

The view that is emerging regarding the relationship of the T3–Ti antigen receptor to the IL-2 receptor points to a mechanism whereby external stimuli, i.e. antigen, direct the magnitude and extent of T cell clonal proliferation by means of the IL-2 hormone–receptor system. Immunological specificity is ensured by the antigen dependence of the IL-2 receptor expression, and yet the reciprocal appearance of T3–Ti and IL-2 receptors presumably leaves the cell in a state of responsiveness to either the hormone or antigen ligand. However, the transient expression of the IL-2 receptors themselves serves as a fail-safe system to eliminate any possibility of uncontrolled growth through an IL-2 dependent mechanism. Whether alteration of this mechanism results in tumour growth *in vivo*, as suggested by studies on continuously growing IL-2-producing primate tumour lines of T cell lineage, remains to be determined but appears likely [74].

The present construct clearly implies that T cell proliferation is mediated through an autocrine network in which antigen receptor triggering leads to IL-2 receptor expression, IL-2 production, IL-2 release and subsequent IL-2 receptor occupancy, which ultimately promotes cell division. Note that it does not necessarily follow that all T cells produce and respond to their own IL-2. In fact, one would predict that the failure of certain cells to proliferate to antigen occurs as a consequence of their inability to produce sufficient amounts of endogenous IL-2, although such cells would, in all likelihood, be triggered to express IL-2 receptors. Furthermore, even for IL-2-producing T cells, exogenous sources of IL-2 would amplify clonal proliferation. However, the present results are clearly different than anticipated from the conventional endocrine notion that suggests that one cell type produces IL-2 while a different one responds to it.

### Implications of the Molecular Definition of the T3–Ti T Cell Receptor Complex for Human Disease

The above functional and structural studies indicate that the T3–Ti T cell antigen receptor complex consists of at least five polypeptide chains: the polymorphic $\alpha$ and $\beta$ subunits of the Ti molecule and the invariant glycosylated 20 kDa and 25 kDa and non-glycosylated 20 kDa T3 subunits (Fig. 7). The specificity of the anticlonotypic antibodies for the Ti molecule and unique peptides resulting from proteolysis of different Ti heterodimers implies that variable regions exist within both Ti $\alpha$ and $\beta$ subunits. Hence, the latter presumably serves as a ligand binding site. In contrast, the 20 and 25 kDa T3 molecules are likely necessary in transducing signals from surface into the cytosol following allosteric changes created by ligand–Ti binding. This transduction appears to be associated with sodium- and calcium-dependent plasma membrane mediated depolarization [75]. The observation that the 20 kDa subunit (non-glycosylated) has a long cytoplasmic tail whereas the Ti $\alpha$ and $\beta$ subunits do not further supports the notion that one or more of the T3 subunits functions in signal transduction. Whether the T3 subunits themselves represent ion channels as shown for other receptors (i.e. acetylcholine, etc.) is not known at present. However, since the non-glycosylated 20 kDa subunit of T3 is hydrophobic, this subunit represents a candidate for an ion channel.

Structural and/or functional defects of the T3–Ti T cell activation system could result in a number of immunodeficiency states. Release of cells lacking the T3–Ti complex either prematurely or as a consequence of aberrations of T cell maturation would lead to an immunoincompetent state. The lymphoid cells of most SCID (severe combined immunodeficiency) patients, for example, lack surface T3 expression and hence have not yet acquired T cell antigen receptors [76]. In contrast, other patients demonstrate phenotypically mature (T3+)

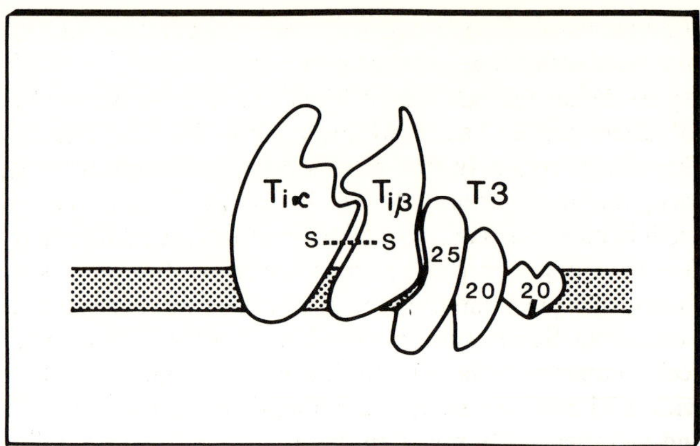


Fig. 7. *Subunit composition of the human T cell receptor*

Ti $\alpha$ and $\beta$ subunits are held together by disulphide bonds and are associated with the 25 kDa chain of the T3 molecule. The $\alpha$ and $\beta$ subunits are anchored in the cell membrane with their transmembrane segments. The T3 complex consists of two additional subunits with molecular masses of 20 kDa.

inducer T cells but fail to respond to antigens or mitogens [77]. These cells are likely to display defects in transmembrane signalling or aberrant function of the IL-2-dependent autocrine pathway. Additional functional and biochemical analyses of such patients should provide insight into the precise level of dysfunction. Whether functional abnormalities of T4+ inducer cells in patients with acquired immunodeficiency syndrome or other viral diseases is related to defects in T cell receptor activation pathway is not known at present, but appears likely, particularly in view of recent reports suggesting that anti-T3 antibody fails to activate T cells of patients with AIDS.

While congenital or acquired defects in the T cell activation system may account for the devastating effects of immunodeficiency disorders, understanding of the molecular basis of T cell activation via the T3–Ti complex could be utilized to design selective immunosuppressive therapies. In this regard, it is already known that cyclosporin A inhibits antigen-specific T cell proliferation [78] but not IL-2-induced proliferation. Thus, it may inhibit T3–Ti signal transduction leading to T cell proliferation. Human T cell clones and anti-receptor monoclonal antibody should provide ideal systems with which to test the effects of various experimental drugs for potency in inhibiting T cell activation.

The ability to produce anticlonotypic antibodies has, in addition, important diagnostic and therapeutic implications for autoimmune disorders. In diseases such as multiple sclerosis, dematomyositis and rheumatoid arthritis, it is likely that a single dominant clone initiates destruction of a particular target organ. Anticlonotypic monoclonal antibodies specific for such effector or regulatory cells could provide a rapid and reliable means to monitor disease activity. Moreover, the anticlonotypes may provide a powerful selective therapeutic strategy for elimination of these cells. Because crossreactive anticlonotypic antibodies exist in many experimental antibody systems, it is also conceivable that anticlonotypic reagents to 'renegade' T cell clones in one multiple sclerosis patient would react with those in other patients.

The ability to define cell surface antigens that appear at specific stages of human T cell differentiation has, in addition, allowed for the orderly dissection of T cell malignant processes. In fact, these T cell diseases reflect the same degree of heterogeneity and maturation as is seen in normal T cell ontogeny. Whereas the tumour cells in most cases of acute T cell lymphoblastic leukaemia arise from an early T3− thymocyte or prothymocyte compartment, the tumours from patients with acute lymphoblastic lymphoma, T cell chronic lymphatic leukaemia, adult T cell leukaemia, Sezary syndrome and mycosis fungoides are derived from a mature T cell compartment bearing the T3 molecule. Since T3 is co-ordinately expressed with a Ti receptor structure, anticlonotypic reagents may be diagnostically and therapeutically useful in these T cell malignancies as well.

## References


1. Benacerraf, B. & McDevitt, H. (1979) *Science* **175**, 273
2. Schlossman, S. F. (1979) *Transplant. Rev.* **10**, 97
3. Zinkernagel, R. M. & Doherty, P. C. (1975) *J. Exp. Med.* **141**, 1427
4. Corradin, G. & Chiller, J. M. (1979) *J. Exp. Med.* **149**, 436
5. Hunig, T. & Bevan, M. (1981) *Nature (London)* **294**, 460
6. Cerottini, J. C. (1980) *Prog. Immunol.* **4**, 622
7. Doherty, P. C. (1980) *Prog. Immunol.* **4**, 563
8. Quinnan, G. V., Kirmani, N., Rook, A. H. *et al.* (1982) *N. Engl. J. Med.* **307**, 7
9. Wallace, L. E., Rickinson, A. B., Rose, M. & Epstein, M. A. (1982) *Nature (London)* **297**, 413
10. Meuer, S. C., Hodgdon, J. C., Cooper, D. A. *et al.* (1983) *J. Immunol.* **131**, 186
11. Gershon, R. K. (1974) *Contemp. Top. Immunol.* **3**, 1
12. Cantor, H. & Boyse, E. A. (1977) *Cold Spring Harbor Symp. Quant. Biol.* **41**, 23
13. Cohen, S., Pick, E. & Oppenheim, J. J. in *Biology of the Lymphokines* (Oppenheim, J. J., ed), pp. 179–195. Academic Press, New York.
14. Waldmann, T. A. (1978) *Ann. Int. Med.* **88**, 226
15. Cantor, H. & Gershon, R. K. (1979) *Fed. Proc. Fed. Am. Soc. Exp. Biol.* **38**, 2051
16. Lipton, J. M., Reinherz, E. L., Kudisch, M. *et al.* (1980) *J. Exp. Med.* **152**, 350
17. Reinherz, E. L. & Schlossman, S. F. (1980) *Cell* **19**, 821
18. Reinherz, E. L. & Schlossman, S. F. (1980) *N. Engl. J. Med.* **303**, 370
19. Reinherz, E. L. & Schlossman, S. F. (1981) *Immunol. Today* **2**, 69
20. Morgan, D. A., Ruscetti, R. W. & Gallo, R. C. (1976) *Science* **193**, 1007
21. Kurnick, J. T., Gronvik, K. O., Kimura, A. K. *et al.* (1979) *J. Immunol.* **122**, 255
22. Bonnard, G. D., Yasaka, K. & Macad, R. D. (1980) *Cell. Immunol.* **51**, 390
23. Sredni, B., Tse, H. Y. & Schwartz, R. H. (1980) *Nature (London)* **283**, 581
24. Meuer, S. C., Schlossman, S. F. & Reinherz, E. L. (1982) *Proc. Natl. Acad. Sci. U.S.A.* **79**, 4395
25. Meuer, S. C., Acuto, O., Hussey, R. E. *et al.* (1983) *Nature (London)* **303**, 808
26. Meuer, S. C., Fitzgerald, K. A., Hussey, R. E. *et al.* (1983) *J. Exp. Med.* **157**, 705
27. Reinherz, E. L., Hussey, R. E. & Schlossman, S. F. (1980) *Eur. J. Immunol.* **10**, 758
28. van Wauwe, F. P., DeMay, J. R. & Goosener, J. G. (1982) *J. Immunol.* **124**, 2708
29. Chang, T. W., Kung, P. C., Gingras, S. & Goldstein, G. (1981) *Proc. Natl. Acad. Sci. U.S.A.* **78**, 1805
30. Burns, G. F., Boyd, A. E. & Beverley, P. C. L. (1982) *J. Immunol.* **124**, 1451
31. Reinherz, E. L., Meuer, S. C., Fitzgerald, K. A. *et al.* (1982) *Cell* **30**, 735
32. Umiel, T., Daley, J. F., Bhan, A. K. *et al.* (1982) *J. Immunol.* **129**, 1054
33. Reinherz, E. L., Morimoto, C., Penta, A. C. & Schlossman, S. F. (1980) *Eur. J. Immunol.* **10**, 570
34. Reinherz, E. L., Kung, P. C., Goldstein, G. & Schlossman, S. F. (1979) *Proc. Natl. Acad. Sci. U.S.A.* **76**, 4061
35. Reinherz, E. L., Kung, P. C., Goldstein, G. & Schlossman, S. F. (1980) *J. Immunol.* **124**, 1301
36. Reinherz, E. L., Kung, P. C., Pesando, J. M. *et al.* (1979) *J. Exp. Med.* **150**, 1472

37. Krensky, A. M., Clayberger, C., Reiss, C. S. *et al.* (1982) *J. Immunol.* **129**, 2001
38. Meuer, S. C., Schlossman, S. F. & Reinherz, E. L. (1982) UCLA Symp. Mol. Cell. Biol. **24**, 127
39. Biddison, W. E., Rao, P. E., Thalle, M. A. *et al.* (1982) *J. Exp. Med.* **156**, 1065
40. Reinherz, E. L., Hussey, R. E., Fitzgerald, K. A. *et al.* (1981) *Nature (London)* **294**, 168
41. Meuer, S. C., Hussey, R. E., Hodgdon, J. C. *et al.* (1982) *Science* **218**, 471
42. Meuer, S. C., Cooper, D. A., Hodgdon, J. C. *et al.* (1983) *Science* **222**, 1239
43. Meuer, S. C., Hodgdon, J. C., Hussey, R. E. *et al.* (1983) *J. Exp. Med.* **158**, 988
44. Acuto, O., Meuer, S. C., Hodgdon, J. C. *et al.* (1983) *J. Exp. Med.* **158**, 1368
45. Acuto, O., Hussey, R. E., Fitzgerald, K. A. *et al.* (1983) *Cell* **34**, 717
46. Acuto, O., Fabbi, M., Smart, J. *et al.* (1984) *Proc. Natl. Acad. Sci. U.S.A.* **81**, 3851
47. Yanagi, Y., Yoshikai, Y., Legget, K. *et al.* (1984) *Nature (London)* **308**, 145
48. Hedrick, S. M., Cohen, D. I., Nielsen, E. A. & Davis, M. M. (1984) *Nature (London)* **308**, 149
49. Hedrick, S. M., Nielsen, E. A., Kavaler, J. *et al.* (1984) *Nature (London)* **308**, 153
50. Kyte, J. & Doolittle, R. F. (1982) *J. Mol. Biol.* **157**, 133
51. Kabat, E. A., Wu, T. T. & Bilofsky, H. (1983) *Sequences of Proteins of Immunological Interest*, US Department of Health and Human Services, Bethesda
52. Saito, H., Kranz, D. M., Takagaki, D. *et al.* (1984) *Nature (London)* **312**, 36
53. Chien, Y., Becker, D. M., Lindsten, T. *et al.* (1984) *Nature (London)* **312**, 31
54. Hannum, C. H., Kappler, J. W., Trowbridge, E. A. *et al.* (1984) *Nature (London)* **312**, 65
55. Fabbi, M., Acuto, O., Smart, J. E. & Reinherz, E. L. (1984) *Nature (London)* **312**, 269
56. Royer, H. D., Bensussan, A., Acuto, O. & Reinherz, E. L. (1984) *J. Exp. Med.* **160**, 947
57. Siu, G., Clark, S. P., Yoshikai, Y., Malissen, M., Yanagi, Y., Strauss, E. & Mak, T. W. (1984) *Cell* **37**, 393
58. Yoshikai, Y., Anatoniou, D., Clark, S. P., Yanagi, Y., Sangster, R., van den Elsen, P., Terhorst, C. & Mak, T. W. (1984) *Nature (London)* **312**, 521
59. Siu, G., Kronenberg, M., Strauss, E., Haars, R., Mak, T. W. & Hood, L. (1984) *Nature (London)* **311**, 344
60. Early, P., Huang, H., Davis, M., Calame, K. & Hood, L. (1980) *Cell* **19**, 981
61. Meuer, S. C., Hussey, R. E., Cantrell, D. A. *et al.* (1984) *Proc. Natl. Acad. Sci. U.S.A.* **81**, 1509
62. Spits, H., Igssel, H., Thomas, A. & deVries, J. E. (1983) *J. Immunol.* **131**, 678
63. Royer, H. D., Acuto, O., Fabbi, M. *et al.* (1984) *Cell* **39**, 261
64. Reinherz, E. L., Kung, P. C., Goldstein, G. *et al.* (1980) *Proc. Natl. Acad. Sci. U.S.A.* **77**, 1588
65. Early, P., Huang, H., Davis, M. *et al.* (1980) *Cell* **19**, 981
66. Allison, J. P., McIntyre, R. W. & Bloch, D. (1982) *J. Immunol.* **129**, 2293
67. Haskins, K., Kubo, R., White, J. *et al.* (1983) *J. Exp. Med.* **157**, 1149
68. Robb, R. J., Munck, A. & Smith, K. A. (1982) *J. Exp. Med.* **154**, 1455
69. Smith, K. A., Lachman, L. B., Oppenheim, J. J. & Favata, M. F. (1980) *J. Exp. Med.* **151**, 1551
70. Smith, K. A., Baker, P. E., Gillis, S. & Ruscetti, R. W. (1980) *Mol. Immunol.* **17**, 579
71. Uchiyama, T., Broder, S. & Waldmann, T. A. (1981) *J. Immunol.* **126**, 1393
72. Leonard, W. J., Depper, J. M., Uchiyama, T. *et al.* (1982) *Nature (London)* **300**, 267
73. Depper, J. M., Leonard, W. J., Robb, R. J. *et al.* (1983) *J. Immunol.* **131**, 690
74. Gootenberg, J. E., Ruscetti, F. W., Mier, J. W. *et al.* (1981) *J. Exp. Med.* **154**, 1403
75. Weiss, M. J., Daley, J. F., Hodgdon, J. C. & Reinherz, E. L. (1984) *Proc. Natl. Acad. Sci. U.S.A.* **81**, 6836
76. Reinherz, E. L., Cooper, M. D., Schlossman, S. F. & Rosen, F. S. (1981) *J. Clin. Invest.* **68**, 699
77. Reinherz, E. L., Geha, R., Wohl, M. E. *et al.* (1981) *N. Engl. J. Med.* **304**, 811
78. Pawelec, G. & Wernet, P. (1983) *Int. J. Immunopharm.* **5**, 315

# Subject Index

First and last page numbers of papers to which entries refer are given.